A Genius for Deception

# A GENIUS
# FOR DECEPTION

*How Cunning Helped the British*
*Win Two World Wars*

NICHOLAS RANKIN

OXFORD
UNIVERSITY PRESS

## OXFORD
UNIVERSITY PRESS

Oxford University Press, Inc., publishes works that further
Oxford University's objective of excellence
in research, scholarship, and education.

Oxford   New York
Auckland   Cape Town   Dar es Salaam   Hong Kong   Karachi
Kuala Lumpur   Madrid   Melbourne   Mexico City   Nairobi
New Delhi   Shanghai   Taipei   Toronto

With offices in
Argentina   Austria   Brazil   Chile   Czech Republic   France   Greece
Guatemala   Hungary   Italy   Japan   Poland   Portugal   Singapore
South Korea   Switzerland   Thailand   Turkey   Ukraine   Vietnam

Copyright © 2008 by Nicholas Rankin

First published in Great Britain as *Churchill's Wizards:*
*The British Genius for Deception, 1914–1945* by Faber and Faber, Ltd.

First published in the United States in 2009 by Oxford University Press, Inc.
198 Madison Avenue, New York, New York 10016

www.oup.com

First issued as an Oxford University Press paperback 2011

Oxford is a registered trademark of Oxford University Press

Library of Congress Cataloging-in-Publication Data
Rankin, Nicholas, 1950–
A genius for deception : how cunning helped the British
win two world wars / Nicholas Rankin.
p.   cm. — (Churchill's wizards)
Includes bibliographical references and index.
ISBN 978-0-19-538704-9
ISBN-13: 978-0-19-976917-9 Paperback
1. Deception (Military science)—History—20th century.
2. World War, 1914–1918—Deception—Great Britain.
3. World War, 1939–1945—Deception—Great Britain.
4. Strategy—History—20th century. I. Title.
U167.5.D37R36 2009
940.4'8641—dc22      2009018155

This book is for my dearest darling wife of twenty-five years
Maggie Gee
who helped so much

War has a way of masking the stage with scenery crudely daubed with fearsome apparitions.

CARL VON CLAUSEWITZ, *On War*


War is a game that is played with a smile. If you can't smile, grin.

WINSTON S. CHURCHILL, in the trenches near Ploegsteert


'Then why have you been so hard to find?'
'Isn't this what the twentieth century is all about?'
'What?'
'People go into hiding even when no one is looking for them.'

DON DELILLO, *White Noise*


There is nothing more deceptive than an obvious fact.

MR SHERLOCK HOLMES in *The Boscombe Valley Mystery*
by Sir Arthur Conan Doyle

# Contents

# CONTENTS

# Illustrations

Cartoon Museum, London; © Lawrence Pollinger Ltd, London. Reproduced with permission.

15 The 'Prop-Shop' at the Special Operations Executive's Station XV in the Thatched Barn roadhouse near Elstree film-studios. 99© The National Archives, Richmond, Surrey. HS 7/49.

16 Dummy British aircraft at El-Adem airbase near Tobruk, part of an 'A Force' deception scheme in the Mediterranean. © The National Archives, CN 26/1.

17 Dummy landing craft moored in North Africa, purportedly for the invasion of Greece in 1943. © The National Archives, CN 26/1.

18 Dummy tank to fool German and Italian observers in North Africa in 1943. © IWM, MH20759.

19 Lt Col. David Stirling, founder of the SAS, with patrol commander Lt Edward McDonald (with Fairbairn Sykes Commando dagger) and Cpl Bill Kennedy. © IWM, E21339.

20 Lt Col. Dudley Clarke, head of 'A Force' in Cairo and Britain's top deceiver, in drag in Madrid in October 1941. Churchill Papers, CHAR 20/25/52. © Churchill Archives Centre, Cambridge.

21 Fluent German-speaking *Daily Express* journalist Sefton Delmer, the maestro of British 'black' propaganda. Reproduced by permission of Felix Delmer.

22 The genuine corpse of 'The Man Who Never Was', before shipping by submarine to Spain in April 1943. © The National Archives, WO 106/5921.

23 A still from the film *I Was Monty's Double*. © IWM, HU47556.

24 Barcelona-born Juan Pujol García, the most successful double agent of WW2. © The National Archives, KV 2/70.

25 Juan Pujol's crucial message as received by teleprinter at German HQ on 9 June 1944. © The National Archives.

26 Prime Minister Winston Churchill sets foot on liberated France on 12 June 1944, six days after D-Day. © IWM, B5357.

# Preface

The British enjoy deceiving their enemies. When the Prussian strategist Carl von Clausewitz defined war in 1833 as 'those acts of force to compel our enemy to do our will', he missed out the dimension that the British political philosopher Thomas Hobbes had spotted nearly two centuries earlier: 'Force and fraud are in war the two cardinal virtues.'

Sir Alan Lascelles called Winston Churchill 'the Arch-Mountebank' and he certainly had a penchant for display. No other British twentieth-century politician was photographed in so many different kinds of headgear as Churchill: whether in boater, bowler, cocked hat, flying helmet, homburg, peaked cap, sombrero, sola topi, sou'wester, Stetson, topper or Tommy's tin hat, he always dressed the part.

Acting is a long-established area of British talent. 'The British like to pretend,' observes a former US Ambassador, Raymond Seitz. 'They seem to prize few things so much as a good performance.' And the theatre director Richard Eyre notes the national 'love of ritual, procession . . . and dressing-up'. 'On the surface they are so open,' writes novelist Geoffrey Household of his countrymen, 'and yet so naturally and unconsciously secretive about anything which is of real importance to them.' British self-deprecation, wit and irony are also forms of concealment. The British do not say what they mean, or mean what they say, and often mask seriousness with jokes as a cover for shyness or sentiment. Jorge Luis Borges says of Herbert Ashe in *Ficciones*: 'He suffered from unreality, like so many of the British.'

On 10 February 1910, a party of six well-bred young people conned their way on to the flagship of the Royal Navy's Home Fleet, HMS *Dreadnought*, by impersonating the Emperor of Abyssinia and his suite. A guard of honour met the train at Weymouth, and an admiral and a commander showed them over the ship. Among these impostors was a 'Prince Mendax', blacked up and false-bearded by the famous theatrical costumier Willie Clarkson of Wardour Street, and complete with turban, caftan and heavy gold chain. Prince Mendax was in fact the future modernist novelist and literary heroine Virginia Woolf. It was a good hoax to play on the Royal Navy that sustained the British Empire at its height. But aristocratic bluff was also a deceptive performance skill that helped the people of a small island nation to rule a vast worldwide empire.

Of course, military deception (MILDEC), which the US Joint Chiefs define as 'actions executed to deliberately mislead adversary decision makers as to friendly military capabilities, intentions and operations', has been used all over the world. In China 2,400 years ago, Sun Tzu said in *The Art of War* that 'All warfare is based on deception'. The *hadith* or proverb, *'al-harb khuda'*, attributed to the Prophet Muhammad – peace be upon him – also means 'war is deception'. The famous stratagem that toppled Troy was the Greek gift of a wooden horse – with a special force, including the wily Odysseus, hidden inside.

The Trojan War was fought over a sexually attractive woman, and deception has deep roots in biology. Vladimir Nabokov observed that 'Everything is deception . . . from the insect that mimics a leaf to the popular enticements of procreation.' Deception also marks predator–prey relations. Weaker animals evolve disguises or camouflages to protect themselves against more powerful ones. Human animals, however, are uneasy with the idea of deception because it confers unfair advantage and destroys cooperation. In *The Republic*, Plato said that only the rulers of the city are entitled to tell lies, and then only in order to benefit the city and in direct response to the actions of enemies or troublesome citizens.

The British developed deception in both World Wars as a response to dangers represented by new military technologies on land, at sea and in the air. This was especially true in WW2, when the weakened nation had its back to the wall after the rest of Europe had fallen into what Churchill called 'the grip of the Gestapo and all the odious apparatus of Nazi rule'. It is against Nazi Germany that Britain developed some of the most sophisticated and brilliant deceptions in history. Winston Churchill's interest in secrecy and deception, however, goes back to the beginning of WW1.

In July 1911, during the Agadir crisis, Churchill was Home Secretary, and was talking to the Chief Commissioner of Police at a Downing Street garden party. Germany was flexing some naval muscle in Morocco. The Chief Commissioner mentioned that the Home Office was responsible for guarding two magazines where the Royal Navy stored its explosives, and that only a few constables were on the job. Churchill asked what might happen if 'twenty determined Germans in two or three motor cars arrived well armed upon the scene one night', and was told such a force could not be held off. Churchill promptly 'quitted the garden party', armed and reinforced the Metropolitan Police and called out the British Army to help secure the cordite reserves. This set him thinking about espionage and counter-espionage, and he signed new warrants to allow the mail of suspected German agents to be opened. As First Lord of the Admiralty in 1914, one of his earliest 'Most Secret' memos of war-time gave instructions to build a dummy fleet of ten large merchant vessels mocked up in wood and canvas to look like far bigger battleships in silhouette, so as to baffle and distract enemy aeroplanes and submarines. Three of these 'battleships' were sent to the Dardanelles in February 1915 to lure the German fleet out into the North Sea.

As a young man, he scorned dishonesty – 'I had no idea in those days of the enormous and unquestionably helpful part that humbug plays in the social life of great peoples dwelling in a state of democratic freedom' – but Churchill came thoroughly to approve of deception in warfare. This book argues that British twentieth-century military deception has four pillars: camouflage, propaganda, secret intelligence and special forces. Churchill was excited by T. E. Lawrence's ideas about guerrilla warfare, based on disguise and surprise rather than frontal assault, and it was in Churchill's prime ministership that the Commandos and the SAS were founded. He also became a master of propaganda who, as the broadcaster Edward R. Murrow said, 'mobilized the English language and sent it into battle'.

Native cunning links all Churchill's wizards, creative people using their skills to help their country in a struggle for survival. The two World Wars recruited widely from the nation's pool of talent, not just from the narrow caste of professional soldiers. This book is about artists and scientists, film and theatre people, novelists and naturalists, as well as daredevils, commandos and the Home Guard who disguised machine-gun posts as gentlemen's toilets or genteel tea-rooms.

The first half of the book is about WW1, in which the first British explorations of camouflage brought together the skills of the big-game hunter Hesketh Prichard, the theatre-hand Oliver Bernard and the society portrait painter Solomon J. Solomon. The Oxford archaeologist T. E. Lawrence dressed up in white robes and became 'Lawrence of Arabia', George Bernard Shaw (in khaki and tin hat) sent polemical dispatches from the trenches, and the popular author John Buchan became the head of British propaganda. The second half of the book is about WW2, and introduces two little-known geniuses, the larger-than-life, German-speaking *Daily Express* journalist Sefton Delmer and a hyper-observant, cinema-loving regular soldier called Dudley Clarke. Delmer, Clarke and their cohorts wove an ever more complex web of tactical trickery, strategic deception and black propaganda to help the Allies win the war.

But our story begins in the summer of 1914.

# PART I

# I

# The War of Nerves

When Winston Churchill read the newspapers in Portsmouth he had a sudden, vivid feeling that something 'sinister and measureless' had occurred. On 28 June 1914, the Emperor Franz Joseph I's nephew and heir-presumptive, the Archduke Franz Ferdinand, and his pregnant wife, Sophie, had taken the wrong turning in Sarajevo. A faraway cloud no bigger than a man's hand was about to become a great storm, embroiling millions of people from scores of nations.

The Archduke Ferdinand was the hated symbol of the Austro-Hungarian Empire that had annexed Bosnia and Herzegovina in October 1908, tearing it away from Greater Serbia. The Black Hand, a Serb nationalist terrorist cell, intended to kill the archduke as he drove in his motorcade through Sarajevo. One of the conspirators threw a bomb from the crowd, but the chauffeur floored the accelerator and the black car shot over the device, which exploded behind, injuring dignitaries in the following vehicle as well as some bystanders. Hours later, driving back from a hospital visit to the injured, the chauffeur took his fateful wrong turn.

As the car reversed slowly back up Gebet Street, it passed a tubercular and weedy-looking youth called Gavrilo Princip, consoling himself with a sandwich in Moritz Schiller's cafe. The 19-year-old Bosnian Serb could hardly believe his luck, for he was one of the seven-strong gang disappointed by the failure of the earlier bomb. In one pocket Princip had a cyanide capsule and in the other a Belgian-made Browning 9 mm semi-automatic pistol. The open-topped Austrian car offered him a second opportunity for his cause to make its mark on history, and he shot the Archduke and his wife at close range.

Throughout July 1914, the widening reverberations of this incident in the Balkans tipped other nations towards war. In London's Fleet Street, where all Britain's national newspapers were edited and printed, Philip Gibbs's sensitive, well-bred face was a familiar sight. At 37, he had fingers yellow from chain-smoking, but he was a star journalist of many scoops who had written the first best-selling novel about newspaper reporters, *The Street of Adventure*. As events unfolded, Gibbs reported 'dazed incredibility' in middle England, uncertainty in Whitehall's corridors of power, and 'profound ignorance' behind all the feverish activity of Fleet Street newspaper offices. In Paris, too, where Gibbs arrived on assignment for the last days of July, the word was '*Incroyable!*'

In England, *Much Ado About Nothing* was opening the summer festival at Stratford-upon-Avon; Wimbledon was under way; there was racing at Goodwood and eights training for Henley. War seemed as stunningly unlikely as the heat on that August Bank Holiday weekend. The Scottish writer John Buchan was moving among the leading lights of the Liberal government, and breakfasted with the Foreign Secretary, Sir Edward Grey, on Saturday, 1 August, finding him 'pale and a little haggard but steadfast as a rock'. Buchan also recalled 'Mr Churchill's high spirits, which sobered now and then when he remembered the desperate issues'.

Winston Churchill, approaching his fortieth birthday, had been the First Lord of the Admiralty since 1911. Britain would not be un-prepared for war on his watch: ammunition dumps and oil depots were guarded, coastal patrols instituted, the First Fleet quietly sent from Portland to the North Sea in case of a sudden German attack. Churchill was playing bridge with F. E. Smith and Max Aitken at the Admiralty around 10 p.m. that Saturday, when a large red Foreign Office despatch box arrived with a small sheet of paper inside bearing a single line of news: Imperial Germany has declared war on Imperial Russia. Churchill rang a bell for a servant, changed out of his dinner jacket and left the room to go and see the Prime Minister. Aitken (the Canadian adventurer who later became Lord Beaverbrook, owner of the *Daily* and *Sunday Express*) remembered him as oddly calm and businesslike. Churchill entered 10 Downing Street 'by the garden gate' and found Asquith with Grey and Haldane and Lord Crewe. He told them he was going to decree full mobilisation of the Grand Fleet of the Royal Navy. From the Admiralty, Churchill wrote to his wife at one in the morning:

Cat – dear – it is all up. Germany has quenched the last hope of peace by declaring war on Russia, & the declaration against France is momentarily expected . . . the world is gone mad . . .

Lord Haldane, deputising at the War Office for the Prime Minister, was described by John Buchan as displaying 'uncanny placidity'; this was exactly what he had been preparing for. As Secretary of State for War from 1905 to 1912, Haldane had created the General Staff, the Territorial Force, the Special Reserve, the Officers' Training Corps in schools and universities, and, in 1907, the British Expeditionary Force (BEF). Britain was the only country in Europe that did not have conscription. Its small yet professional army had already been on 'Precautionary Measures' for a few days, with all regular soldiers recalled from leave.

The first shots were fired between French and Germans at Petit-Croix, near Belfort, on Sunday, 2 August. Imperial Germany declared war on France on the following day. This meant the Germans had campaigns on two fronts: east against Russia, west against France. Because the Germans knew that backward Russia would mobilise more slowly, seven of the eight German armies were dedicated to attacking France first. General von Moltke followed the plan of Count von Schlieffen for his main attack, which was to strike at the heart of France by encircling and seizing Paris. The best way to do this was to drive through the neutral kingdom of Belgium and then wheel most of his armed forces left, to the west of Paris. On 3 August, the Germans demanded free passage through Belgium's territory. King Albert I and his government refused 'to sacrifice the honour of their nation and betray their duty towards Europe'. Germany then declared war on Belgium.

Gunfire in Brussels acted as the starting pistol for the UK. The British government now requested an assurance from the German government that Belgium's wishes be respected. Britain was a signatory to the 1839 Treaty of London that had guaranteed the independence and neutrality of Belgium, and a hostile power just over the Channel in Belgium directly threatened British interests and British shipping. To Bethmann-Hollweg, the German Imperial Chancellor, a treaty about Belgium was just 'a scrap of paper'; but the British said that their word was binding.

'The die is cast,' pronounced *The Times* first leader on Monday, 3 August: 'Europe is to be the scene of the most terrible war that she has witnessed since the fall of the Roman Empire.' John Buchan thought that Bank Holiday Monday was 'the strangest in the memory of man':

An air of great and terrible things impending impressed the most casual visitor. Crowds hung about telegraph offices and railway stations; men stood in the street in little groups; there was not much talking but many spells of tense silence. The country was uneasy.

Sir Edward Grey, the Foreign Secretary, put the case in Parliament and only five MPs voted against war in defence of Belgium. Among them was the dissenting aristocrat and peace campaigner Arthur Ponsonby. The opening chapter of his 1928 bestseller, *Falsehood in War-Time*, concerns what was not talked about in the parliamentary debate: the secret military arrangements that Britain had made with France in 1911 for seven divisions of the BEF to support the French left and for the Royal Navy to protect the French north coast in the event of a German attack. 'This commitment was not known to the people; it was not known to Parliament; it was not even known to all the members of the Cabinet.' Ponsonby argued that Sir Edward Grey's statement was disingenuous. If these contingency plans had been made public, Imperial Germany might have hesitated instead of precipitately declaring war. For Ponsonby, it was 'a deplorable subterfuge' for Grey to insist that Parliament was free to decide.

'What happens now?' Churchill asked the Foreign Secretary as they left the chamber. 'Now,' replied Grey, 'we shall send them an ultimatum.' Grey and Asquith hand-wrote the demand to Imperial Germany between them on the Cabinet table in No. 10 Downing Street. Unless German troops withdrew from Belgium by midnight German time (11 p.m. GMT) on Tuesday, 4 August, Britain would declare war.

Five German armies violated Belgian neutrality around dawn on that day. Although the invasion force of a million men was one of the largest ever seen, Belgian soldiers and the Garde Civique started shooting back. It took eight German divisions finally to reduce Liège by 16 August. Panicky and sometimes drunk German soldiers were so angered by the brave Belgian resistance, so afraid of irregulars or guerrillas without uniforms known as *francs-tireurs* or free-shooters, and so upset by rumours that captured Germans were being mutilated, that they began burning buildings, using Belgian civilians as 'human shields', and bayoneting or shooting them out of hand.

In London, Lord Haldane gave the order to go to war at 4 p.m. on Tuesday, 4 August. The 'War Book' was opened. From the chaotic hive of the War Office buzzed out many terse telegrams: MOBILISE, signed Troopers. As this message cascaded from army to corps to division to battalion, all British army reservists were sent further individual telegrams ordering them to report back to their old regimental depots early the following day. Every soldier and staff-officer worth his salt wanted to be in the BEF and see some action before it was all over.

Many remembered the oddly festive mood at the outbreak of war, with patriotic mafficking and crowds singing 'God Save the King' outside Buckingham Palace. The society portrait painter Solomon J. Solomon, who had recently been in the Palace doing studies of the royal family for a huge painting for the Guildhall, was in a dull committee meeting of the Royal Academy in Burlington House, Piccadilly, late on the evening of 4 August. A dozen silver candelabra with lighted candles shone on the few council members round the table, under portraits by Thomas Gainsborough and Sir Joshua Reynolds. They discussed gallery closures and what to do now that the porters were being mobilised. 'Towards the end of our meeting,' Solomon recalled, 'an eerie distant shouting was heard from a howling mass of presumably young people swaying up St James's Street. It had been announced from Buckingham Palace that we were also to take up arms against Germany.'

In France, Philip Gibbs was chafing at the bit. The *Daily Chronicle* had sent him abroad as a war correspondent, but the French military authorities stopped him getting to the front. He went west to Nancy and saw French lancers trotting through dust and the horse teams pulling batteries of guns along tree-lined avenues. He watched the French infantry marching off towards the Alsace frontier, wearing kepis and bright horizon-blue coats and baggy red trousers, led by their officers with swords and white gloves. Foch's staff ordered him back to Paris. Gibbs was not allowed to see the French army being blown away by German howitzers, nor their conspicuous uniforms riddled by machine-gun bullets.

The previous Monday, a grim Lord Kitchener had been on the return journey to Egypt when he was called back to London. 'Lord Kitchener was more than a national hero,' wrote Violet Bonham Carter, Asquith's daughter. 'He was a national institution.' Herbert Horatio Kitchener was the general who industrialised British imperial warfare. He was summoned to No. 10 for a Council of War on the Wednesday. His view was that the war would not be won by sea-power alone, but by great battles on the Continent. It would last three years, and take manpower in the millions. Prime Minister Asquith asked him to take on the job of Secretary of State for War. Three days later Kitchener made his first appeal for men to join his 'New Armies'. Up went vast posters in places like Trafalgar Square, emblazoned with his moustached face and pointing finger: 'Your Country Needs YOU.'

On Thursday, 6 August 1914, the British Cabinet agreed to send the 100,000 men of the BEF, with Field Marshal Sir John French as commander-in-chief, to the Franco–Belgian border to support the left of the eight French armies, and to face the advancing German right. Protected by the Royal Navy's warships and wireless-fitted aircraft, packed troopships sailed from Southampton to Rouen and Boulogne over the weekend of 8–9 August. Most of the BEF was safely in position in northern France and southern Belgium by the 20th.

They went with no publicity and no press coverage, because, under Kitchener, British censorship became total. The Committee of Imperial Defence drafted the first Defence of the Realm Act (DoRA) giving the government extra coercive and censorship powers ('to prevent persons communicating with the enemy or obtaining information for that purpose'). It became law on 8 August 1914 and was extended or 'consolidated', as the term was, several times during the war. New standing orders forbade servicemen 'to give any military information to press correspondents, military attachés or civilians'.

This news blackout enabled the BEF to move in secrecy without German intelligence also reading about it in the newspapers, but Philip Gibbs recalled how the draconian censorship 'throttled' journalism at a stroke: he was actually on the telephone from Paris to the London office when the line was cut off in mid-sentence. Staff journalists now lived the lives of desperate harried freelances, without accreditation or support, ingeniously improvising ways to get their dispatches through, while trying to evade arrest by both French and British military authorities.

Out in the field, Gibbs palled up with two other correspondents, W. T. Massey, whom he called 'the Strategist', and H. M. Tomlinson, nicknamed 'the Philosopher'. In the first two months of the war, these three covered thousands of miles in France and Belgium by train, bus, taxi, and on foot, grasping at straws, or contemplating defeat:

Yet we went on, mixed up always in refugee rushes, in masses of troops moving forward to the front or backwards in retreat, getting brief glimpses of the real happenings behind the screen of secrecy.

Philip Gibbs, *The Soul of the War* (1915)

Sometimes the reporters had to carry their own copy back to Fleet Street, staying a few hours before crossing the Channel again to France. They wore civilian clothes, had no military passports, and carried bags of money to hire cars at exorbitant prices, to live in hotels and to bribe doorkeepers in the ante-chambers of war. They were still threatened with being shot as traitors. 'Many [journalists] were arrested, put into prison, let out, caught again in forbidden places, re-arrested and expelled from France.' Gibbs was himself arrested five times.

The forward thrust of the German forces caused the Belgian, French and British armies to retreat in the first four weeks. German brutalities panicked thousands of civilians into becoming refugees on the roads, or fighting their way on to trains. H. R. Knickerbocker commented:

Whenever you find hundreds and thousands of sane people trying to get out of a place and a little bunch of madmen trying to get in, you know the latter are reporters.

But the journalists did good work. Lloyd George says in his *War Memoirs* that Kitchener's military briefings were terse almost to the point of unintelligibility, so the first clear news the British Cabinet itself got about the desperate fighting retreat of the British army was from Arthur Moore's report in a special Sunday edition of *The Times* on 30 August 'that had escaped the censor'. The contagion of fear from the war zone was palpable. 'The shadow of its looming terror crept across the fields of France,' wrote Gibbs in 1914, 'though they lay golden in the sunlight of the harvest month.'

In contrast the Germans wanted both their own and foreign newspapers to trumpet their awesome advance. When the veteran American war correspondent Richard Harding Davis witnessed the

German army marching unopposed into Brussels on 20 August 1914, he noticed the disconcerting power of their uniforms to deceive and disguise. *Feldgrau* ('field grey') 'held the mystery and menace of fog rolling toward you across the sea':

All moved under a cloak of invisibility. Only after the most numerous and severe tests at all distances, with all materials and combinations of colors that give forth no color, could this gray have been discovered . . .

After you have seen this service uniform . . . you are convinced that for the German soldier it is his strongest weapon. Even the most expert marksman cannot hit a target he cannot see . . . It is the gray of the hour just before daybreak, the gray of unpolished steel . . . Like a river of steel it flowed, gray and ghostlike.

The *News Chronicle*, London, 23 August 1914

Nothing galvanises the British quite like being despised by Germans. A canny staff officer at British GHQ put the British Expeditionary Force on their mettle by telling them (untruthfully) that Kaiser Wilhelm II had called them 'General French's contemptible little army'. The title 'Old Contemptible' became a badge of honour.

My mother's father, Geoffrey Page, was in that BEF and went off to France with them in August 1914. Second Lieutenant Page, son of the vicar of Mountfield in Sussex and not long out of Sandhurst, was proud, at the age of 20, to be leading fifty or so men of No. 3 platoon, A company, 2nd Battalion Lancashire Fusiliers. When he disembarked in France in 1914 it was his first time abroad and he had never heard a shot fired in anger. For a modern reader, two themes emerge from the diary that he should not have kept: the shock of the new, powerful technologies of twentieth-century warfare – aeroplanes, artillery and machine guns – and the men's need to find shelter from them.

Early on the morning of 26 August, his platoon had barely scraped lying-down trenches in the stubble and stooks of a wheat field north of Longsart Farm near Esnes when German shrapnel broke over their heads, and machine-gun bullets began chopping up their parapets of dirt and straw. His platoon made their stand at the extreme left of the five-mile British line at the Battle of Le Cateau, from which only nine escaped. In the continuing retreat from Mons to the Marne, Geoffrey Page's diary shows an obsession with spies, because being seen and spotted brings down violent retribution.

--

John Buchan's fiction caught the zeitgeist of 1914, the spy mania and invasion paranoia that marked the start of the Great War, especially along the North Sea coastline between Cromer and Dover. About 120 miles away from the gunfire of Le Cateau, across the English Channel in the seaside town of Broadstairs in Kent, Wednesday, 26 August 1914 was also Buchan's thirty-ninth birthday. He was recuperating from an attack of duodenal ulcers and writing a fast-paced yarn about the secret forces and hidden hands behind political events, a book which incorporated his own age into its title, *The Thirty-Nine Steps*.

John Buchan had entered *Who's Who* as an undergraduate at Oxford, been an elite administrator in South Africa, written short stories, essays, poems, history and biography, deputy-edited *The Spectator*, and was now the prospective Unionist candidate for the Parliamentary seat of Peeblesshire and Selkirk. Since 1907 he had also been the literary adviser to the Scottish publishing house of Thomas Nelson and Sons. To keep the presses running, he had agreed to edit and write a weekly illustrated magazine called *The War* (which folded after six months) as well as almost single-handedly researching and writing a monthly partwork, *Nelson's History of the War*, more than a million words of contemporaneous narrative history which ended up as twenty-four red volumes. Buchan (who gave all his profits and royalties from it to war charities) compared himself to Thucydides writing the *History of the Peloponnesian War* in which he himself was taking part.

In this new thriller, his eighteenth book, the real Broadstairs was transmogrified into 'Bradgate' for the dramatic climax. Further up the east coast of England, genteel Frinton-on-Sea in Essex was also in a state of high excitement in early August 1914 when a 15-year-old public-schoolboy called Dudley Wrangel Clarke came back home from the Charterhouse Officer Training Corps summer camp in Staffordshire. Already determined to become a professional soldier, he had not yet developed the talents he would later show as the genius of British deception in WW2. For now the boy was delighted to see soldiers digging up the front for defences and naval destroyers aggressively patrolling the sea, in face of a supposed 'threatened hostile landing' in East Anglia. Further north, Great Yarmouth was full of journalists eager to scoop the story of barges imminently expected from the Frisian Islands, packed with pointy-helmeted Huns, grinding on to British summer holiday beaches.

The Royal Academician Solomon J. Solomon, author of *The Practice of Oil Painting and Drawing*, and a still vigorous and enthusiastic man in his fifties, was spending the summer on the outskirts of St Albans in Hertfordshire. The war fever of that hot August prompted him to consider how art and painting could help hide things from the eyes of enemy Zeppelins in the skies. In his mother-in-law's large garden Solomon began furiously experimenting. He used paints and dyes he had bought as well as mud-pies and crushed leaves, staining sheets of butter muslin that he dried on the lawn and the tennis net. Then he hung the results between plants and shrubs or draped them over bamboo canes by trees and hedges, looking down from upstairs at their colours and shadows as the light slowly changed.

--

In September 1914 Winston Churchill was caught up in his own espionage drama in north-west Scotland. In 'My Spy-Story', published in *Thoughts and Adventures* in 1932, Churchill relates how he went north by special train to the Highlands and was travelling west by car to visit the fleet when the flotilla commodore, who was in the back with the director of Naval Intelligence, pointed out a large searchlight mounted on the turreted roof of a Scottish baronial castle in a deer-forest near Achnasheen. As the car sped on into Wester Ross, they all tried to puzzle out what the device might actually be used for.

At last the road went winding downwards round a purple hill, and before us far below there gleamed a bay of blue water in which rode at anchor, outlined in miniature as in a plan, the twenty Dreadnoughts and Super-Dreadnoughts on which the command of the seas depended. Around them and darting about between them were many scores of small craft. The vessels themselves were painted for the first time in the queer mottled fashion which marked the early beginnings of the science of Camouflage. The whole scene bursting thus suddenly upon the eye and with all its immense significance filling the mind, was one which I shall never forget . . .*

'What would the German Emperor give,' I said to my companions, 'to see this?'

* Is Churchill's vivid image of the ships, written down eighteen years later, an accurate memory, or is it coloured by intervening developments? Mottled or dispersive naval camouflage is not usually recorded as appearing until 1915.

The Admiralty officials discussed the means by which such intelligence might get back to Germany. Submarines were causing much anxiety. After the light cruiser HMS *Birmingham* had sliced through the German submarine *U-15* on 9 August, hundreds of miles from the nearest German naval base, the Royal Navy began to understand these vessels had greater range than anyone had realised. Churchill was anxious that U-boats could be picking up wireless messages:

' . . . suppose a submarine flotilla were lurking about behind some of the islands and suppose a Zeppelin came over and saw the Fleet, couldn't she tell them and lay them on at once?' . . . 'Suppose there was a spy on shore who signalled to the Zeppelin, and the Zeppelin without coming near the bay signalled to the submarines' . . . 'Suppose, for instance, . . . someone had a *searchlight* . . . '

At lunch on Admiral Sir John Jellicoe's flagship HMS *Iron Duke* the subject came up again. Rumours about that shooting estate, involving foreigners and aeroplanes, made Churchill even more determined to investigate. He requisitioned four pistols from the battleship's armoury, just in case 'the searchlight was an enemy signal and a Scotch shooting-lodge a nest of desperate German spies'.

Churchill led an armed and uniformed naval party back to Lochrosque Lodge, and summoned its owner. The former Liberal Unionist MP and Carlton Club member Sir Arthur Bignold was surprised to be called from his dinner to explain that he actually kept a 24-inch searchlight on the battlements only because its beam picked up the gleaming green eyes of the deer on the braes at night so the ghillies and stalkers could locate them more easily the next day for the shooting parties. Churchill found this hard to believe, although it was quite true.

Whatever the cause, the anxiety was quite understandable. No one knew the U-boats' range, and no protective defences against them were ready in harbours like Scapa Flow in Orkney. If anyone had any lingering doubts about the vulnerability of very big ships to torpedoes, the Germans scotched them dramatically in the North Sea on 22 September 1914 when Kapitänleutnant Otto Weddingen in *U-9* sank three British armoured cruisers, *Aboukir*, *Hogue* and *Cressey*, with the loss of 1,460 lives.

From November 1914, Churchill encouraged the development of armed Decoy Ships, 'mystery ships' or 'Q-Boats', to help counter the menace of German submarines on the high seas. The deception was based on observation of their practice. Because the Germans thriftily saved their torpedoes for larger armed vessels, U-boats attacking merchant shipping would usually surface, force the merchant seamen crew to evacuate in their lifeboats and then sink the ship by holing the water line with 37 mm gunfire. Churchill's 'Q-ships' accordingly looked like merchant navy vessels, dingy cargo ships, coasters, colliers or trawlers. They had scruffy civilian crews and they flew the Red (merchant navy) Ensign, but they also carried concealed guns 'which by a pantomime trick of trap doors and shutters could suddenly come into action', as Churchill wrote in *The World Crisis*.

Over 300 Q-ships had gone out by November 1918, and although they claimed only eleven of the 182 German submarines sunk in WWI, their dramatic exploits stirred the British imagination that liked pirate stories. *The Wonder Book of Daring Deeds*, a typical 1930s volume of cheery British imperial propaganda, told how Lieutenant Stuart and Seaman Williams won VCs to match the one their captain, Gordon Campbell, had already won, when their Q-ship *Pargust* was torpedoed by a U-boat off Ireland on 2 June 1917. The 'panic party' rowed away from the ship, but when the enemy submarine surfaced, screens dropped to reveal hidden guns that opened fire.

'Deception, however, was not a British monopoly' says Edwyn A. Gray, completing the story in his book *The U-Boat War 1914–1918*:

[Kapitänleutnant Ernst] Rosenow replied by sending some of his crew on deck with their hands raised in surrender. Campbell immediately ordered the guns to hold fire but suddenly realised that UC-29 was trying to escape under cover of the truce. Once again *Pargust*'s guns blazed, and this time no quarter was asked or given.

--

The twentieth-century surge in camouflage and deception was not just a response to the machinery of new weapons on land, at sea and in the air, but also to the new information technologies which in time of war became dangerous. The secret war therefore aimed from the beginning to destroy or disrupt enemy communications.

Early on 5 August 1914, the crew of the British cable ship *Telconia*, offshore from Emden on the German–Dutch coastline, dealt with five German telegraph cables that ran down the English Channel to France, Spain, Africa and the Americas, grappling them, hauling them up, and chopping the bright wires through their slimy gutta-percha sheathing.

The electromagnetic telegraph had been born in the USA in 1846, but the British Empire was the first to get wired. In 1866, Brunel's ship the *Great Eastern* laid the successful transAtlantic cable that made use of Samuel Morse's code of dots and dashes. By 1870 the UK was linked to Bombay, and the line was extended via Dutch Java to Australia in 1871. The Pacific Cable Board was set up at the beginning of the twentieth century by the governments of Britain, Canada, Australia and New Zealand to provide telegraphy within the British Empire.

Thirty years after Morse's telegraph, Edison's improved telephone arrived, which still needed wires; another twenty years later came literally 'wire-less' communication: the magic of radio. The German physicist Heinrich Hertz had worked out the principle of radiation and was the first to produce electromagnetic waves artificially, but an Italian physicist, Guglielmo Marconi, continued his experiments in Britain, aiming to set up the world's first permanent wireless station. By July 1900, he had installed radio apparatuses on a Royal Navy cruiser and a battleship so successfully that the British Admiralty soon contracted for dozens more warships to be fitted.

Radio/wireless signalling – and interception – play an important role in WW1. The first British Empire soldier to fire a shot against the Germans, on 12 August 1914, was Regimental Sergeant Major Alhaji Grunshi of the Gold Coast Regiment, during the campaign to silence the German wireless station at Kamina in Togoland. This station linked Germany to Windhoek in German South-West Africa, to Dar-es-Salaam in German East Africa, to German shipping in the South Atlantic and to German agents all around South America. All the German radio stations across the Pacific – Yap, Apia, Rabaul, Nauru – were silenced by British Empire forces in August and September 1914.

Germany's response to this communications war was to slice through Britain's global cables wherever their longest stretches were hardest to repair. On 7 September 1914, a three-funnelled warship, flying a friendly French flag, dropped anchor just off the north-west corner of Fanning Island, a low coral atoll in the middle of the Pacific Ocean. Two boatloads of men rowed swiftly ashore. The German sailors and marines of the Imperial German Navy light cruiser *Nürnberg* pulled out pistols and machine guns and began to wreck the cable station under the coconut palms. They harmed no one, but a demolition crew blew up the generators and accumulators. Axes smashed the control-room instruments and liberated ammonium chloride from the cells. The landing party also looted all the gold sovereigns from the superintendent's safe, where they found Alfred Smith's treasure map showing where he had hidden the spare instruments and the Fanning Island Volunteer Reserve's arms and ammunition. These were duly dug up and destroyed. Meanwhile, *Nürnberg*'s companion ship, the Bremen class light cruiser *Leipzig*, had hauled up and severed the Fanning–Canada cable and dragged the long end out to sink in deep water. Then they cut the Fanning–Fiji cable, but dragged it out only as far as the shallow reef where, luckily, it could quite soon be dredged up and reconnected by the British.

Another German light cruiser from China, the *Emden*, headed west into the Indian Ocean and became the most famous raider of the war, causing millions of pounds worth of havoc by shelling the oil tanks at Madras, attacking Penang harbour and capturing and sinking merchant ships. Her captain, Karl von Müller, was adept at deception. By fitting a fake fourth smokestack and flying a Royal Navy ensign, *Emden* sometimes passed as the British cruiser HMS *Yarmouth* until she got within range and revealed her true colours.

On 9 November 1914, *Emden* carried out another cable-cutting raid. This time the mission was to sever the Indian Ocean telegraph connection between South Africa and Australia at one of its junctions, Direction Island in the Cocos (Keeling) Islands. Again, a landing party came ashore, harmed no one, smashed all the Morse machinery with axes and severed two cables. But there was an additional task in the Cocos: to blow up the wireless tower, because by this time shore-to-ship and ship-to-ship radio was helping the Allies to coordinate their tracking of the enemy. (*Emden* always kept radio silence, but listened carefully to all other wireless traffic, judging distance by strength of signal only.) The wireless demolition job was done remarkably politely; the Germans agreed not to drop the tower across the only tennis court. But they were just too late. The men of the Eastern Telegraph Extension Company had managed to transmit a signal that John Keegan has described as 'perhaps the earliest ever piece of real-time intelligence of the electronic age'. This wireless message, 'strange ship in entrance', reached an Australian convoy two hours away, which sent the cruiser HMAS *Sydney* to investigate. *Emden* had to up anchor and run for it; outgunned by *Sydney*, she was scuttled and shelled on a reef, with half her crew killed or wounded. The other two cable-cutters, *Nürnberg* and *Leipzig*, were both sunk on 8 December 1914 in the Battle of the Falkland Islands.

Britain was determined to rule the airwaves. The German navy or *Kaiserliche Marine* talked to its vessels at sea by radio, using codes. German U-boats communicated with their control centre at Wilhelms-haven, hundreds of miles away, on the 400 metre waveband. On 2 August 1914, the British government had taken 'control over the transmission of messages by wireless telegraphy', closing down amateur and merchant marine use. In the late summer of 1914, the Royal Navy's Director of the Intelligence Division of the Naval Staff (DID), Rear Admiral Henry Oliver, raised to fourteen the number of radio intercept stations along Britain's east coast. Their task was to monitor all the German *Hochseeflotte* signals traffic and to supply useful information to the Admiralty. These stations were staffed by ex-General Post Office (GPO) engineers and by 'ham' radio operators whose private sets had been banned. From 1915 onwards, the British employed the new technique of wireless direction finding (DF) to locate German transmitters and to intercept their radio transmissions. The principal site for this was Hunstanton Coast Guard Station on the Wash in Norfolk, which could 'tap the air' in Flanders and northern France as well as the North Sea, and sent non-naval military infor-mation about the Western Front to the War Office and thence to the Intelligence Section of GHQ France.

Rear Admiral Oliver also asked the director of Naval Education, Sir Alfred Ewing, to set up a department to break the codes and ciphers that the German navy were using. Ewing hired the cleverest teachers and academics he knew. One of them was the hockey-playing Scotsman Alastair Denniston, who had been teaching German at the Royal Naval College at Osborne on the Isle of Wight. Denniston became the great cryptanalyst who later headed the Government Code and Cipher School (GC&CS, nicknamed the Golf, Cheese and Chess Society, the forerunner of GCHQ) and who started Bletchley Park.

In the decoding and deciphering game, the British got very lucky very soon. By the end of 1914, the Royal Navy possessed the three principal codebooks of the Imperial German Navy, one obtained in Australia, one captured by Russian allies, and one trawled up in a

British fishing net from a sunken German destroyer off the Dutch coast. Foolishly, the Germans did not change their signal books, and so in Room 40 of the Old Building of the Admiralty in London, British Naval Intelligence, by stops and starts, began reading the enemy messages that would help them to win the war. It was as First Lord of the Admiralty in WW1 that Winston Churchill first got the taste for Signals Intelligence (SIGINT) that would be so important to him as Prime Minister in WW2.

--

The Duke of Wellington once said: 'All the business of war, and indeed all the business of life, is to endeavour to find out what you don't know by what you do; that's what I called "guessing what was at the other side of the hill".'

If you get high enough, you can see right over the other side of the hill. Since 1783, when Frenchmen pioneered the new technology, balloons had offered the tantalising chance of doing this. British military ballooning began in 1878, first used in the field by the Royal Engineers Balloon Company, who kept an observer aloft for seven hours in the 1885 campaign against the Mahdi in the Sudan. In the Boer War, sappers took reconnaissance photographs from balloons, and in May 1904 they first transmitted and received wireless communications while aloft. By 1914, all kite-balloons that were fastened to the ground belonged to the Royal Flying Corps (RFC) and the ones tethered to ships to the Royal Navy Air Service (RNAS). Spotters in the navy kite-balloons at the Dardanelles in 1915, for example, helped to direct naval gunfire and to keep an eye out for enemy submarines.

Captive 'sausage' balloons used for observation were a familiar sight along all the front lines from Macedonia to Belgium. In November 1914, the historic cloth hall and cathedral of Ypres were destroyed by explosive shellfire directed from German observation balloons. British observers in balloons were linked by telephone to field batteries below, to HQ and to the wider world. In a quiet spell in 1917, one observer gave his private number to the operator and found himself, a quarter of an hour later, high above France talking to his wife in north London. Naturally, observation balloons became targets too. When the official war artist C. R. W Nevinson went up in a kite-balloon above the Western front one night in 1917 to observe and sketch the flashes of the guns in the darkness, he was strapped to a bulky parachute, ready to jump should incendiary bullets ignite the hydrogen in the 'gas-bag' above him. (Nevinson was the first British artist to paint the aerial view. His lithographs of 1917, *In The Air* and *Banking at 4000 Feet*, catch the queasy yaw of looking down on patchwork fields from a small open aeroplane.)

The Germans took balloon technology forward by investing heavily in Count Zeppelin's rigid inflatable airships, but the French also led the way in the development of aeroplanes for use in war. By the summer of 1911, the French air corps had over 200 aeroplanes, and their coordination with the cavalry, infantry and artillery on the ground in the military manoeuvres at the Camp de Châlons impressed foreign observers. By 1914, the Imperial German Air Service, the *Luftstreitkräfte*, had 246 aircraft and eleven airships compared to Britain's 110 aircraft and six airships. The British may have come later to the game, but then they forged ahead. The RFC was formed in spring 1912 with only eighteen machines, but with the aim of creating seven 'squadrons' of a dozen planes, plus an airship/kite squadron in its Military Wing. The earliest, flimsiest aeroplanes of the RFC had no integral armaments because their purpose was reconnaissance. The leader of the RFC in 1914 was Brigadier General Sir David Henderson, who had been staff captain (Intelligence) in the Sudan campaign and director of Military Intelligence in the second Boer War. David Henderson had himself learned to fly in 1911, the year when aeroplanes first proved their value in both the French and German Army manoeuvres, and became, at 49, one of the oldest pilots in the world. He wrote an important primer on information gathering, *Field Intelligence*, in 1903 and published *The Art of Reconnaissance* in 1907.

During the British Army military exercises or manoeuvres of September 1912, aircraft of the RFC proved invaluable in reconnaissance over Norfolk. Jimmy Grierson, defending Thetford, had the army airship *Gamma* communicating to him (by wireless, from up to thirty-five miles away) all the daylight movements of Douglas Haig's division, attacking from the east, trying to move under the cover of roadside hedges. They became more rather than less conspicuous to the aerial spotters when they tried a primitive sort of Birnam Wood camouflage, covering wagons and guns with branches of trees.

The third edition of Henderson's *Art of Reconnaissance*, published in May 1914, contained a whole new chapter on 'Aerial

Reconnaissance'. Henderson predicted that the new aircraft would make it impossible to prevent enemy surveillance. Aerial spotting would lift the fog of secrecy from strategic moves and make commanders more cautious, because surprise would be harder to achieve. Henderson could foresee the air arm completely superseding the cavalry. Churchill too saw the appeal of the air and its freedom. He first flew as a passenger in 1912, and many hundreds of flights would follow. He encouraged the Naval Wing of the RFC to pioneer wireless telegraphy in airships, and to detect submarines from the air. Relishing the Royal Navy tradition of attack, Churchill foresaw a far more aggressive role for aircraft than Henderson's idea of intelligence gathering and reporting.

When early pilots on opposing sides met in the air, they either ignored each other, or saluted in a display of *Brüderschaft*. Then manners broke down. According to John Masters, they first began throwing objects like bricks at each other, then using small arms. Once this started, clearly they had to kill each other, and fitting machine guns with an interrupter gear to shoot through the revolving propeller was a logical development.

Quite soon, single combat in the air became epic. In June 1915, a week after the Germans first dropped bombs on London, a monoplane flown by Flight Sub Lieutenant Reginald Warneford of the RNAS almost collided with Zeppelin airship LZ37 over Bruges, as it was returning to base after fog had prevented it from raiding England. In the ensuing fight, 'Rex' Warneford used a technique that had first been recommended by Churchill. Driven off by the Zeppelin commander's gondola guns, Warneford forced his Morane-Saulnier Parasol higher and higher, then dived down from 11,000 feet into a hail of bullets from the Zeppelin's roof gun until he was a hundred feet above the grey back of the dirigible. Straightening out, he jerked his bomb-releases and planted all six of his twenty-pound bombs along the Zeppelin's length. The ensuing explosions blew his plane, upside down, hundreds of feet up in the air (where only his safety straps prevented him falling out), and silenced his engine. He managed to glide down, land in enemy territory, fix his fuel line with a cigarette holder and a handkerchief, take off again and coast with an empty tank until he could crash-land near Cap Gris Nez. Meanwhile, the blazing German airship had crashed on to the convent of St Elizabeth at Ghent, killing two nuns and an orphan. The crew's only survivor was Steuermann Alfred Mühler who jumped from the flaming gondola and smashed through an attic skylight on to a feather bed. Fragments of the first wrecked enemy airship became secret souvenirs for patriotic Belgians. However, one lump of scorched metal ended up as a paperweight on the desk of the First Lord of the Admiralty, Mr Winston Churchill.

Warneford, the first pilot to 'spike-bozzle' (completely destroy) a Zeppelin in the air, became the hero of the hour. He received a telegram from Buckingham Palace: 'I most heartily congratulate you upon your splendid achievement of yesterday in which you single handed destroyed an enemy Zeppelin. I have much satisfaction in conferring on you the Victoria Cross for this gallant act. GEORGE R.I.'

But ten days after his great victory, feted and honoured by Paris society and competed over by French women, 24-year-old 'Rex' Warneford VC fell to his death near Versailles when the Henri Farman F-27 two-seater reconnaissance bomber he was flying (without safety harness) rolled over in a steep turn and broke up in mid-air. Warneford was still alive when they reached his face-down body. The enamelled insignia of his *Chevalier de la Légion d'Honneur* was driven deep into his left side.

--

Even reconnaissance was not as easy as it sounds. On their first wartime recce in France, on Wednesday 19 August 1914, two RFC pilots lost their way, and each other, in cloud. One flew over Brussels without knowing what city it was; the other had to land twice and get directions from a gendarme. Tangible evidence of the enemy first came back on 22 August in the form of a rifle bullet in the bloody leg of an airborne observer called Jillings. New observers were not always sure what it was they were seeing. At Ypres in 1914, some reported tarmac as troops on the move, and gravestones as bivouacs. By trial and error, it was found that 6,000 feet gave a good view, almost at the limit of rifle fire from the ground.

From the 1890s the ballistic charge of artillery ammunition had been changing, so the reconnaissance fliers of 1914 no longer saw the distinctive and conspicuous clouds of white smoke that used to billow from big guns. The greater power and range that this smokeless propellant gave to the field guns meant that by 1914 they did not have to be fired 'over open sights', directly looking at the target they were shooting at, but could be kept miles back under cover, their 'indirect fire' guided by forward observation officers linked by telephone. This technique had begun in the 1860s in the American Civil War, when spotters with flags in fixed Union balloons overhead had helped gunners below to shell Confederate positions accurately.

The introduction of mobile reconnaissance aircraft now put many more guns at risk. An airman could see an enemy battery and then drop a smoke bomb to mark the place. Friendly spotters would then work out the range with a telemeter and direct the artillery to shell the place. If the aeroplane was fitted with wireless, map grid-references could be radioed back. Here are the wireless messages sent down from a Royal Flying Corps plane spotting for British gunners on the ground shelling an enemy gun battery on 24 September 1914:

4.2 p.m. A very little short. Fire. Fire.

4.4 p.m. Fire again. Fire again.

4.12 p.m. A little short; line O.K.

4.15 p.m. Short. Over, over and a little left.

4.20 p.m. You were just between two batteries. Search two hundred yards each side of your last shot. Range O.K.

4.22 p.m. You have them.

4.26 p.m. Hit. Hit. Hit.

4.32 p.m. About 50 yards short and to the right.

4.37 p.m. Your last shot in the middle of three batteries in action; search all round within 300 yards of your last shot and you have them.

4.42 p.m. I am coming home now.

The fear of spotters is a theme of John Buchan's WW1 novel *Mr Standfast* (1919). Buchan's old Afrikaaner scout, Peter Pienaar, joined the RFC and developed a genius for air combat. 'He apparently knew how to hide in the empty air as cleverly as in the long grass of the Lebombo Flats.' The climax of the book, during the huge offensive by the Germans in April 1918, is Pienaar's desperate and glorious flight in a little Shark-Gladas to try and stop the enemy air ace's aeroplane getting back to the German HQ with the news of a glaring weakness in the Allied line, 'the knowledge which for us was death'.

More alarming than the spotter was the camera. Aerial

Photographic Reconnaissance (APR) yielded harder evidence than observers' impressions and anecdotes. APR helped construct a scientific record of terrain from overlapping oblique and vertical pictures, which could be scrutinised in detail and matched to the map. The French, first to take photographs from a balloon in 1858, were soon photographing German positions in WW1. Rudyard Kipling, visiting French troops in his 1915 report *France at War*, noted 'the Intelligence with its stupefying photo-plans of the enemy's trenches'.

In the 1914 edition of *The Art of Reconnaissance*, Henderson did not mention photography because it was then a new and secret application of the technology of flight but, in a single day in 1914, British officers had managed to photograph all the defences of the Isle of Wight, Portsmouth and the Solent from a height of 5,600 feet, and to develop the negatives in the air so they were ready for printing upon landing.

The RFC's Air Photograph section was officially founded in January 1915. When Douglas Haig launched the first British offensive at Neuve Chapelle two months later, the artillery, infantry and aircrews were all working to the same map of the German trenches, prepared from the 4" x 5" aerial photographic plates taken by the Thornton-Pickard box cameras specially built for the RFC biplanes. This attack failed like so many others, not because people did not know where they were, but because they could not talk to each other; telephony had not managed to keep up with telemetry. By the end of the war, hundreds of aerial cameras had made nearly six million black and white prints for distribution, and the Allies reckoned no German could deepen a trench without it being known about.

It was the paramount need to deceive eyes in the skies that led to the rise of camouflage.

# 2

# The Nature of Camouflage

Camouflage does not feature in the famous Eleventh Edition of the *Encylopaedia Britannica* in 1910–11, but the cataclysm of the Great War taught everyone about it. The Twelfth *Britannica* in 1922 had illustrated articles on the subject, including one by the marine artist Norman Wilkinson, who had devised a startling way of deceiving the eye about ships at sea. The word 'camouflage' itself is French, and was said by Eric Partridge to derive from the Parisian slang verb *camoufler* meaning 'to disguise', or perhaps from the Italian *camuffare*, derived from *capo muffare*, 'to muffle the head'. 'Camouflage' entered the English language during WW1, and the *Oxford English Dictionary*'s first example of published usage is from the *Daily Mail* in May 1917: 'The act of hiding anything from your enemy is termed "camouflage".'

There are two stories about the first use of camouflage in 1914, and both are linked to artillery, artists and aircraft. A 43-year-old Parisian portrait painter called Lucien-Victor Guirand de Scévola, serving as a second-class gunner, was said to be the first person to think of covering artillery with painted sheets so the enemy could not spot them. His motive was to save lives; after a German shell hit his battery in the open and wounded his companions badly he thought he might be able to stop it happening again by blurring the shape and colour of the guns. In another story, a German aeroplane dropped a bomb on a battery of the 6th Regiment of Foot Artillery at Toul, west of Nancy, killing and injuring some friends of the painter Louis Guingot and Sergeant Eugène Courbin. Courbin had some canvas sheets, edged with eyelets, run up in the workshops of the Associated Stores in Nancy where he used to work as an administrator. These splotch-painted tarpaulins were then stretched over the guns and lashed to poles. Meanwhile, Guingot got hold of some capes or cagoules which he dyed yellowish and then daubed with green splashes outlined in black for the artillerymen to don over their bright blue uniforms.

All this was approved by their commanding officer, Colonel Fetter, who put it to the test by having his aviator son fly over a newly 'camouflaged' battery at a height of 300 metres. The pilot dropped a message saying he could only spot the five men deliberately left wearing their regular bright blue uniforms. Colonel Fetter then got Courbin and Guingot, together with Henri Roger and Eugène Renain, to start making covers and capes to help conceal the 120 and 150 mm guns and their crews, as they moved up to Metz.

From October 1914 onwards the war on the western front became bogged down in trenches. The war of the future became a grotesque return to medieval siege warfare, and as the armies rooted into the mud the first shoots of true camouflage began to show. You could, in the traditional way, stack wicker fences around the guns to hide them from view but that was not really camouflage, just screening, like those 'masked batteries' of field guns that shelled the Martian machines in H. G. Wells's 1898 thriller, *The War of the Worlds*. Camouflage was a new art that painters would help pioneer.

The French Army's *section de camouflage* at Amiens received official status on 12 February 1915, with the immediate priority of disguising guns and gunners from enemy view. Guirand de Scévola was promoted to lead it, with the 63-year-old Impressionist Jean-Louis Forain as its first Inspector General. General Joffre gradually expanded camouflage by attaching workshops to each army corps, not only in Amiens, but also in places like Arras, Bourget, Châlons-sur-Marne, Chantilly, Epernay, Nancy and at 34 rue du Plateau, Paris.

The *section de camouflage* started with thirty officers and seven men drawn from the worlds of theatre, painting, sculpture and design. All the artists wore a white-and-red armband or brassard embroidered in wire with their unit's badge: a silver chameleon, the slow-moving African lizard whose pebbly-looking skin can mottle and alter colour. Chameleons' swivelling, binocular eyes, set on each side of the head, help them to calculate the range of their insect prey and to shoot out a lightning-fast sticky tongue that snatches it out of the air. Chameleons seem to combine the skills of *camoufleurs*, spotters and gunners.

The art of camouflage in WW1 follows the basic principles of survival of the fittest in nature. Camouflage confers strategic and tactical advantage in the arms race between seeing creatures. Both predators and prey can deceive each other by colouring that blends them into the background, or by patterning that disrupts conventional outline. Seeing and not being seen are matters of life and death. When a Cabbage White butterfly settles on a striped green-and-white cornus bush, its folded wings perfectly match the ragged pale edges of the leaves. In *My Early Life*, Winston Churchill says that of the magic-lantern lectures that he could remember from Harrow School, two were on battles, one on empire, one on geography, but one was 'about how butterflies protect themselves by their colouring'. Although many animals and birds deceive, *Homo sapiens* succeeded as a species because human individuals used their intelligence to outwit each other, as well as other species, in competition for resources. The decoys and disguises of the early hunters helped to deceive their prey. An early student of this was Abbott H. Thayer (1849–1921), an eccentric autocratic artist from Monadnock, New Hampshire. His article 'The Law Which Underlies Protective Coloration' in the ornithological magazine *The Auk* in April 1896 stated that camouflage was a matter of depth perception, as well as surface colouring. Thayer painted naturalistic *trompe l'oeil* pictures which made people think a two-dimensional canvas was solid and three-dimensional. He came to realise that nature sometimes did just the opposite.

In the real world, the contours of an object in relief are shown by brightness on the side facing the light source, and shadow on the other side. Thayer was the first to observe that many animals had what he called 'countershading': darker colours above, on their backs facing the sun, grading to lighter colours below, on the belly. Because these arrangements worked against the usual visual expectation – lighter above, darker below – they made a round-bodied creature appear flat against its background. The British biologist Edward Bagnall Poulton was the first to note in the 1880s how the white spots stippling the darker shaded side of a Purple Emperor butterfly chrysalis made it seem as flat as a leaf, though he did not then grasp that it was a general principle right across the animal kingdom.

In 1909, Thayer's son, Gerald, published a groundbreaking book, *Concealing-Coloration in the Animal Kingdom, An Exposition of the Laws of Disguise through Color and Pattern: Being a Summary of Abbott H. Thayer's Discoveries*. Abbott Thayer's introduction was grandiose and combative. Zoologists and naturalists could not understand what painters saw: 'The disguising patterns worn by animals . . . are, in the best sense of the word, triumphs of art.' Both Thayers were over-insistent that all animal camouflage was about the organism's 'obliteration' in its environment, and neither had read or understood *The Origin of Species* where Darwin specifically linked bright plumage to sexual selection rather than disguise. Nevertheless, it is still a remarkable work, an important urtext for *camoufleurs* before the actual word existed, and a manifestation of ideas that were also expressed in contemporaneous new movements in art. Thayer's pointing out of what he called 'ruptive marks', for example the black and white bands that break up and disguise the outline of a ringed plover nesting among pebbles, occurred at just the same time as artists like Georges Braque across the Atlantic in Paris were beginning to disrupt surface resemblance and the single viewpoint with cubism.

Avant-garde painters certainly influenced the French military during the Great War. Writing not long after the armistice, Guirand de Scévola said it was the Cubists who sprang to mind when he first thought about disguising the form of the guns. Violent techniques that Braque and Picasso had used to distort figure and ground, and jam together different viewpoints and perspectives to show things in new lights, could also be used to alter the look of objects so they could not be recognised.

A photo in *The War Illustrated* (3 July 1915) headlined 'Hide and Seek with Heavy Artillery' shows disruptive patterns on artillery, and is captioned: 'The latest *ruse de guerre* of our ingenious ally. French gunners painting '75s' the colour of the landscape, to form an effective disguise from inquisitive aircraft.'

Picasso, who once defined cubism as not painting what you see but what you know to be there, recognised camouflage as his bastard child. Gertrude Stein recalled being with Picasso on the Boulevard Raspail in Paris one night in 1915 when one of the first camouflage-painted heavy artillery pieces was hauled past them. Picasso looked amazed at the cannon with its blocks of disruptive pattern and then cried out, '*C'est nous qui avons fait ça!*' ('We invented that!) Later, in Alsace, Stein noted how culturally specific camouflage was:

Another thing that interested us enormously was how different the camouflage of the french [*sic*] looked from the camouflage of the germans, and then once we came across some very neat camouflage and it was american . . . The colour schemes were different, the designs were different, the way of placing them was different, it made plain the whole theory of art and its inevitability . . .

In *Technics and Civilization* (1934) Lewis Mumford put forward the idea that the invention of the photographic camera had made humans more extrovert, so people began posing for the shot or acting for the motion picture, as if they were all constantly on stage or up on the screen. He thought this technological shift, from the self-examination of the mirror to the self-exposure of the camera, put a premium on presentation. Apply this to war, and the front you need to present to face the enemy, and one can see the new possibilities of a false presentation, of deception.

This was exactly how the British understood the idea of camouflage by the end of the Great War: '*Deception*, not *concealment*, is the object of camouflage' stated the official pamphlet on its principles and practice. The 1921 *Manual of Field Works (All Arms)* defined camouflage as 'the art of concealing that something is being concealed. Its keynote is deception.' Popular usage during the later war years came to reflect this: a half-lie or prevarication would be 'camouflaged truth'; a doubtful patriot a 'camouflaged Hun', and so on. In journalism, an article placed with false information for the enemy was referred to as 'a camouflage'.

Although the British practice of camouflage got official status in 1915, it took until 1917 for the word 'camouflage' to begin entering the public domain. Because it was 'official' in wartime it was therefore an official secret – protected by DoRA – and so the art of hiding things had to be censored to keep it from all those who might be interested in what was being hidden.

When George Bernard Shaw was visited for the last time by G. K. Chesterton's younger brother, the pugnacious political journalist Cecil Chesterton, he was dressed in khaki, 'a deeply sunburnt, hopelessly unsoldierlike figure':

The word camouflage was in everyone's mouth then; and . . . my unruly imagination instantly presented me with a picture of Cecil camouflaging himself as a beetroot on a sack of potatoes by simply standing stock still.

--

The British military took the word camouflage from the French painters, and then British painters helped their own military to enact the idea. Foremost among them was Solomon J. Solomon, the portly Royal Academician whom we last saw spreading coloured muslin sheets over the shrubbery of his mother-in-law's garden at the outbreak of war. Solomon got a persistent bee in his bonnet about the subject and became indefatigable in proselytising for it.

Solomon Joseph Solomon was born in the Paragon, an elegant part of Blackheath, south London, on 16 September 1860, fifth of the twelve children of Joseph Solomon and his cultured wife Helena Lichtenstadt from Prague. He was a lively and brilliant boy who believed in numerology: his luck apparently ran in sixes. Solomon's great-grandfather had been a silversmith in Amsterdam and they were a respected Anglo-Jewish family who went to the synagogue but were only moderately orthodox. Because of the second commandment prohibiting 'graven images', it was not usual at that time for a Jew to become a painter by vocation, but at 17, Solomon was enrolled in the Royal Academy Schools in the basement of Burlington House. Among his teachers, John Everett Millais was particularly kind to him. There had only ever been one Jewish Royal Academician: Solomon Hart of Plymouth, elected in 1840. But Solomon was determined to be another. And why not, if he were good enough? After all, Benjamin Disraeli was Jewish-born, and he had risen to become Prime Minister, twice. Being Anglo-Jewish meant putting the emphasis on the 'Anglo', not being incongruous. Integration was a kind of social camouflage.*

After travelling through Europe and to Tangier, studying and looking at pictures, S. J. Solomon started painting noble and romantic works of art himself, first making his name in 1887 with the dramatic

---

* When Sir Samuel Montagu, a patron of the arts and benefactor of the Jewish community, received a letter signed 'S. Solomon' asking for his help because the writer was in some distress, he hurried to help the young student. He was received, however, by Simeon Solomon, not young Solomon Solomon. Simeon was a superb pre-Raphaelite artist but he was also a gay man who had been arrested in a public lavatory in 1873 and charged with committing buggery, and finally died as an alcoholic indigent in St Giles's workhouse. After this visit, Montagu sharply advised Solomon to sign his letters Solomon J. Solomon. (Many years later, Solomon's daughter would marry Montagu's grandson.)

painting *Samson* (now in the Walker Art Gallery, Liverpool) which depicts a wild-eyed muscle man being restrained by brawny Philistines as semi-naked Delilah brandishes his chopped-off hair. Solomon J. Solomon's obituary in *The Times* said that his work in this vein 'suffered from the two tendencies, the sensational and the sentimental'. His obituarist thought portrait painting was his real gift, because of Solomon's 'wide knowledge of humanity'.

He was certainly jovial and clubbable. A stalwart of the convivial Savage Club, he was also a member of the New English Art Club, the Royal Institute of Oil Painters, the Art-Workers' Guild and the Royal Society of Portrait Painters, and later became President of the Royal Society of British Artists, saying: 'I feel I ought to accept because I am a Jew.' He was a good singer and dancer, a canoeist and an avid horseman.

A convinced Zionist, he was the first president of the Order of Maccabeans, an association of English Jewish professional men, as well as a member of the Jewish Territorial Organization, which was dedicated to finding a homeland for the Jewish people after the first scheme to settle them in Uganda had failed. And yet Solomon embraced Englishness. In 1906, one of his lucky years with a six in it, the artistic establishment of the Royal Academy elected him Academician. Solomon was 'R.A.'d in all his glory' after painting a picture of England's national saint, St George.

Approaching his fifty-fourth birthday at the time WW1 broke out, Solomon was a well-established and ambitious painter beginning to get portrait commissions from the rich and powerful, and from large institutions like the Houses of Parliament. He had recently been to Buckingham Palace in July 1914 to paint oil-on-panel preliminary portraits of King George V, Queen Mary and Edward, Duke of Windsor for a huge canvas depicting the 1910 Coronation Luncheon held at the Guildhall.

When war broke out, Solomon signed up as a private soldier in one of the first volunteer corps for home defence, formed from older men in the arts world who could not join the Regulars or the Territorials. The United Arts Rifles had the playwright Sir Arthur Pinero as their chairman and were nicknamed 'The Unshrinkables' from the white jerseys that were their first drill uniforms. Solomon designed their badge, a dove flaring on to a sabre (which the irreverent dubbed 'the duck and skewer'), and got them permission to drill in the courtyard at Burlington House in Piccadilly, as well as the right to use half of the galleries at the Royal Academy. The refreshment room doubled as the United Arts Rifles' mess and the store for their elderly Japanese rifles, although they were forbidden to leave ammunition on site.

Solomon first staked a public claim to the as-yet-unnamed field of camouflage by the very English expedient of writing a pompous letter to *The Times*. It appeared on Wednesday, 27 January 1915, under the headline 'Uniform and colour':

Sir, – The protection afforded animate creatures by Nature's gift of colour assimilation to their environment might provide a lesson to those who equip an army; seeing that invisibility is an essential in modern strategy. To be invisible to the enemy is to be non-existent for him. Our attempts in this direction might well be a little more scientific. A knowledge of light and shade and its effect on the landscape is a necessary aid to the imagination of a designer of the uniform in particular, and the appurtenances of war in general.

Solomon had clearly read Abbott Thayer on counter-shading because his letter criticises the sameness of uniforms: '. . . the khaki tunic is good in summer – in winter it is too yellow – but the same colour cloth clads the whole man. Here a knowledge of light and shade comes in.'

Solomon suggested darkening soldiers' caps and shoulders and lightening their trousers and gaiters, and questioned why uniforms had to be so uniform. If in each section the colour of the tunic or coat varied between the excellent winter blue of the Guards' greatcoat, a grey-green, and the present khaki, a broken effect of colouring would be obtained with advantage. He warned of the danger of shape and shadow: 'The cap now worn detaches the men in this way from almost any setting and affords a most excellent target for the enemy marksman.'

He suggested new forms of colour assimilation:

The artillery officer is covering his gun with grey tarpaulin, but with a team of six or eight horses in front of it, the airman is not likely to mistake it for a butcher's cart. The horses have merely to be covered with a thin grey-green stuff to make them equally inconspicuous. Wagons are a leaden grey, unlike anything in nature; a warm dust colour would be more harmonious. A similar observation applies to warships. The North Sea is almost invariably a pearly green, and experiments with models should evolve something more subtle than their metallic hue.

He ended by proposing that painters and artists, 'the makers of the arts of peace', could be useful to 'the designers of the munitions of war'.

--

From the earliest days of organised human fighting, elaborate head-dresses, shiny armour and shields, garish warpaint and costumes were designed to alarm the enemy, like the threat displays of other non-human animals. The advance to close quarters of massed units – Roman legionaries in *testudo* (tortoise) formation, covering them-selves in a hard shell of shields, or deep-singing, shield-clashing Zulu impis or British heavy infantry in red coats and bearskin busbies – was intended to strike fear into enemies or panic them into running away. According to Philip Mansel's book *Dressed to Rule*, military uniforms spread through Europe between 1650 and 1720, designed, among other things, to instil

discipline, courage and *esprit de corps* . . . to impress spectators . . . to inspire fear in the enemy; and, as innumerable recruiting posters show, to attract young men to enlist.

It was the development of accurate guns that did most to cause exuberant brightly coloured uniforms to give way to the familiar drab tones of the modern soldier. In the late eighteenth century special forward units of scouts, skirmishers and sharpshooters like the American Rangers, the German *Jägerbataillon* and the British Rifle Corps had already begun to wear some kind of green to maximise their cover. Guns, 'glorious products of science', stirred life in remote places from traditional torpor, according to Winston Churchill in *My Early Life*:

The convenience of the breech-loading, and still more of the magazine, rifle was nowhere more appreciated than in the Indian highlands. A weapon which could kill with accuracy at fifteen hundred yards opened a whole new vista of delights to every family or clan which could acquire it. One could actually remain in one's own house and fire at one's neighbour nearly a mile away. One could lie in wait on some high crag, and at hitherto unheard-of ranges hit a horseman far below.

When the British soldier's white helmet or pipeclay belt became the target of native musketry, 'Tommy Atkins' began to stain his accoutrements with *chai* or tea. Khaki first appeared in the Indian Army and the word is derived from *khak*, the Urdu and Persian word for 'dust-coloured'. Harry Lumsden's famous Corps of Guides, one of the irregular Indian forces raised by the British in the Punjab in 1846 and used for scouting and intelligence gathering, was the first unit to wear khaki-coloured uniforms, though during the hot-weather fighting in India in 1857 many British soldiers began to dye their summer-wear unlined white cotton tunics and trousers with tea, earth and curry powder.

The Indian Mutiny was a key stage in the transformation of European field uniforms from symbolic display to aids to concealment. By 1885 stout twilled cotton khaki drill was universal in the Indian Army and the British Army in India, and in 1896 sandy brown khaki (both in cotton and serge) was approved for British Army foreign service outside Europe. The South African War of 1899–1902 against the Boers, whose own homespun clothing was coloured like the land they fought over, permanently convinced the British that bright pillar-box red was best kept for the parade ground. Muddy field-manoeuvres needed dingier or dungier battledress, though they never got the colours quite right: Kipling described the colour of British WW1 khaki as 'gassed grass'. After colonial wars in Cuba and the Philippines, the US army similarly adopted khaki in 1902, as did the Japanese fighting the Russians in Manchuria in 1905. The entire Imperial German army turned over to *feldgrau*, field grey, in 1910. Their *Tuch* or cloth mixed grey, blue and green fibres.

Solomon took an interest not just in colours for clothing, for in the early days of the war he was carrying on his experiments with screens of dyed muslin and bamboo poles to cover trenches, according to his undated diary:

I sent some of these screens, with drawings, to the War Office – they caught on, and I was asked to make fifty yards of them at Woolwich Dockyard, where materials would be found me as well as a little assistance in preparing them . . .

Fifty yards of trenches were accordingly dug, and in the presence of large group of officers, including generals, Solomon fixed the screens over one section. An airman was detailed to fly over the scene, and reported he could see the uncovered trenches, but not the one that Solomon had camouflaged. According to Solomon, the officers present were enthusiastic, and his drawings of covered trenches were sent to France. At this stage, the commander-in-chief, Sir John French, turned down Solomon's ideas. But his time would come.

# 3

# Engineering Opinion

Just as camouflage brought painters, designers and artists into WWI, so the propaganda effort required authors, critics, poets and playwrights to lend a hand. Like 'camouflage', the word 'propaganda' did not have an entry in the eleventh edition of *Encylopaedia Britannica*, but everybody knew about it by the end of WWI when the twelfth edition came out. Of course the concept was not wholly new. As Samuel Johnson observed in the eighteenth century:

Among the calamities of war may be justly numbered the diminution of the love of truth, by the falsehoods which interest dictates and credulity encourages . . . I know not whether more is to be dreaded from streets filled with soldiers accustomed to plunder, or from garrets filled with scribblers accustomed to lie.

Arthur Ponsonby, the author of *Falsehood in War-Time*, recognised that the lie was an extremely useful weapon in warfare, deliberately employed by every country 'to deceive its own people, to attract neutrals, and to mislead the enemy'. He wrote this book because he thought the 'authoritative organization of lying' in wartime was not sufficiently recognised: 'The deception of whole peoples is not a matter which can be lightly regarded.' Ponsonby knew that famous writers were better able 'to clothe the rough tissues of falsehood with phrases of literary merit' than statesmen.

As early as 2 September 1914, Charles Masterman, a member of Asquith's cabinet, called a meeting of senior British writers to get together a response to German propaganda leaflets and manifestos. In one room were gathered some impressive names, among them J. M. Barrie, Arnold Bennett, G. K. Chesterton, Arthur Conan Doyle, John Galsworthy, Thomas Hardy, Gilbert Murray, George Trevelyan, H. G. Wells and Israel Zangwill. Also invited but not able to attend were Arthur Quiller-Couch and Rudyard Kipling.

After a second meeting on 7 September 1914 with writers and editors from the respectable British press (no pacifists or socialists were invited), Charles Masterman set up a War Propaganda Bureau at Wellington House, Buckingham Gate, in London. Its mission was to sell the British line and counter the arguments of 'The Unspeakable Prussian' to educated elites in Allied and neutral nations, rather than in Britain or Germany and Austria. By June 1915, this discreet clearing house had distributed 2.5 million copies of speeches, booklets and official publications in seventeen different languages. A year later, it was distributing a million illustrated newspapers every fortnight, and had helped publish 300 books and pamphlets.

Anthony Hope Hawkins, author of *The Prisoner of Zenda*, was the War Propaganda Bureau's literary adviser. Arnold Toynbee and Lewis Namier were among the historical consultants. William Archer, translator of Ibsen, headed the Scandinavian department. G. K. Chesterton wrote a tract called *The Barbarism of Berlin*. Arthur Conan Doyle tackled a history of the campaigns in France and Flanders. John Galsworthy wrote articles. The historian G. M. Trevelyan wrote and lectured on the Serbs and the Austrians before leaving for Italy. John Masefield wrote one book on Gallipoli, and another on the Somme. Popular novelist Mrs Humphrey Ward promoted her 1916 paean of praise to the war workers, *England's Effort: Letters to an American Friend*, on tour in the United States.

The USA was considered the most crucial country to get on side, and so the War Propaganda Bureau put the Canadian-born romantic novelist Sir Gilbert Parker in charge of the public relations campaign aimed across the Atlantic. The basic propaganda message was that the decent British and their allies were honourably muddling through against the *Schrecklichkeit* or 'frightfulness' of the belligerent Kaiser and his ruthless Huns. The rape of plucky little Belgium was the first atrocity to be cited. The Imperial German army certainly killed at least 5,500 civilians in Belgium, but some of the more imaginative bestialities they were accused of probably owed more to fantasy than truth. 'War is fought in this fog of falsehood,' wrote Ponsonby. 'The fog arises from fear and is fed by panic.'

1915 brought a rich harvest of war atrocity stories from Belgium, most notably the execution of the British nurse Edith Louisa Cavell in October 1915. The matron of the Berkendael Institute in Brussels who stayed at her post when it became a Red Cross hospital after war broke out, Miss Cavell, the 49-year-old unmarried daughter of a Norfolk vicar, was formally tried and shot by German firing-squad in Brussels for the crime of helping Belgian, British and French soldiers escape from German-occupied territory into neutral Holland. The British never denied that she had done this. The Germans incurred a propaganda disaster by prosecuting and executing Edith Cavell for treacherously undermining the German war effort, without pausing for merciful gestures and without considering the publicity it would generate.

Her execution duly caused outrage in the UK, and in the USA. Killing a nurse in wartime hardly wins public approval, and Edith Cavell's death was milked by British propagandists as the murder of an angel of mercy. She became the perfect symbol of Belgian martyrdom, and a justification for the war. *The War Illustrated* (30 October 1915) has a drawing of a glaring-eyed prognathous Prussian approaching a figure lying on the ground. Headlined 'The murder of Nurse Cavell', the caption reads:

The ill-fated woman had no strength to face the firing party, and swooned away, whereupon the officer in charge approached the prostrate form, and, drawing a heavy Service pistol, took his murderous aim, while the firing-party looked on.

In March 1920, Queen Alexandra unveiled Cavell's memorial statue in St Martin's Place in central London, just north of Trafalgar Square, the heart of the British Empire, between the National Portrait Gallery and the church of St Martin-in-the-Fields.

M. R. D. Foot, once a wartime intelligence officer, pointed out in his book *MI9: Escape and Evasion 1939–1945* (written with J. M. Langley) that Norman Crockatt, the head of this secret organisation founded in WW2 to help servicemen get out of enemy territory, traced the rivalry between different British secret services back to Edith Cavell. She had in fact been working for the Secret Intelligence Service (SIS or MI6), but had been exposed through helping prisoners of war to escape. This was why the older set of spooks, SIS, wanted nothing to do with MI9, because SIS 'were determined to prevent evaders and escapers from involving them in any way'.

Her secret role was also revealed in Paul Routledge's *Public Servant,*

*Secret Agent: the enigmatic life and violent death of Airey Neave.*
Foot, reviewing it in the *TLS* in May 2002, noted

> a story on which I have had to sit for a generation: that Edith Cavell, shot by
> the Germans in Brussels in 1915 for having helped scores of British soldiers to
> escape into Holland, had, in fact, been an exceptionally well placed spy,
> despised in the Secret Service for having turned aside from her duty as a spy
> to perform a work of mercy.

Cavell's work could not be acknowledged for the usual reason: the
secret services have to stay secret in order to be effective. She probably
also suffered because of her sex and the popular view of it in the media.
Women did not have the vote then and they did not serve in the armed
forces; feminine heroism was mostly framed in terms of self-sacrifice.
Thus to call nurse Edith Cavell anything like a 'spy' (with all its lurid
connotations then) would mean sliding her down the scale of female
achievement, away from worthies like Florence Nightingale towards
houris like Mata Hari. Compromising her virtue might have diminished
her propaganda value. When the British Prime Minister, Gordon
Brown, included Edith Cavell in his book *Courage: Eight Portraits* in
June 2007, he also made no mention of her secret service activities.

---

Perhaps the major *cause célèbre* of WW1 propaganda was the sinking
of the Cunard passenger liner RMS *Lusitania*, torpedoed by *U-20* off
Ireland on 7 May 1915. One hundred and twenty-eight American lives
were lost, and the incident outraged the USA, whose government
protested that such an attack on a passenger ship was a flagrant
breach of the rules of war and, as Assistant Secretary of the Navy F. D.
Roosevelt put it, 'piracy on a vaster scale of murder than old-time
pirates ever practised'. In short, it was an act of 'terrorism'.

The German government's defence was that the *Lusitania* was an
armed merchantman built with British government funds, mounted
with hidden guns and quite prepared to ram submarines, and that she
was carrying Canadian troops for the Western Front as well as
thousands of crates of illicit munitions (which, they said, the torpedo
caused to explode, thus sinking the ship in eighteen minutes). They
added that this was a war zone in wartime and that the Imperial
German ambassador in Washington DC, Count Bernstorff, had placed

notices in US newspapers stating that British and allied vessels might be attacked, so there was fair warning.

Some of these arguments are not true. The *Lusitania* was unarmed and had no hidden guns, and there was only one Canadian soldier on board, running off with his mistress. Others are half true: the *Lusitania* was indeed carrying four million rounds of .303 rifle ammunition and 5,000 3.3-inch Bethlehem Steel shrapnel shells not yet filled with explosive, but marine archaeology does not suggest the ammunition blew up. Certainly the British government was not anxious to publicise the existence of munitions on a passenger ship, which would have undermined their righteous indignation. In any case, the German justifications could never carry as much emotional weight in world public opinion as the distressingly horrible deaths of 1,200 innocent people, including many women and nearly a hundred children, a third of them babies. This was one of the great shock-horror stories for newspaper front-page headlines: 'The Huns Sink the *Lusitania*' said *The Daily Sketch* on 8 May; 'Full Story of the Great Murder', '*Lusitania* Survivors' Terrible Stories'.

When the *Lusitania* sailed on her last voyage the passenger list included a small, deaf, angry designer called Oliver Percy Bernard. The immediate outlet for his rage on board the Cunard liner was British caste and class snobbery, but the fires of his anger had been long stoked by the frustrations of life.

Born among 'vague and violent people' in Lambeth, where his father boxed with bare knuckles, Oliver 'Bunny' Bernard had been sent as a 13-year-old orphan to learn backstage theatrical arts in Manchester, where he taught himself to draw by paying attention to trees, carefully drawing their boles, branches and bark. From a lonely adolescence, Oliver Bernard grew into an outsider who liked the theatre but was cold-eyed about

the tiresome vanity of successful actors, the emotional insincerity of favourite actresses . . . those who practise deception are most deeply deceived; those who excel in the simulations of grief are most early reduced to tears; the liar falls most completely for the lie.

By 1915 he was a successful stage architect and scenic artist. Oliver Bernard loved the effects that music and drama could achieve but loathed the 'consecrated humbug' of grand opera in London, Boston and New York, so often a world of 'beasts and bitches', charlatans and frauds. Unloved, unhappy in love, resentful of the rich lording it on board, ashamed to be a non-combatant in wartime, and remembering how 'deafness and discriminating methods of muddled recruitment had prevented him from becoming cannon fodder in 1914', it was a rather disgruntled and acerbic 'Bunny' Bernard who paced the deck of the *Lusitania* as her sirens hooted into the Atlantic fog.

On the sixth day out, the sun was shining off the south-west of Ireland, and the passengers' mood on board the floating luxury hotel brightened. After lunch, around 2.15 p.m., Bernard went up on deck. The smooth, still sea was like 'an opaque sheet of polished indigo' and the horizon was undisturbed by the smoke or sails of any other vessel. Bernard's reverie was interrupted by 'a frothy track snaking up . . . like an express'. The torpedo was nearly seven metres long and weighed over a ton: it carried 160 kilos of high explosive in its nose, and was travelling at over 80 kph towards the ship.

Oliver Bernard felt a slight shock through the deck, as though a tug-boat had run into the giant liner. Then there was a terrific explosion. A column of white water rose high in the air, followed by an eruption of debris. Lumps of coal bounced on the deck. He was no longer alone. Fellow passengers in the floating hotel appeared from every-where in a rush of trampling feet, wails and cries. Bernard dutifully went down to B deck to fetch his lifebelt from his cabin. The lights were all out. He fell down tilting stairs; could not balance; reeled in darkened corridors to his cabin. Back on the crowded deck, a woman demented with fear snatched his lifebelt from him. No one knew what to do, and there was no loudspeaker system to tell anyone. Passengers had no lifejackets or put them on wrongly. As the ship canted more to starboard and dipped down forward, Bernard began taking off his clothes, methodically folding his coat, waistcoat, collar and tie, carefully putting his tie-pin in his trouser-pocket like a man about to have a wash. But 'Bunny' could not swim. He slid down the steep sloping deck and in 'a wild lucky splash' scrambled into a lifeboat that had to be hacked away from its bow davit. By rowing frantically they only narrowly missed engulfment by a huge smokestack as the ship slid sideways under the waters. From the boat they watched the triumphant sea pouring into the funnel's steaming black maw, and then the *Lusitania*'s mastheads disappearing.

'All that remained was a boiling wilderness that rose up as if a volcanic disturbance had occurred beneath a placid sea.'

--

Public anger burned long on the fuel of the *Lusitania* story. Gruesome horrors lasted for weeks: the morgues and mass graves at Queenstown; the bloated corpses with seagull-pecked faces washing up on Irish beaches; the pathetic stories; the private griefs. The propaganda press feasted on it in words and graphics. Rioting mobs in Liverpool and London sacked shops with Germanic names. That emotional barometer, D. H. Lawrence, said, 'I am mad with rage myself. I would like to kill a million Germans – two millions.' The Liberal government in Britain ordered the arrest and internment of up to 30,000 'enemy alien' males.

The *Lusitania* incident not only destroyed German propaganda hopes in America, but fitted right into the War Propaganda Bureau aim of demonising the Germans. There was no shortage of material that month. On 15 May 1915, *The Times* added more details to a completely untrue story it had run on 10 May about a Canadian soldier being crucified by German bayonets on a barn wall in Belgium. This was just a gobbet of tainted meat to add to the ghoulish feast of the official Bryce Report into the Alleged German Outrages in Belgium, published on 13 May 1915 and distributed by Wellington House to almost every important newspaper in America and in twenty-seven languages to many countries around the world. Its author, James Bryce, was a distinguished jurist, member of the House of Lords and former ambassador to Washington DC, who had helped Roger Casement to expose the involvement of British-owned companies in atrocious exploitation of rubber-tappers in the Amazon in 1907. But his Royal Commission report on Belgium is naïvely credulous, luridly recounting 'witness' stories of mass rape, amputation and baby-bayoneting, collected without any cross-examination or corroboration.

'Your report has swept America,' Charles Masterman wrote to Lord Bryce, 'As you probably know even the most sceptical declare themselves converted, just because it is signed by you!' War Propaganda Bureau operatives in America told Masterman: 'Even in papers hostile to the Allies, there is not the slightest attempt to impugn the correctness of the facts alleged. Lord Bryce's prestige in America puts scepticism out of the question.'

Some sceptics did want to spoil the horror stories, including a furious Roger Casement, but he was just a cranky, homosexual Irish nationalist who would soon be hanged for high treason in Pentonville prison on 3 August 1916. The US lawyer Clarence Darrow went to France later in 1915 and could not find any of Bryce's eyewitnesses, though he offered $1,000 to meet any Belgian child amputee. The Pope, the Italian Prime Minister and David Lloyd George also had diligent inquiries made, but no one ever found the supposed handless kiddies. The atrocity stories were designed to unite people against the foe.

But not everyone in Britain shared these views. The brilliant, gentle cartoons of William Heath Robinson, born into a family of illustrators in 1872, are a wonderful deflation of both sides in the combat. He said that 'the much advertised frightfulness of the German army' gave him one of his best opportunities as an artist, and in such books as *Some 'Frightful' War Pictures* (1915), *Hunlikely!* (1916) and *The Saintly Hun: a Book of German Virtues* (1917), he ridiculed the demonisation of the enemy by accusing Germans of minute failures of sporting etiquette but also showing them in improbable acts of saintliness. German aeronauts protect the modesty of a young Englishwoman in her attic; an enormously fat, be-helmeted Prussian general withstands the tempting aroma of a pie carried by a starving child, and another 'benignant Boche returning good for evil' offers a cigar to a British soldier as the latter impales him with a bayonet. Heath Robinson was a good advertisement for British amateurishness and larkishness, and an antidote to the over-serious simplicities of propaganda.

---

They are masters of propaganda, you know. Dick, have you ever considered what a diabolical weapon that can be – using all the channels of modern publicity to poison and warp men's minds? It is the most dangerous thing on earth. You can use it cleanly – as I think on the whole we did in the War – but you can use it to establish the most damnable lies.

John Buchan, *The Three Hostages* (1924)

John Buchan was not well known enough to attend Charles Masterman's first meeting of writers in Whitehall on 2 September 1914, but he later became the master of propaganda in journalism, fiction and history. Buchan wrote many books for Masterman's War Propaganda Bureau, and in February 1917 he became Masterman's boss when the Prime Minister appointed him director of the Department of Information, charged with coordinating all British propaganda.

Buchan was the son of a Church of Scotland minister and understood that effective propaganda was linked to deep belief. The word 'propaganda' is religious in origin, coming from the Roman Catholic Church's *congregatio de propaganda fide*, 'congregation for propagation of the faith', a body set up to aid the missionary work of the Church. But Buchan links propaganda to less orthodox spirituality in his novel *The Three Hostages*, published in 1924, the era when Lenin, Stalin and Hitler emerged:

The true wizard is the man who works by spirit on spirit. We are only beginning to realize the strange crannies of the human soul. The real magician, if he turned up today, wouldn't bother about drugs and dopes . . . The great offensives of the future would be psychological, and . . . the most deadly weapon in the world was the power of mass-persuasion . . .

In March 1918, Lord Beaverbrook took over the Ministry of Information, and John Buchan was renamed director of Intelligence for the last eight months of the war. Anthony Masters, in *Literary Agents: The Novelist as Spy*, says Buchan's work then is 'shrouded in mystery', but some idea may be gathered from Anthony Clayton's *Forearmed: A History of the Intelligence Corps* (1993):

John Buchan, later Lord Tweedsmuir, was commissioned as a lieutenant in the Intelligence Corps in 1915 to assist with the communiqués for, and later for an official account of, the Battle of the Somme . . . Deception plans and misleading information were used by GHQ Intelligence on occasions – false reports being given to the Press or drafted into carefully prepared political speeches.

Other clues are in his fiction. John Buchan was a classical scholar who energised a new literary genre, the paranoid spy-thriller, for popular consumption in the early twentieth century. The story is often a sinister plot that threatens England. John Buchan was fascinated by deception and 'the veiled prophets who are behind the scenes in a crisis'. His adventures often involve joining up disconnected pieces of information to reveal a picture of the problem or danger which then has to be resolved by decisive, heroic action. Such popular books have upbeat endings because the hero always prevails and restores order, but they also articulate in an interesting way the anxieties and prejudices of the author's group.

Buchan's new hero first appeared in October 1915 in his 'shocker', *The Thirty-Nine Steps*, which sold 25,000 copies by Christmas. This hero, Richard 'Dick' Hannay, is first encountered as a rough-and-ready mining engineer from Rhodesia, bored in London in May 1914 until he gets caught up in a fast-moving adventure of murder and escape that eventually unravels a German spy ring called the Black Stone, *Der Schwarzestein*. In chapter V, Hannay remembers Peter Pienaar, an old Boer scout in Rhodesia, telling him that the secret of playing a part was to think yourself into it. 'You could never keep it up, he said, unless you could manage to convince yourself you were it.' In chapter X, Hannay recalls Pienaar's advice that the secret of effective disguise was to blend fully into your surroundings. Hannay then remembers hunting a dun-coloured rhebok with his dog in the Pali Hills in Rhodesia:

That buck simply leaked out of the landscape . . . Against the grey rocks of the kopjes it showed no more than a crow against a thundercloud. It didn't need to run away, all it had to do was to stand still and melt into the background.

The leader of the German Black Stone spy ring is a master of disguise who successfully impersonates the British First Sea Lord in front of his military colleagues, precisely because they are expecting to see him and so take him for granted. 'If it had been anybody else you might have looked more closely, but it was natural for him to be here and that put you all to sleep.'

In the final chapter, Hannay realises the ruthless German spy ring has also managed to camouflage itself into 'the great, comfortable, satisfied middle-class world, the folk that live in villas and suburbs'. Hannay remembers the old scout's theory of 'atmosphere' in matching your surroundings: 'A fool tries to look different: a clever man looks the same and is different.'

In Buchan's second Richard Hannay adventure novel, *Greenmantle*, Hannay pretends to be an anti-British, pro-German Boer called Cornelius Brand in order to travel deep into the Kaiser's Germany. This exploit is modelled on the true story of John Buchan's friend and fellow Scot, Edmund Ironside, the future Chief of the Imperial General Staff (CIGS). As a young officer in 1903, Ironside went undercover in

German South-West Africa (now Namibia) to investigate German activities during the revolt by the Herero people. The Intelligence Department helped disguise the burly Ironside as a Boer ox-cart driver in battered hat and veldskoens. He grew a beard, smoked Boer tobacco in a foul pipe, and spoke authentic colloquial Cape Dutch. He was soon accepted, but was horrified one day to see his white bull terrier proudly trotting alongside his wagon in a bright collar proclaiming his owner's name: 'Lt. Ironside: Royal Artillery'. Nevertheless, Ironside managed to bluff his way through and even got a German medal (which he later displayed to Adolf Hitler).

*Greenmantle* was Buchan's tenth novel and thirtieth book and remains one of the finest novels of the imperial 'Great Game', perhaps second only to Rudyard Kipling's 1901 novel *Kim*. There is an allusion to the fact that Kim, the boy spy, worked with a red-bearded Afghan horse-trader called Mahbub Ali when the fictional head of the Secret Service in *Greenmantle*, Sir Walter Bullivant, says:

I have reports from agents everywhere – pedlars in South Russia, Afghan horse-dealers, Turcoman merchants, pilgrims on the road to Mecca, sheikhs in North Africa, sailors on the Black Sea coasters, sheep-skinned Mongols, Hindu fakirs, Greek traders in the Gulf, as well as respectable Consuls who use cyphers.

The classic opening chapter of *Greenmantle*, 'A Mission Is Proposed', was chosen by Graham Greene and Hugh Greene to open their 1957 anthology, *The Spy's Bedside Book*, in tribute to the author whose memoirs *Memory Hold-The-Door* recorded that one side of his WWI duties 'brought me into touch with the queer subterranean world of the Secret Service'.

'You Britishers haven't any notion how wide-awake your Intelligence Service is,' the American agent John S. Blenkiron flatteringly says in *Greenmantle*, adding, 'If I had a big proposition to handle and could have my pick of helpers I'd plump for the Intelligence Department of the British Admiralty.' From November 1914 on, British Naval Intelligence had as its director Admiral W. Reginald Hall, who had commanded the battle cruiser HMS *Queen Mary* at the Battle of Heligoland Bight, and inherited OB40, the cryptographic department led by Sir Alfred Ewing, which cracked German military and diplomatic codes. 'Hall is one genius the war has developed,' the American ambassador in London wrote to US President Wilson. 'Neither in fiction nor in fact can you find any such man to match him.'

John Buchan's character Sir Walter Bullivant, the spymaster in *Greenmantle*, was very like Admiral Sir Reginald Hall. Though small, Hall was the archetypal forceful naval officer, from the dome of his bald head to the cleft in his clean-shaven chin. His eyes glared out under bushy eyebrows above a great hooked beak of a nose. This look of an alert peregrine falcon, with a disconcerting eyelid twitch, earned Hall the nickname 'Blinker'. Hall wielded his power 'vigorously', according to F. H. Hinsley, the historian of British Intelligence, 'building up his own espionage system, deciding for himself when and how to release intelligence to other departments, and acting on intelligence independently of other departments in matters of policy that lay beyond the concerns of the Admiralty'. Translating from the bureaucratic, that means he was a ruthless and cunning rogue elephant. His biographer, Admiral Sir William James, said: 'There was nothing Hall enjoyed more than planning ruses to deceive the Germans.'

'Blinker' Hall had a genius for picking people. He hired civilians whose professional work was analytical, like academics, bankers, lawyers, scientists, and mixed them with the artistic: actors, authors, designers, dilettantes, etc. He also employed clever women at a time when that was unusual, like the formidable, cigar-smoking Lady Hambro who marshalled the secretaries.

John Buchan knew Reginald Hall well, and *Greenmantle* can be read as a novel about an imaginary British intelligence operation involving disguise and deception, that uses insider knowledge of other operations. It begins a year on from the end of *The Thirty-Nine Steps*: Major Richard Hannay of the (fictional) Lennox Highlanders is back in England recuperating from wounds received in the real Battle of Loos in late September 1915. 'Loos was no picnic,' says Hannay, in a typical stiff-upper-lip understatement of the catastrophe which left 8,000 dead. Loos was the big attack in the grimy Belgian colliery district where the British first used their own chlorine gas, 140 tons of it, five months after the Germans used gas at Ypres. The 6th battalion of the Royal Scots Fusiliers, which lost three-quarters of its officers and half its other ranks there, got its new commanding officer in France early in 1916: Lieutenant Colonel Winston S. Churchill.

John Buchan begins *Greenmantle* with Richard Hannay convalescing from Loos in the same Hampshire country house as his friend and brother officer who has just saved his life, 'Sandy' Arbuthnot, the second son of Lord Clanroyden. Arbuthnot is a man with a 'passion for queer company' – in the old sense. In London, we learn, you get news of Sandy Arbuthnot from 'lean brown men from the ends of the earth . . . in creased clothes, walking with the light outland step, slinking into clubs as if they could not remember whether or not they belonged to them'. Sandy Arbuthnot is a creature of romantic imperial fantasy:

> He rode through Yemen, which no white man ever did before. The Arabs let him pass, for they thought him stark mad and argued that the hand of Allah was heavy enough on him without their efforts. He's blood-brother to every kind of Albanian bandit. Also he used to take a hand in Turkish politics, and got a huge reputation . . . We call ourselves insular, but the truth is that we are the only race on earth that can produce men capable of getting inside the skin of remote peoples.

Buchan based Sandy Arbuthnot on a real-life crusader for small nations, the Honourable Aubrey Herbert, second son of the Earl of Carnarvon. Semi-blind at Eton, reckless at Oxford, Herbert had nevertheless got a First in History, joined the diplomatic service as an honorary attaché and was an MP for seven years. At the start of the war he had had an officer's uniform made by a military tailor and slipped into the ranks of the Irish Guards as they left for France. Smuggled on to a troopship in Southampton by officer friends, he went off to war with the BEF as an interpreter. Within a month he was wounded, captured, freed, sent home.

In Salonika, Aubrey Herbert had acquired a ferocious Albanian bodyguard called Kiazim who sprouted daggers and pistols and took him to hashish dens. He learned fluent Turkish in Constantinople, and like his fictional counterpart travelled widely. Herbert's only known comment on the character that Buchan based on him was 'He brings in my nerves all right, doesn't he?'

*Greenmantle* ends at the fall of Erzerum, in Turkey, in 1916. This is where the daring deception is finally revealed. In real life, Britain and Russia were fighting Germany and Turkey, and in Buchan's novel, a British deceiver manages to infiltrate the German-inspired Islamist revolt. Richard Hannay, the Boer scout Peter Pienaar and the

American John Blenkiron help their Russian allies to find the weak link in the Turkish defences, and join the grey-clad Cossack cavalry in the final ride across the snow. Ahead of them, in the van of the charge, is one man . . .

He was turbaned and rode like one possessed, and against the snow I caught the dark sheen of emerald. As he rode it seemed that the fleeing Turks were stricken still, and sank by the roadside with eyes strained after his unheeding figure . . . Then I knew that the prophecy had been true, and that their prophet had not failed them. The long-looked for revelation had come. Greenmantle had appeared at last to an awaiting people.

The radical 'Islamic' prophet, Greenmantle, turns out to be a British intelligence officer in camouflage: Sandy Arbuthnot. But he also looks and sounds exactly like Lawrence of Arabia, pursuing British policy in native disguise. T. E. Lawrence, that master of dressing up, was impressed by Buchan's 'clean-lined, speedy, breathless' books. In 1933, he wrote perceptively to Edward Garnett about John Buchan's novels:

For our age they mean nothing: they are sport, only: but will a century hence disinter them and proclaim him the great romancer of our blind and undeserving generation?

John Buchan's adventures are the premier novels of twentieth-century camouflage and deception. Their villains pass as fine gentlemen at ease in society, and their heroes are also disguised. 'For men who live so dangerously, they are oddly conventional,' observed Graham Greene. Buchan's constant theme is shamming, pretence, tactical deception.

'I found out in the war that it didn't do to underrate your opponent's brains. He's pretty certain to expect a feint and not to be taken in. I'm for something a little subtler.'
'Meaning?'
'Meaning that you feint in one place, so that your opponent believes it to be a feint and pays no attention – and then you sail in and get to work in that very place.'

John Buchan, *John Macnab* (1924)

The subliminal effect of the Richard Hannay adventures on the generation that fought in WW2 was immense. As boys or young men, all

47

of them had read Buchan. Their coeval George Orwell speaks for them:

Personally I believe that most people are influenced far more than they would care to admit by novels, serial stories, films and so forth . . . It is probable that many people who would consider themselves extremely sophisticated and 'advanced' are actually carrying through life an imaginative background which they acquired in childhood from (for instance) Sapper and Ian Hay.

Richard Usborne wrote *Clubland Heroes*, a study of the fictions of John Buchan, Dornford Yates and Sapper, and was himself in the Special Operations Executive (SOE), set up by Winston Churchill on 19 July 1940 'to co-ordinate all action by way of subversion and sabotage, against the enemy overseas'. Usborne said that almost every single SOE officer he ever met in WW2 pictured himself as Richard Hannay or Sandy Arbuthnot.

John Buchan's Dick Hannay novels also exemplify the British success in deception in both World Wars:

'See here, Dick. How do we want to treat the Boche? Why, to fill him up with all the cunningest lies and get him to act on them.'

John Buchan, *Mr Standfast* (1919)

# 4

# Hiding and Sniping

When he first saw it from a distance in May 1915, John Buchan thought that the Flanders wool town Ypres 'looked a gracious and delicate little city in its cincture of green'. By day, in the spring sunshine with the birds singing, the wartime world was not immediately apparent. But as he walked into its dusty pale centre, he thought an earthquake had hit the place and driven everyone away. Ypres was being hard fought-over. The houses of a once-rich town were skeletal, disembowelled by shells that had ripped open their fronts, exposing middle-class furniture and fancy fittings to raw weather.

The British section of the front line bulged eastward from Boesinghe round to St Eloi to protect Ypres in its low basin. The bag of soggy marshland this bulge enclosed was known as the Ypres Salient, and from the winter of 1914 through to the spring of 1915 the sodden earth was often too wet to dig deep slit-trenches, so soldiers had to build parapets and revetments up in order to get what shelter they could from enemy artillery fire.

In Second Army, Archibald Wavell of the Black Watch was brigade major of the 9th Infantry Brigade at Hooge. He was the best kind of staff officer, who insisted on visiting every part of his front line when he was in the Ypres Salient from November 1914 to June 1915, and he deplored the irrational way in which the command automatically valued trenches, the holes in the ground that someone had dug somewhere, without considering whether the place where they happened to be was any good or not, whether the ground was wet or dry, protected or concealed. It might make better tactical sense to move the line yards eastwards to more solid chalk, but the obsessive refusal to yield an inch meant it could not be done, or not officially. So men died because ordered to hold, at all costs, unsheltered and enfiladed terrain, which became their own muddy grave.

Winston Churchill showed his practical good sense when he wrote to the Prime Minister Herbert Asquith on 7 January 1915: 'Ought we not to get into a more comfortable, dry, habitable line, even if we have to retire a few miles? Our troops are rotting.' Wavell once found the whole garrison of a trench perched on the parapet above it, preferring to risk bullets rather than endure the freezing swampy gruel. His brigade soon started a factory for duckboards, mudscoops, sandbags and revetting material. In the months and years to come, this construction became industrial.

On 16 June 1915 Wavell's brigade attacked three lines of trenches near Bellewarde Lake at the eastern tip of the Ypres Salient. It was one of those minor attacks on a narrow front that was meant to 'support' other British and French attacks miles away to the south, but the whole assault could be easily observed by the enemy and fired on from three sides. Despite meticulous plans, on the day it turned into a muddle, and the division lost 3,500 men. Continuous accurate German shelling turned two-thirds of the 9th Brigade into casualties, including 73 of the 96 officers. One of them was big Archie Wavell. A piece of shrapnel or a bullet destroyed his left eye. When his un-bandaged right eye also closed and he could only open it with both hands he thought he ought to walk back to the dressing station. He was given morphine and woke up in a ward of the Rawalpindi General Hospital at Wimereux, near Boulogne. The stupidity of orthodox military tactics in WW1 impressed on Wavell the need for new ways of waging war, including deceptive stratagems, to avoid mass slaughter. We will meet him again, with a glass eye, in the Middle East.

---

By June 1915 the position of journalists at the front had finally been regularised by authority. Pressure from the three principal newspaper proprietors in London had forced Lord Kitchener to concede a role to selected war correspondents, replacing the single official spokesman Major Ernest Swinton, whose hundred or so articles, bylined 'Eye-witness' ('Eyewash' to critics), had been vetted by Kitchener himself. From March 1915, these newly accredited journalists wore the khaki of British Army officers with a green band on the right arm and held the honorary rank of captain, but they were still civilians at heart, trying to tell the people back home what their menfolk were going through.

Philip Gibbs, representing the *Daily Chronicle* and the *Daily Telegraph*, was one of the six British correspondents, along with Percival Phillips of the *Morning Post*, William Beach Thomas of the *Daily Mail*, H. Perry Robinson of *The Times*, Herbert Russell of Reuters news agency and Basil Clarke of the Amalgamated Press. The reporters started work in a spiral-staircased chateau at Tatinghem, not far from GHQ at St Omer, where the stuffier staff officers resented their presence and tried to waste their time. The correspondents always had to be escorted by military officers and every word they wrote had to pass a military censor.

One of the first escort/censors was a tall Assistant Press Officer called Hesketh Prichard. He was the Indian-born son of a popular officer in the 24th Punjabis, who had died of typhoid six weeks before his son was born. By the time war came in 1914, 37-year-old Prichard had written books about his travels in Haiti (*Where Black Rules White*), South America (*Through the Heart of Patagonia*, seventy years before Bruce Chatwin) and Canada (*Through Trackless Labrador*). His big-game-shooting adventures in *Hunting Camps in Wood and Wilderness* (1910) were admired by Teddy Roosevelt.

Hesketh Prichard was best known as a cricketer, a notable right-arm fast bowler for Hampshire from 1900–13, good enough to be picked three times for the 'Gentlemen' team of Marylebone Cricket Club or MCC. He was also a good marksman. A hunting companion once said that he never missed a crucial shot: 'H.P.' would always get the last hope of food, even at distance in fading light. As others became shaky, he grew steadier.

When John Buchan came out to write a series of articles for *The Times* in May 1915, his escort at the front was Hesketh Prichard, who told him about the enemy snipers, 'forest rangers from South Germany'. Buchan's first piece, on Monday, 17 May, points up the contrast between the calm-looking countryside and 'this secret warfare, hidden in the earth and the crooks of hills. There is something desperate in its secrecy, something deadly in its silence.'

Later the next month, Prichard escorted H. M. Tomlinson of the *Daily News*, not the kind of man he was used to, a self-described working-class Socialist with trade union connections, agreeing with the radical views expressed by George Bernard Shaw in November 1914 in the *New Statesman*: 'Both armies should shoot their officers and go home to gather in their harvests in the villages and make a revolution in the towns.'

But Prichard and Tomlinson liked each other. In cautious walks very near the front, tall Prichard confided to tiny Tomlinson 'that he thought he was not doing enough. He was not killing Germans. But the German snipers were doing our men a lot of harm. He wanted rifles with telescopic sights and men trained in stalking and sniping.' The escort officer could not do much about the horrors of shelling or chlorine gas, but he decided to use his personal skills and his hunter's craft to stop other friends of his from being shot through the head. Hesketh Prichard determined to take on the snipers.

Prichard had already brought over from England several hunting rifles with telescopic sights, and regularly carried them with him on his duty trips or lent them to units in the line. He was also spending hours watching the German lines through a telescope and making his earliest attempts at shooting back. It was fitting that a sportsman take on the job. 'Sniping' had begun as an amateur sport in the colonies, the skilful shooting of long-billed bleating marsh birds that fly fast in zigzags.

The Germans dominated the sniping war between the trenches from the beginning of 1915. The poet Robert Graves and the courtier Alan Lascelles both had their trench periscopes neatly drilled through by a sniper's bullet; the poet Siegfried Sassoon was shot through the chest by one, and the future BBC leader John Reith through the cheek; the famous white hunter Frederick Courtney Selous and the sardonic writer 'Saki' (H. H. Munro) were both killed by snipers. 'Take me over the sea,' the soldiers sang in the 1915 song *I Want To Go Home*, 'Where the snipers they can't snipe at me. Oh my, I don't want to die, I want to go home.' Sniping helped drive soldiers below ground and made them troglodytic, cautious, and eventually camouflage-minded.

German snipers were better equipped because scientific Germany was more advanced than Britain in making optical instruments. By the end of 1914 the Germans already had 20,000 telescope sights, made by companies like Carl Zeiss, for their Mauser Gewehr 98 rifles. German and Austrian *Scharfschützen* or sharpshooters, who came from a long tradition of hunting, also carried two-foot by three-foot steel plates, strong enough to resist a British .303 bullet, with a loophole in the middle through which they shot. At this stage of the war, the Germans were also ahead in the camouflage game. They had better-disguised loopholes set in their deliberately 'untidy', irregular trenches. These had revetments made of differently coloured sandbags, blue, yellow, red, green, black, pink, striped, amid a litter of corrugated iron, old biscuit tins, drainpipes and dummy plates, all of which confused the eye, so hidden riflemen were harder to spot.

It took time and many deaths for the tidy-minded British regulars to begin to understand that excessive smartness could be fatal. Neatly flattening the top row of sandbags by banging them with a spade in order to make a crisp revetment or straight parapet was only making the enemy sniper's job easier by underlining the gleaming half-moon of a soldier's hat, which as Solomon J. Solomon pointed out, made 'a most excellent target'.

Snipers had a bad effect on morale. Sudden violent death with horrible injuries induced many emotions in onlookers, including fear and shock. An angry reaction of leaping straight up to shoot back would all too often just produce another casualty. Tall men were regularly picked off because they forgot to stoop. Others were shot as they shat. 'Many a poor Tommy met his end in a latrine sap,' wrote George Coppard. In early 1915, one battalion lost eighteen men in a single day to snipers, but lower daily figures, week in, week out, month after month, all along the line, added up to thousands of Allied casualties.

This attrition continued the lesson that the British had been taught by the Boers in the South African War. Henry Charles Bosman's classic short story *The Rooinek* opens with two ragged Boers lying in hiding, shooting with smokeless Mausers at smart British officers riding out openly on horseback. The last British regiment ever raised by a Highland clan chief, the Lovat Scouts, were formed in response to such guerrilla marksmen in South Africa.

The original draft of the Lovat Scouts was formed in 1900 by Simon Fraser, 14th Lord Lovat and 41st chief of Clan Fraser. It comprised 150 stalkers and ghillies recruited from game estates in Scotland (where Fraser's family owned over 180,000 acres), all men with the fieldcraft to take on the Boer Commandos. These crack shots who had spent their lives outdoors spotting stags for rich Victorians adapted well to the job of scouting, observing, signalling, guiding and sniping in South Africa, the only drawback being that few of their Anglicised officers spoke the Gaelic.

Hesketh Prichard considered the Lovat Scouts Sharpshooters were the best battle observers and scouts in the British Army in WW1, saying that:

... behind the lines the Major-General, the Corps Commander, the Army Commander and the Commander-in-Chief himself are all blind. Their brains direct the battle, but it is with the eyes of Sandy McTosh that they see.

The Lovat Scouts were superb 'glassmen'. Like Hesketh Prichard, they used brass sliding telescopes made by Ross of London that they brought out with them (always preferring a spyglass to the binoculars that became so popular with officers on both sides in WW1). With a telescope they could read a cap badge or cockade at 140 yards or even identify a shoulder flash seen upside down through the enemy's own trench-top periscope. Years of crawling through heather, gauging wind, observing tiny movements, and counting the points on distant stags, paid off. 'The Lovats never let one down,' wrote Hesketh Prichard. 'If they reported a thing, it was as they reported it.'

Lord Lovat's Scouts were a re-invention, exactly a hundred years later, of the British Colonel Coote Manningham's experimental Corps of Riflemen which later evolved into the famous 95th Rifles. The Riflemen dressed in bottle green and black, used ground and cover to best advantage, and went forward as scouts to observe and report but also to inflict serious damage on the enemy with their short, grooved Ezekiel Baker rifles. These mobile marksmen were trained by humane officers like Sir John Moore to be 'intelligent, handy and active', rather than just mindless cannon-fodder.

The sharpshooting rifleman evolved into the sniper. Visible signs of status began disappearing from military officers in the field after keen-eyed, well-hidden riflemen began singling out badges of rank. Germans shot at 'thin legs' during WW1 advances because officers had better-cut trousers. The job of military snipers, who often work in pairs, alternating spotting and shooting, is precisely to pick off enemy leaders, officers and NCOs, artillery and mortar crews, forward observation officers, reconnaissance scouts and other snipers, to cause maximum damage for minimum effort, and to be 'force-multipliers'.

After the Boer War and defeats like Magersfontein in 1899, British infantry shooting was radically improved by the Musketry School at Hythe in Kent. Its chief instructor from 1901–3 and commandant from 1903–7 was the South African veteran Sir Charles Carmichael

Monro. Thanks to his ideas of training troops for battle conditions rather than competitive target-shooting, the Germans mistook British infantry rifle fire for massed machine guns at Mons in 1914. Monro linked firepower to tactical movement forward. New tactics, making use of ground while shooting fast and accurately, seemed more interesting and competitive to junior leaders. By 1911, 'the Monro doctrine' was accepted throughout the British Army.

In August 1914, 54-year-old Major General Monro led the 2nd Division; by January 1915 he was in charge of the whole of 1st Corps, and on 15th July 1915 he was appointed to command the newly formed Third Army. This was Hesketh Prichard's chance. Monro was not only interested in new ideas about shooting, he also liked, and played, cricket. Monro appreciated what Prichard was trying to do, and secured him a roving commission with Third Army to develop the art of counter-sniping. Sniping appealed to men in the field because it was something more skilful and individual than group fire. Aubrey Herbert also found it more sporting:

At one place on the way, we ran like deer, dodging. The General, when he had had a number of bullets at him, also ran. Sniping is better fun than shrapnel; it's more human. You pit your wits against the enemy in a rather friendly sort of way.

Alan Lascelles visited a Yorkshire regiment in September 1915:

A little lower down the trench we came on a sergeant perched just under the top of the parapet, with a telescope and rifle, in a little eyrie of sandbags that he had built with extreme cunning. 'This man,' said the Captain who was showing us round, 'is our crack sniper. He has mopped up eleven of them since we came in five days ago.' 'Twelve sir,' said the Sergeant. 'They've just pulled him into yon dug-out.'

The counter-sniping project got under way in June 1915 when Major Hesketh Prichard went to see his old friend Captain Alfred Gathorne-Hardy, further south down the line at Neuve Chapelle with the 9th Scottish Rifles (he was killed a few months later at Loos). Together they crawled out across no-man's-land to steal some of the large protective iron plates through whose loopholes the German snipers used to shoot. Prichard took them home on leave in July to test against different rifles and ammunition. He found that big bullets (.577 or .470 Nitro Express) from a double-barrelled elephant gun, or even the

55

smaller but high-velocity Jeffreys .333, punched through the sheet-metal as if it were chocolate.

Hesketh Prichard approached John Buchan in London about raising money to buy more such guns. The *Spectator* ran an appeal and Buchan got Lord Haldane and other wealthy men to assist. Meanwhile, Prichard visited Willie Clarkson (the famous London costumier who had dressed Virginia Woolf for the Abyssinian Dreadnought hoax), from whom he obtained a supply of the model heads used to display wigs. In September 1915, H.P. managed to escape from GHQ escort duties and began teaching 'Sniping, Observation and Scouting' to officers and men of Third Army. In the summer of 1916, he started an innovatory sniping school at Linghem in Belgium for First Army. By then, his mentor Charles Monro was back from Gallipoli and commanding First Army.

Prichard had to overcome inertia above and ignorance below. He started out as a lone individual with no 'Establishment', authority or charge code. He had to step down in rank from major on the staff to infantry captain, and received no pay for eight months. Telescopic sights for rifles were in short supply, and 80 per cent of them were useless because improperly aligned and maintained, and no one knew anything about concealment or observation. But slowly, as he moved from brigade to brigade, Major Hesketh Prichard found allies and converts ('Who is this blighter who's coming?' . . . 'Plays cricket, doesn't he?') as he demonstrated old ruses and new tricks to counter German sharpshooters, helping sniper/scouts to earn their fleur-de-lys badge.

The theatrical heads from Clarkson's the costumiers could be used as decoys to help locate hidden snipers. The head, set on a stick that slid up and down a grooved board, would be pushed cautiously above the parapet like someone taking a look; if hit by a sniper's bullet, it was swiftly lowered. By inserting a rifle-cleaning rod through the bullet's entry and exit holes in the dummy head you could get the exact angle and alignment of the shooter. Or you could slide a periscope up the groove in place of the head, spot the sniper, and then get counter-snipers to fix him in their sights.

When Prichard visited the French Camouflage Works at Amiens in 1916 he fell on the *camoufleur* Henri Bouchard with joy. Here was a sculptor already making brilliantly realistic heads and shoulders of French and British soldiers out of papier mâché. They were more readily available than Clarkson's models from London, and so well done that they were impossible to tell from the real thing at 300 yards. H.P. got Bouchard modelling Gurkha and Sikh individual heads too, to vary the target and to worry German intelligence compiling an 'order of battle' or inventory of enemy troops. Some dummies even had a slot in the mouth for a lighted cigarette which could be puffed from below through a rubber tube. Prichard wrote: 'It is a curious sensation to have the head through which you are smoking a cigarette suddenly shot with a Mauser bullet.'

*Camoufleurs* helped snipers in the field by making realistic hides and observation posts which fitted seamlessly into no-man's-land or the trenches: shattered brickwork, a French milestone, shorn-off poplars, a swollen dead horse, even the corpse of a Prussian or a French soldier. *Camoufleurs* also painted special full-length 'sniper's robes' in the appropriate earth and vegetation colours.

German soldiers grasped sooner than British ones that sticking out like a sore thumb was no good. Philip Gibbs describes a wooded section of the line between Vaux-sur-Somme and Curlu where a kind of warfare more like violent paintballing went on in the summer of 1915. Raiding parties of thirty to forty men stole into the thickets of no-man's-land where they met

a party of Germans . . . creeping forward from the other direction, in just the same way, disguised in parti-coloured clothes splashed with greens and reds and browns to make them invisible between the trees, with brown masks over their faces. Then suddenly contact was made.

Into the silence of the wood came the sharp crack of rifles and the zip-zip of bullets, the shouts of men who had given up the game of invisibility.

*Realities of War* (1920)

Major Underhill of the King's Shropshire Light Infantry recreated no-man's-land for Allied trainees to crawl around in at night, on a realistic site made by blowing craters in an old cornfield and littering the zone with wire and other authentic detritus. The (dummy) corpses had German *Soldbücher* or pass-books in their pockets and other useful identification on their sleeves. While defenders fired flares, attackers had to crawl as close as possible and hammer in a peg to prove in the morning where they had got to.

Making use of cover, stalking, hiding, blending, waiting, concealment, careful aiming: it was precisely the world of rough shooting and big-game hunting, but with quarry that could fire back. Prichard's book *Sniping in France* is like African or Indian *shikar* or hunting literature where, as in the classic stories by Jim Corbett, Colonel Patterson *et al*, the hunter has to offer himself rather than a tethered goat as the bait for a man-eater. Drawing the sniper's shots by pretending to be an over-eager duffer, blazing away carelessly from a loophole, while other judiciously sited spotters on your own side pin-point through telescopes the enemy's flicker of muzzle flash or the wisp of smoke that lingered longer on chilly days, was all part of the deadly sport.

Hesketh Prichard's quest for what he called 'the hunter spirit' in the army was what first led him to the Lovat Scouts. They contributed the 'ghillie suit' to the art of camouflage from their deer-stalking origins. Modern British Army snipers still make 'ghillie suits' themselves for field use: a shrubby overcoat and trousers hung with long ragged strips of frayed nylon and dyed hessian, topped with dreadlocks of greenery, a camouflage suit that makes them look like vegetating yetis or sloths when they move, but turns them into bushy undergrowth wherever they settle to kill.

Hunters understood camouflage because it was part of their regular practice. Two days after Solomon J. Solomon's letter about camouflage appeared in January 1915, *The Times* printed a response from Walter Winans, a trotting-horse fanatic and Olympic pistol-shooting champion. Born in St Petersburg in 1852, where his father was US Consul, fabulously wealthy from Baltimore railway and engineering money, domiciled in England, but with a large boar-hunting estate in his ancestral Belgium, Walter Winans indulged his sporting passions to the limit.

### Uniforms and colour

To the Editor of *The Times*

Sir, – There is one point 'S.J.S.' has left out of his letter, with which I entirely agree otherwise. That is the importance of breaking up the outline. However well the tone of the clothing of a man is made to agree with its surroundings, the outline of the man is apt to show.

Now, as an artist and big-game shot, I have found that if the waistcoat is one colour, the coat another, the leg coverings another, &c., its outline is less easy to make out. For instance, if lying down on a Scotch deer-forest waiting for deer – if the cap is the colour of a stone, the coat a peat hag, the knickerbockers grass colour, the stockings and boots black, to represent the exposed black peat, if the man keeps still he looks, not like one object, but an agglomeration of a small stone, peat hag, patch of grass, and a piece of exposed peat. The face is the difficulty, and that can be got over by wearing a veil, green or grey. I have walked close up to a man dressed as I have described and his face covered with a long bag veil of grey without noticing him, although he was the very man I was trying to find, when out deer stalking. The great thing, next to protective colouring, is breaking up the outline. I suppose rifle barrels get rusty, or else it would be as well to paint them grey or green, as they are apt to flash. W.W.

Other sportsmen too saw what Hesketh Prichard was trying to do. When George A. B. Dewar, editor of the *Saturday Review* and author of several books on fishing and wildlife, came out on one of many visits to the Western Front in the summer of 1917, he wrote a piece on the sniper schools for *The Times*, extolling the virtues of hunting and shooting as preparation for war.

The best natural training for sniping in warfare lies in 'rough' sport . . . The best sniper in war is he who can not only hit his game but discover it himself, and at the same time hide himself from it . . .

The snipers we need today to put against the cunning enemy are men who can not only shoot true, but who, besides, can 'creep and crawl' . . . for hours unseen; who knows how to avail himself of every plant stem and grass patch as cover; and who – perhaps above all – can spy between the lines of the landscape and read its tiniest types.

Lord Lovat himself visited Hesketh Prichard's school and was impressed enough to loan him his head stalker, Corporal Donald Cameron, to teach detailed observation, compass work and intelligent use of the telescope. Once, when students reported 'soldiers in blue uniforms' at 6,000 yards, Cameron looked through the glass and was able to pronounce them Portuguese. The uniforms could have been French-style, but the shape of their British headgear marked them clearly as 'our oldest ally', the Portuguese.

Often a subtler, more deductive kind of intelligence was required to make sense of the information that came from close observation. Thinking about why a tortoiseshell cat should be strolling or sunning itself regularly unmolested on one particular section of the rat-haunted trenches led to the identification of a German officers' front-line mess by photoreconnaissance, and its eventual destruction by shelling.

# 5

# Deception in the Dardanelles

Chlorine gas at Ypres had no impact on the agreeably privileged life of Duff Cooper at the Foreign Office in late April 1915. He was more upset at being hit in the mouth by the beautiful Lady Diana Manners in a weekend tiff at the Cavendish Hotel. (They later married.) What really shocked him in Monday's newspaper was seeing that the 'good poet' and 'beautiful man' Rupert Brooke had died 'from sun-stroke' (in fact an infected mosquito-bite on his lip) on a Greek island in the Mediterranean, on his way to fight the Turks. Brooke's obituary tribute in *The Times* was written by Winston Churchill, for the poet had been serving in the Royal Naval Division that Churchill had founded, and died at the beginning of the strategic Dardanelles Campaign which Churchill had inspired.

If you unrolled a map centred on the Mediterranean Sea in 1915, you would have seen only three entrances for ships. Two of them, Gibraltar and Suez, were controlled by the British, but the third was held by the Ottoman Turks, then an enemy in alliance with the Central Powers, Germany and Austria-Hungary. In the north-east corner of the map, the narrow passage called the Dardanelles runs out of the Aegean Sea, past the peninsula of Gallipoli (Gelibolu in Turkish) to the Sea of Marmara and then the Black Sea. The importance of the Dardanelles for the Entente Allies in WW1 was as a lifeline to Imperial Russia, the only sea route from the Mediterranean to the Black Sea ports of Odessa and Sebastopol. After Ottoman Turkey joined forces with the Central Powers, Russia's way out from the Black Sea was blocked. Northern ports were frozen in winter so the Ukraine grain harvest could not be exported, nor military supplies imported. While fending off Germany in the west, Russia was also being attacked by Turkish troops in the Caucasus. The Tsar appealed to his allies, Britain and France, for a demonstration of force to draw off the Ottoman Turks.

From the very start of 1915, Winston Churchill, as First Lord of the Admiralty, wanted to support the Russians by attacking the Dardanelles. Lord Kitchener and Jackie Fisher, the First Sea Lord, seemed to support him. But the idea grew in Churchill's mind from a diversion to a grand vision: a daring thrust through the Dardanelles in order to take Constantinople, which the Turks called Istanbul, and to knock Ottoman Turkey out of the war. Larger hopes (or what Sir Ian Hamilton called 'a bagful of hallucinations') also rode upon this coup: Germany would thus be cut off from meddling in the East, Greece sustained, Serbia saved, Egypt and the Persian Gulf protected, the Balkans rallied, the mouth of the Danube seized, Russia rescued and her granaries freed, and the Central Powers enclosed in the Allies' ring of iron. There was also a non-secular vision, something like the one in Ernest Raymond's 1922 best-selling novel, *Tell England*: 'It's the Cross against the Crescent again, my lads. By Jove, it's splendid, perfectly splendid! And an English cross too!'

The British and French navies tried to force a passage through the Dardanelles on 18 March 1915 with ten battleships (mostly expendable old ones destined for the scrapyard). First they bombarded the shore forts and the Turkish guns sited along a dozen miles of both the Gallipoli peninsula and the Asiatic mainland. The kite-balloon ship HMS *Manica* sent spotters aloft to report how shells were falling up to seven miles away. The newest Dreadnought battleship *Queen Elizabeth* also tested her 15-inch guns with tremendous sound and fury, signifying effectively nothing. Assuming the Turkish guns had been silenced, the civilian trawlers were then supposed to clear a 900-yard passage through the minefields. But after a string of twenty Turkish sea mines, spotted neither by seaplanes nor picket boats, managed to sink the French battleship *Bouvet* and two British battleships, *Irresistible* and *Ocean*, with over 600 (mainly French) dead, naval operations were halted by Rear Admiral John de Robeck.

On the actual day of the Allied attack, Churchill was visiting French trenches among the sand dunes of the Belgian coast where barbed wire ran right down into the North Sea, snagging corpses covered in seaweed and washed to and fro by the tides. He says he tried not to think about what was happening in the Dardanelles, though he knew that if they succeeded there, this stagnant deadlock in France and Flanders could be broken. Back in London the next day, the politicians and top brass seemed intent on persevering. Fisher said the Navy could safely lose a dozen battleships.

But by 23 March, Rear Admiral de Robeck had lost his nerve and sent Churchill a cable saying that the Dardanelles could not be taken by the navy without the army first destroying the artillery on shore. Some Turkish guns were mobile, others hidden; they could not be spotted from the sea or from the air by the *Ark Royal*'s weedy Sopwith seaplanes. This telegram filled Churchill with consternation. Delays only gave the enemy time to reinforce. Because Captain Hall's Room 40 had decrypted messages between Berlin and the German commander of the Ottoman Navy, Churchill composed a telegram to send to de Robeck:

We know the forts are short of ammunition and supply of mines is limited. We do not think the time has yet come to give up the plan of forcing Dardanelles by a purely naval operation.

But the three senior admirals in London backed the judgement of de Robeck as the man on the spot; Churchill's cable was written but never sent, and the purely naval attack never resumed. Churchill bowed to the sea lords' decision 'with regret and anxiety'.

This was a crux of history. What might have happened had the Allied ships kept pressing on through the Dardanelles in the third week of March 1915? Believers, like the daring attacking submariner Commodore Roger Keyes, tried to revive Churchill's idea of the naval-only assault through the straits to seize Constantinople, but were overruled. Ten years afterwards, commanding the Mediterranean fleet, Keyes steamed through the Narrows and was overcome with emotion. 'My God,' he said at last, 'it would have been even easier than I thought; we simply couldn't have failed . . . and because we didn't try, another million lives were thrown away and the war went on for another three years.'

For the rest of his life, Winston Churchill would have the dead of the Dardanelles and the disasters of Gallipoli laid at his charge. What he hoped would be 'one of the great events in the history of the world' did not happen. The truth is that the amphibious landing on the peninsula in April 1915 was neither Churchill's plan nor his original concept: he had put his faith in the ships alone forcing their way through the straits to capture Constantinople.

When Earl Kitchener of Khartoum pronounced that the army would carry through the operations, it was easier said than done. The 'Incomparable' 29th Division of British infantry who joined the Expeditionary Force had been sent more as garrison troops than as an amphibious invasion force. The organisation required for each of the two roles was quite different. The 'Constantinople Expeditionary Force' planning was ad hoc, because not they but the Royal Navy was meant to force the Dardanelles. Ships had been loaded haphazardly, with units separated from their equipment, guns from their ammunition, and not much thought given to what was going to be needed first. There were too few engineers, and no provision for building piers, jetties and cranes at the beachhead. There were not enough smaller boats to ferry supplies and people ashore. Medical stores and personnel were inadequate. No one had thought about the water supply. There wasn't even a big base where all this could be sorted out, because the nearest Greek islands, Imbros and Lemnos (where there was a big harbour at Mudros) did not have enough water. So the ships had to go 800 miles to Alexandria in Egypt. 'There are three islands here,' soldiers said, 'Lemnos, Imbros and Chaos.'

All this gave General Liman von Sanders of the German Military Mission four weeks to organise the Fifth Turkish Army to defend the Dardanelles. There was no secret about the British intentions, but von Sanders did not know exactly where they would land and so split his forces into three equal parts of 20,000 men, to cover the northern neck of the peninsula at Bulair, the southern foot of Gallipoli and the Asiatic side. Obvious landing-beaches in the south of the peninsula were mined, wired and lightly garrisoned, with forces held in reserve to move where needed. Commanding the 10,000 men of the Turkish 19th Division in reserve at Bigali down south in the peninsula was a lieutenant colonel called Mustafa Kemal, the future Atatürk, founder of the Turkish republic and president of Turkey from 1923 to 1938.

A fleet of 200 ships carried the British Mediterranean Expeditionary Force from Lemnos across the wine-dark sea towards the coast where the city of Troy had once stood. General Sir Ian Hamilton deployed six divisions. The two French ones landed at Kum Kale, on the Asiatic side of the straits, but this was merely a feint, as was the diversion by the Royal Naval Division at Bulair at the neck of the peninsula in the north. The main attack was on the southern peninsula, by 30,000 men of the British 29th Division and the two divisions of Anzacs (Australian and New Zealand Army Corps).

It was planned that the British would land at five beaches around Cape Helles to take the high ground at Achi Baba; the Anzacs were to land a dozen miles away at Gaba Tepe to seize the Sari Bair ridge. From there they would advance together on the Pasha Dagh plateau that dominated the 'Narrows' section of the Dardanelles. They had maps but as yet no photoreconnaissance. Kitchener, thinking that the Turks would run away, said there was no need for aeroplanes.

The first man on the peninsula was one of the last men out, nine months later, and his initial job was deception. Lieutenant Commander Bernard Freyberg of the Royal Naval Division was born in London but raised in New Zealand. He had been with Pancho Villa's revolutionary forces in Mexico at the outbreak of war, made his way to England and got into the Royal Naval Division by badgering Churchill on Horse Guards Parade. Not twenty-four hours after burying his brother officer, the poet Rupert Brooke, in a moonlit olive grove on the island of Skyros, he swam two miles to the beach below Bulair. It was Saturday night, 24 April 1915, and he was semi-naked, thickly oiled with engine-room grease, with only the whites of his eyes visible in a mask of brown. He was towing a bag of seven flares to set off on shore to make the Turks think a landing was already underway there, a long way north of the real landing-sites. This action won Freyberg the first of his four DSOs.

The Sunday was a serenely beautiful spring day, with the blue Aegean calm and smooth, ideal sea conditions for a landing just before dawn. At Gaba Tepe in the north, the first of 15,000 Anzac troops were landed under cliffs, over a mile north of beach 'Z', where they should have been. It was not disastrous, just another muddle. John Buchan describes the Australians dropping their packs to scramble a hundred feet up through myrtle scrub and a yellowy rock-garden of spring flowers, purple cistus, grape hyacinth, anemone, asphodel and amaryllis, to entrench under fire at the top of the cliffs, staring straight into the rising sun.

'Now you have only to dig, dig, dig until you are safe,' General Hamilton told the Anzacs. This is where the Australians started earning the nickname 'Diggers', only putting down their spades to resist counter-attacks with their bayonets. In months to come Anzac

Cove would resemble a mad mining camp, quarried from apricot-coloured rock and dirt by half-naked, sun-bronzed men.

At 'S', 'Y' and 'X' beaches around Cape Helles, other landings were relatively easy or unopposed to begin with, but the invaders did little with them. Troops from 'Y' wandered to within yards of Krithia village, which was deserted, then returned to the beach. At that moment, they outnumbered all the Turkish defenders of Cape Helles. No Allied soldier ever got that close to Krithia again.

At 'W' beach the thundering naval bombardment stopped. The 1st Battalion Lancashire Fusiliers were still packed tight in two dozen clinkered ship's boats with their tow-ropes cast, each being rowed ashore by four naval ratings, when accurate Mauser rifle fire from the Turkish redoubts started hitting them. Commander Charles Samson of the RNAS flew overhead in a Maurice Farman. 'I saw Hell let loose,' he wrote. 'The sea was literally whipped into foam by the hail of bullets and small shells.' Only two boats got to shore. Tipped overboard from boats which could not land, weighed down with up to 70 pounds of kit, pith-helmeted men struggled in four feet of water to get towards the barbed wire and mines on the beach. If wet sand jammed their rifles, bayonets were their only weapon. Over 500 were killed and wounded; the dead included 63 of the 80 naval ratings. 'Why are they resting?' said people looking through ship's binoculars at the still bodies on the beach.

The plan at 'V' beach was to run a kind of Trojan horse of a ship ashore into the crescent between a Turkish fort and a crenellated castle, and then to disembark troops through the square sally ports cut in her port and starboard bow, over gangways and across a pontoon of lighters towed into place by a steam hopper. SS *River Clyde* was a 4,000-ton collier or coal transport ship which bore mottled sandy-yellow and black camouflage and the soldiers' nickname 'the Dun Cow'. The *River Clyde*, with 2,000 soldiers aboard and a dozen machine guns sandbagged in the bow, was accompanied by open cutters full of 'Bluecaps', 1st Royal Dublin Fusiliers, towed by steam pinnaces. From above at Seddülbahir the Turkish soldiery opened withering fire from rifles and machine guns, plus shrapnel from two 'pom-poms' or quick-firing cannons. This massacred the Dublins in their boats together with the first companies of 2nd Hampshires and 1st Royal Munster Fusiliers, who raced out of the *Clyde*, falling over each other across the lighter barges. When the shore pontoon drifted away, soldiers jumped into the water and drowned, dragged down by their kit. The naval airman Samson who flew over reported fifty yards of sea 'absolutely red with blood'.

Men still inside the *River Clyde* heard nightmarish noises. They at least had food and water, which the hundreds trapped on the beach did not have. As Royal Navy big ships ventured closer to shell Seddülbahir, small boats tried to collect the wounded in the afternoon and evening. After dark, one of the staff officers on board volunteered to go ashore to assess the situation.

Lieutenant Colonel Richard 'Dick' Doughty-Wylie was 46 years old, a nephew of the Arabian explorer Charles Doughty and the married lover of the Arabist Gertrude Bell. He had spent twenty years as an active professional soldier in Asia and Africa before his wounds pushed him towards the job of military consul in Turkey and Abyssinia. Old soldiers never die, however. In 1909, before the Great War brought enmity with Turkey, Doughty-Wylie used Turkish regular troops to prevent a massacre of Armenians at Adama. He spoke Turkish, and his knowledge of the Ottoman Empire was seen by the Egyptian command as so useful to the Mediterranean Expeditionary Force that he was taken on to Sir Ian Hamilton's staff.

Doughty-Wylie found men still alive sheltering under the bank on the beach below Seddülbahir and in the morning he was among the officers who rallied them. Collecting the remnants of three battalions, the Hampshires, the Dublins and the Munsters, he encouraged the attack that captured the Turkish fort on the left and the ruined village. Doughty-Wylie then came back to arrange for *Queen Elizabeth*'s massive guns to shell the remaining Turkish redoubt on Hill 141 overlooking the beach. When the naval bombardment finished at two in the afternoon, he personally led the infantry attack on the last fort. He had lost one puttee and was carrying only a walking stick because he did not want to bear arms against his old friends the Turks. He was buried where he fell, at the summit of the hill in the moment of victory, when a sniper's bullet blew away the side of his face, killing him instantly. His entry in the *Oxford Dictionary of National Biography* ends: 'Doughty-Wylie was posthumously awarded the Victoria Cross. He was the highest ranking officer to win the award during the Gallipoli campaign.'

It was a cruel spring on the Peninsula. So many soldiers were killed in the first month – more than in the three years of the Boer War, says John Buchan – that the British and the Turks organised a one-day truce on 24 May to bury their dead. Aubrey Herbert came across 'entire companies annihilated – not wounded, but killed' by machine-gun fire. The reek of dead men and mules going bad in the sun made soldiers vomit. Staff officer Compton Mackenzie jumped up on a parapet at Quinn's Post that day: 'Looking down I saw squelching up from the ground on either side of my boot like a rotten mangold the deliquescent green and black flesh of a Turk's head.' The smell of death and putrefaction was 'tangible . . . clammy as the membrane of a bat's wing', and Mackenzie said it took two weeks to get just two hours' exposure to it out of his nostrils. At night, the offshore wind carried the stench out to the ships at sea.

Now camouflage became important. From the moment they set foot on the charnel-house peninsula, Allied soldiers were under accurate rifle fire. They looked for shelter or dug it for themselves with trenching tools. Being neither camouflaged nor concealed put them at a severe disadvantage vis-à-vis their opponents, as the Australian Albert Facey says in his extraordinary autobiography, *A Fortunate Life*, which calls his time on the peninsula 'the worst four months of my whole life'. He found himself with other Anzacs crawling about in small groups, with NCOs having to make the plans because all the officers had been picked off:

We lost many of our chaps to snipers and found that some of these had been shot from behind. This was puzzling so several of us went back to investigate, and what we found put us wise to one of the Turks' tricks. They were sitting and standing in bushes dressed all in green – their hands, faces, boots, rifles and bayonets were all the same colour as the bushes and scrub. You could walk close to them and not know. We had to find a way to flush these snipers out. What we did was fire several shots into every clump of bush that was big enough to hold a man. Many times that we did this Turks jumped out and surrendered or fell out dead.

The *War Illustrated* for 21 August 1915 has a photograph of a captured Turkish sniper between two Australian soldiers in shorts with shouldered rifles. Only his bald head is visible above the mass of leafy shrubbery that hides him. The caption reads:

The Turk, wily as are all Orientals, is quick to assimilate the ideas of his temporary masters. Sniping, which has been such a feature of the Great War in Europe, is also very much in vogue at the Dardanelles. This captive Turkish sniper seems to have found an effective disguise, but not so sufficiently as to escape the vigilance of his foes.

Aubrey Herbert heard about it:

The first convincing proof of treachery which we had was the story of the Turkish girl who had painted her face green in order to look like a tree, and had shot several people at Helles from the boughs of an oak.

'Wiliness', 'treachery': those were the names for camouflage and deception when enemy foreigners used them and we did not. Kangaroo-hunters of the Australian outback and deer-hunters from New Zealand's mountains adapted to new forms of shooting. Billy Sing of the 5th Light Horse was the best-known Australian sniper. Half Indian, very dark with a thick moustache and goatee beard, Sing specialised in snap-shooting on his spotter's commands; his record kill was nine men in one day. When there weren't enough hand grenades, men manufactured their own from dynamite in jam tins packed with rusty metal scrap and snippets of barbed wire. Opposing trenches were sometimes no more than a cricket pitch apart.

Yet these soldiers who were fighting each other tooth and nail to the death did not, on the whole, hate their enemy. They seem to have respected each other's cheerfulness and bravery amid shared squalor. Alan Moorehead compared this to the cruel friendliness of the very poor. 'Abdul', 'Johnny Turk' or 'Jacko', as the Anzacs called the Turkish soldier, was a character highly regarded by soldiers like Private Henry Barnes:

I never heard him decried, he was always a clean fighter and one of the most courageous men in the world. When they came there was no beating about the bush, they faced up to the heaviest rifle fire that you could put up and nothing would stop them, they were almost fanatical. When we met them at the armistice [24 May] we came to the conclusion that he was a very good bloke indeed. We had a lot of time for him.

--

Larks sang above hills dotted with blue cornflowers and scarlet poppies that only exacerbated the hay fever of A. P. Herbert, the future humorist, lawyer and parliamentarian. When he sneezed on patrol one night in no-man's-land, he alerted a Turkish sniper who shot Herbert's fellow scout through the femoral artery. Herbert had to carry the bleeding, dying man back to his own trenches, and put the incident into *The Secret Battle* (1919) which Winston Churchill rightly described as 'one of the most moving of the novels produced by the war . . . a soldier's tale cut in stone'.

Herbert shows in the book how different conditions were in the two theatres where he served, France and the Peninsula. No one ever went home on leave from Gallipoli: you left on a stretcher or sewn up in a blanket. In some French sectors, the line could be quiet for months, but on the Peninsula 'from dawn to dawn it was genuine infantry warfare':

But in those hill-trenches of Gallipoli the Turk and the Gentile fought with each other all day with rifle and bomb, and in the evening crept out and stabbed each other in the dark . . . The Turk was always on higher ground; he knew every inch of all those valleys and vineyards and scrub-strewn slopes; and he had an uncanny accuracy of aim. Moreover, many of his men had the devotion of fanatics . . . content to lie there and pick off the infidels till they too died. They were very brave men. But the Turkish snipers were not confined to the madmen who were caught disguised as trees in the broad daylight and found their way into the picture papers. Every trench was full of snipers, less theatrical but no less effective. And in the night they crept out with inimitable stealth and lay close in to our lines, killing our sentries, and chipping away at our crumbling parapets.

Through the gruesome Gallipoli campaign, Herbert's loathing of Winston Churchill grew. Waiting to invade in April he had written in his diary, 'Winston's name fills everyone with rage. Roman emperors killed slaves to make themselves popular, he is killing free men to make himself famous.'

Three thousand miles away, in London, the dramatic resignation of ageing Admiral Jackie Fisher led to a political crisis. The Liberal Prime Minister Herbert Asquith (reeling from the news that his mistress was marrying another member of his cabinet) was forced to form a wartime coalition government with the opposition, Bonar Law's Conservatives, who were described by Compton Mackenzie as 'barren of policy yet greedy of place and patronage'. The Tories demanded two Liberal scalps as the price of coalition: Lord Haldane's and Winston Churchill's.

On 26 May 1915, Churchill was ousted from the Admiralty, although he kept a seat in Cabinet and on the Dardanelles Committee to try and see the enterprise through. This was a tremendous shock for him – Violet Bonham Carter believed it 'the sharpest and the deepest wound he suffered in his whole career'. Clementine Churchill told her husband's biographer, Martin Gilbert: 'I thought he would die of grief.' Aubrey Herbert's wife, Mary Vesey, dined at No. 10, Downing Street in June 1915, and was seated next to Winston Churchill:

He was in a curious state, really rather dignified, but so bitter. He and Clemmy look very broken. He told me that if he was Prime Minister for 20 years it wouldn't make up for this fall.

Aubrey Herbert wrote back angrily from the peninsula, 'As for Winston, I would like him to die in some of the torments I have seen so many die in here. But his only 'agony' you say is missing being PM.'

The Tories in the government coalition blocked Churchill from going out in person to ginger up Gallipoli in July. Churchill could do nothing. The scheme was out of his hands. But the setback did give him something manual to occupy his mind – he started painting.

Like a sea-beast fetched up from the depths, or a diver too suddenly hoisted, my veins threatened to burst from the fall in pressure. . . I had to watch the unhappy casting-away of great opportunities, and the feeble execution of plans which I had launched and in which I heartily believed . . . And then it was the Muse of Painting came to my rescue . . .

Churchill started with his children's paintbox one Sunday in July 1915. The next day he procured easel, canvas, oil paints, palette, brushes, and a long white dustcoat. For Churchill, painting a picture was a mixture of fighting a battle (but with 'no evil fate' to avenge 'the jaunty violence') and a sort of enchantment. His daughter Mary Soames said: 'When he picked up a paint brush it was like picking up a magic wand.'

Back in the Aegean, in August 1915, another painter, Norman Wilkinson, had climbed to the foretop of HMS *Jonquil* to observe, from a safe distance, the landings of the British 9th Corps at lightly defended Suvla Bay, in the last big push of the Dardanelles campaign. He called it 'the living cinema of battle':

Glasses were necessary to distinguish the light khaki of our men against the scrub and sand. The troops marching in open order across the salt lake . . . crossed the unbroken surface of silver-white. Overhead shrapnel burst unceasingly, leaving small crumpled forms on the ground, one or more of which would slowly rise and walk shoreward, while others lay where they fell . . .

Alan Moorehead's *Gallipoli*, which is a masterpiece of narrative history, includes Norman Wilkinson's painting of soldiers crossing the salt lake. Moorehead, writing forty years later, saw a new principle slowly being revealed in the Gallipoli campaign:

Everything that was done by stealth and imagination was a success, while everything that was done by means of the headlong frontal attack was foredoomed to failure.

This was true of the flank landings at Helles in April 1915, and of the way Gurkha Bluff was taken by the British on 12 May. It was true, too, of Compton Mackenzie's deception initiative on Lesbos in July. The writer was sent in his Royal Marines uniform from GHQ to Mytilene on the island of Lesbos. His orders were to make plans for establishing a 'secret' military base there, in readiness for a forthcoming big Allied attack on Smyrna on the Turkish mainland. No such attack was planned: the whole thing was a diversion. But Mackenzie told various people – the British Consul, *The Times'* correspondent, the Civil Governor – about the plans, 'in confidence', and numerous Greek small businessmen soon came rushing forward with drachma bribes in the hope of future contracts with military forces. Mackenzie, of course, suavely but unconvincingly denied they were coming. This three-week deception operation, planned and set in motion by the staff officer Guy Dawnay, was effective and historically important.

At German-assisted Ottoman HQ, rumours garnered from various sources about British movements were alembicated into hard intelligence about the fictional attack on Smyrna. Enemy troops were braced and reinforced in the wrong places. Thus no U-boats were ready to stop the invaders when they eventually came to Suvla Bay in August. Meanwhile, 25,000 men were landed secretly, by night, at Anzac Cove, and then packed into concealed trenches. Australian tunnellers, many of them ex-goldminers, dug new saps towards the Turkish lines for a lightning surprise attack at Lone Pine on 6th August. Hopes were high for a coordinated breakthrough.

71

Tragically, the Suvla landings and attacks fell apart in exhaustion and indecision because the chain of command did not join up the dots of what they were supposed to be doing. At Suvla, the elderly General Stopford thought his only job was to make camp by the bay and wait for the howitzers to be supplied. At Anzac, a desperate plan involving some of the best soldiers took terrible casualties because the mass of new English troops who were meant to sweep round from Suvla and help the Australians and New Zealanders capture the heights were still bathing on the beach.*

It was all a great disappointment. Forty thousand men were killed and wounded on the peninsula in August to gain only a few square miles. September 1915 was made miserable by dysentery and diarrhoea, 'between the Devil and the W.C.' The Dardanelles project was slowly strangling, but the legs still madly kicked. In October, Churchill proposed using poison gas on the Turks, and Roger Keyes wanted to crash ships through the Chanak Narrows; in November Kitchener considered attacking Bulair, and there were other wild schemes.

The Aegean was still glorious to look at that autumn. Norman Wilkinson's watercolour sketches from Gallipoli make thirty full-page plates in his 1915 book *The Dardanelles*. These elegant sea- and landscapes depict no horrors. By the white tents of a dressing station on the white sands of 'A' beach at Suvla you see netting draped over poles, more for shade than for camouflage, not bloody bandages. You sense a large human enterprise nonetheless dwarfed by vast spaces of sea, sky and intractable land. There are long views of balloon ships, submarines, seaplanes, the *Aquitania* converted into a huge floating hospital. The bright stillness and depth of the paintings is like the slow daze of a holiday afternoon.

And then, suddenly, the whole fleet sailed away. In a flash of fireworks, they vanished. At Troy, the departure of the Greek fleet was the prelude to the deception: the Trojan Horse was left behind on the

---

* This is the climax of the famous 1981 Australian feature film *Gallipoli*, written by David Williamson and directed by Peter Weir. The film – which broadly suggests that decent Australians were sacrificed for English toffs – is mythic because, as L. A. Carlyon rightly observes, 'Gallipoli has become Australia's Homeric tale.'

shore with a special forces unit hiding inside. At Gallipoli, by contrast, the sailing away was the climax of the deception, leaving the Turks only empty trenches on the hills and burnt offerings on the beach.

The evacuation of Gallipoli was the best thing about the campaign. A. J. P. Taylor called it 'the successful end to a sad adventure' and Brigadier John Monash, the Jewish engineer who fought in the Peninsula and went on to become Australia's greatest WW1 general, said it was 'a most brilliant conception, brilliantly organised, and brilliantly executed – and will, I am sure, rank as the greatest joke in the whole range of military history'.

On 11 October 1915, Kitchener asked Hamilton what losses could be expected if the army pulled out: '50 per cent', was the gloomy reply. So, on 16 October, Sir Ian Hamilton was relieved of his command and replaced by Sir Charles Monro, who visited Suvla, Anzac and Helles in one day and on the next, 31 October, recommended evacuating the peninsula. 'He came, he saw, he capitulated,' said Churchill. (Most commentators call this bon mot unfair.) Lord Kitchener himself visited. He had come out eager to push on, but after seeing the rough terrain he sent a telegram to Asquith saying that 'the country is much more difficult than I imagined'.

Back home the architect of the Dardanelles scheme finally made his farewell speech to the House of Commons, resigning from Asquith's coalition cabinet. Winston Churchill was off to join his regiment, as a major, in France. Kitchener was the only one of Churchill's colleagues who formally visited him when he left the Admiralty, an act of kindness the younger man never forgot. But the age of Kitchener was ending in contradictions. His manly, moustachioed face concealed a love of fine china and furnishings, his decisive speech masked a havering temperament, and he simply could not make up his mind about what had to be done in Gallipoli. 'K' blew hot, 'K' blew cold; aides danced attendance and tried to decipher the icon's intentions. Lloyd George had noted Kitchener's 'concealment of his limitations under a cloak of professional secrecy'; and John Buchan said that Kitchener was a poor administrator protected by 'that air of mystery and taciturnity which the ordinary man loves to associate with a great soldier'. Herbert Asquith's wife Margot dismissed him as 'a great poster', and the Prime Minister was already plotting to get rid of him; Sir William Robertson was rising to become CIGS in Kitchener's place.

73

Winter came hard and fast to Gallipoli on 27 November 1915. Twenty-four hours of heavy rain caused flash floods at Suvla that swept down ravines and through trenches and dugouts, drowning over 200 men and resurrecting the bones of the not-long dead. A south-west gale destroyed jetties, and beached a destroyer. Then the wind veered round to the north and temperatures plummeted for two days and nights; thousands of Anzacs experienced their first-ever snow and ice as severe exposure, frostbite and gangrene. Royal Fusilier sentries were found grey and frozen, dead at gelid parapets. Shivering men in stiffened blankets tried to thaw out over tiny flames, while the cruel potshot at enemies scavenging kindling in the open.

On 7 December the British Government decided to evacuate Suvla and Anzac, but to stand firm in Helles at the toe of the peninsula. Anxieties about losing face before the Muslim world had to be squared with military realism. The next day Monro ordered General Birdwood to execute the plans that Colonel Aspinall-Oglander and Lieutenant Colonel Brudenell White had been carefully working up. After a quarter of a million casualties in eight months, it was time to cut and run.

Complete secrecy now was vital: careless talk could cost thousands of lives. In cabinet in London on 24 November, a fearful Lord Curzon had painted the nightmare of a retreat being shelled to shambles, with awful political repercussions. The fact that Lord Milner and Lord Ribblesdale had loudly debated 'Evacuation of the Peninsula' in the House of Lords in October and November accidentally misdirected the Turks and Germans. They could not credit such stupidity and carelessness among intelligent people, so supposed the debate was just propaganda.

In the days before Christmas 1915, life appeared to be going on as normal in daylight at Suvla and Anzac, with men disembarking, mules going up the line with stores, the occasional sniper-shot. In fact, the same teams of men were disembarking each day, and the panniers and boxes were empty. Every night, according to the plan, thousands of men were slipping away down to the beach and making their way to the boats, the sick and wounded first, then PoWs, then batches of infantry. There was fierce competition to be the last to leave; men who had been in the initial landings on 25 April won that risky honour. As the trenches emptied, the rearguard left behind rifles wedged in the parapet. They were rigged to self-fire by a wire or cord round the trigger, linked to a Heath Robinson mechanism of cans that dripped water or trickled sand till the lower can had enough weight to exert a finger's pressure. Lone individuals tended their friends' graves for the last time, moved from loophole to loophole, setting booby traps and mines and pulling barbed wire across communication trenches, then raced downhill with boots muffled by sacking. Evacuation was risky; the advance estimate of casualties ranged from 25,000 to more than 40,000. In the event, the British managed gradually to drain away in secret, until at 4 a.m. on Monday, 20 December 1915, 83,048 people had gone from Suvla and Anzac, with only a few minor injuries, most caused by alcohol.

Incredibly, the miracle was repeated at Helles, starting on 28 December and ending early on 9 January 1916, two days after the Allies beat off a major Turkish attack. Through the hulk of the *River Clyde* and from other jetties, some 35,000 men, 4,000 animals, 110 guns and 1,000 tons of stores were all evacuated by 3.45 a.m. They tried to destroy everything they left: gun-spiking, sandbag-slashing, pouring away drink, spoiling flour and fuel. Men shot over 500 mules and horses on the beach (though the kinder ones freed their donkeys and left them with fodder for the Turks to find). Finally, they set off the charges under a mountain of oil-soaked kit: volcanic explosions mushroomed flames into the night sky, rained debris on the last boats leaving and started, too late, a terrific firework show of enemy shooting and shelling. When the Turks eventually put out the flames they still found more useful materiel than you could shake a stick at. It took two years to ship it all to Istanbul.

German Intelligence found this magical *coup de théâtre* impossible, and promptly spread the plausible rumour that the British had bribed the Turks to let them slip away. Churchill himself wrote to his wife on 13 January 1916, 'Perhaps a little money changed hands & rendered this scuttle of "imperishable memory" less dangerous than it looked.' Noble-minded Henry Nevinson denied it: 'That malignant depreciation of a most skilful enterprise was a libel both on the enemy and on our own officers and men. There was not a vestige of truth in it.' There is no evidence of a deal. In 1916, however, the British offer of a £2 million bribe or *douceur* to the Ottoman Turks to release their besieged army, trapped in Iraq at Kut-al-Amara, is well attested. Anything is possible. In his *Room 40: British Naval Intelligence 1914–18*, Patrick Beesly tells what he himself calls the 'somewhat unbelievable story' that earlier in 1915 Admiral Reginald Hall had authorised two merchant agents in Turkey to spend up to £4 million to try and buy a passage through the Dardanelles from the Young Turks. If that bribe had worked, there would have been no Gallipoli campaign at all.

# 6

## —Steel Trees

The official historian, Brigadier General Sir James E. Edmonds, in *A Short History of World War I* writes: 'Camouflage, already practised, was officially recognized under that name in June 1915, and by 1 January 1916 the service and the manufacture of material were definitely organized.' In this statement, the former Royal Engineers' director of works in France camouflages a lot of activity under bland bureaucratese.

Major General Sir Robert Porter of the Second Army in France said that towards the end of 1915 one of his officers handed him a large file of papers with drawings and descriptions showing how troops, guns and camps could be concealed from the enemy. This packet came from the painter Solomon J. Solomon, last seen badgering *The Times* and doing experiments at Woolwich, who complained in his accompanying letter that he had been sitting on the steps of the War Office for six weeks trying to get an interview with someone about the importance of camouflage.

Somehow Solomon's papers got up to the Second Army commander, General Sir Herbert Plumer, a man who looked like David Low's choleric cartoon figure Colonel Blimp, red-faced, white-moustached, pot-bellied. But Plumer knew cover was important from his experience of leading irregular troops in Matabeleland under Baden-Powell, fighting the Boers in South Africa. Plumer, known as 'Old Plum-and-Apple' and respected as a conscientious general who looked after the lives of his soldiers, saw that Solomon was summoned to France to get concealment going. Early in December, the War Office arranged for Solomon to visit the front in France, to report on what French *camoufleurs* were up to and give an opinion on whether the British should be doing something similar.

Solomon crossed the English Channel to the war zone on a crowded troopship, wearing an inflatable waistcoat under his fur coat lest he should be torpedoed. Near the harbour, he noticed the string of small craft supporting underwater steel netting that protected troopships against German submarines. In Boulogne he was met by a lieutenant from the Royal Engineers. The British Army had decided to put the new *camoufleurs* in with the sappers, on the grounds that if engineers built things, they could hide them as well. In his biography *Kitchener: the Man behind the Legend*, Philip Warner points out that camouflage and deception fall naturally into the province of the sappers not only because 'they have materials for making replicas and deception targets, but because engineers look at landscapes with a more under-standing eye than most other soldiers do'.

The officer sent to meet Solomon, Lieutenant Malcolm Wingate of 459th Field Company, Royal Engineers, was the younger son of the Governor General of the Sudan and Sirdar (Commander-in-Chief) of the Egyptian Army, Sir Reginald Wingate. Malcolm Wingate went on to win the DSO, the MC and the Croix de Guerre before being killed in action near Arras on 21 March 1918, and Solomon always remembered his kindness. The painter was out of his depth in the military world; his artist's temperament put him at odds with soldiers' codes of rank, form and procedure, which he found meaningless and stupid. Wingate, however, made a gentlemanly escort for him, driving Solomon to GHQ at St Omer to dine with the engineer-in-chief, General Fowke, and the next day on to the French camouflage atelier at Amiens, the hub of invention and experimentation. Here artists were working out the right colours for disguising and screening and also making realistic dummies, including armoured trees for use as observation posts. Solomon hit it off famously with the French painters, some of whom had been, like him, at the Ecole des Beaux Arts, but he somehow also got the incorrect idea that *camoufleurs* were not under military control. Many of Solomon's future frus-trations and disappointments sprang from this misapprehension.

Solomon was shown round GHQ at St Omer and taken to dine at the château of the commander-in-chief, Field Marshal Sir John French. However, French was away in Paris, and his temporary house-guest, Winston Churchill, was dining out that night, so Solomon met neither.

By December 1915, the tectonic plates of British command on the Western Front were shifting. Solomon arrived in French's very last days; he had stayed on after the disastrous Battle of Loos in late September only to be outmanoeuvred in October and November by General Sir Douglas Haig (who was now telling influential people, including King George V, that Sir John had muffed things at Loos). The two soldiers were opposites. Philip Chetwode said: 'French was a man who loved life, laughter and women, whereas Haig was a dour Scotsman and the dullest dog I ever had the happiness to meet.' Haig and French came to loathe and despise each other with all the ardour of former friends.

Churchill, however, was still close to French. Fresh from his resignation, the former government minister, now Major W. L. S. Churchill, arrived at Boulogne on 18 November, and the military landing officer told him that the commander-in-chief had sent a staff car. Churchill used the vehicle to touch base with his Yeomanry regiment, the Queen's Own Oxfordshire Hussars, near Boulogne, before going on to the hospitality of French's château. When asked by French what he wanted to do, Churchill told him that he would like to spend some time in the front line with the elite Guards Division. The prodigal son was back in the army again, but he knew he had to master the special conditions of trench warfare as a regimental officer before other possibilities of command could arise. Two days later, Churchill was driven to the 2nd Battalion Grenadier Guards near Merville. They had not been consulted and were none too happy about having a politician foisted on them, even though his bellicose ancestor the Duke of Marlborough had once commanded the regiment.

It took three wet hours to get into the front line. The officers started out on horses, riding through the icy drizzle of a late November afternoon. Occasional red gun-flashes stabbed the darkening plain. Habitation gave way to ruins; shell holes and rubbish increased; leafless trees stood scarred and split among fields rank with weeds. At a halt in the darkness, orderlies took away the horses, and they walked the final two miles over sopping fields to the battalion HQ, Ebenezer Farm, a thousand yards behind the front line, an edifice of shattered brick propped up inside with sandbags. Churchill's education in trench warfare began here. He wrote to his wife:

Filth & rubbish everywhere, graves built into the defences & scattered about promiscuously, feet and clothing breaking through the soil, water & muck on all sides; & about this scene in the dazzling moonlight troops of enormous rats creep and glide, to the unceasing accompaniment of rifles and machine-guns & the venomous whining & whirring of the bullets which pass overhead.

Around the same time Solomon J. Solomon met the Second Army commander, General Herbert Plumer, who had brought him out to France, and went on to Canadian HQ where the bluff and genial General H. E. Burstall, in charge of Canadian artillery, asked Solomon to make forward Observation Posts or 'OPs', also known as 'Oh Pips', disguised as trees, like the French did.

Near Hill 63, between Ploegsteert Wood and Messines, south of Ypres, Solomon sketched a tree on a hill in the rain. The next day, with borrowed sleeping bag and gas mask, Solomon found himself at 39th Division HQ at Brielen, north-west of Ypres, where General Percival said he needed 'Oh Pips' for artillery spotters so he could more accurately shell the German lines. His gunners had found some willows which, if replaced by imitation trees with steel cores, would serve their purpose. Would Solomon go and look?

Sunshine gleamed on the yellow water of the Yser canal, which was bridged by a dozen military pontoons. The opposite bank, nearer the German lines, was piled up twelve feet high with sticky yellow clay thrown up by excavating two tiers of dugouts. Tall poplar and birch grew up beyond that, some broken and splintered by artillery fire. As Solomon watched, a German shell hit a local landmark known as 'the White Château' in a cloud of black smoke. He sketched the scene on brown paper, thinking white paper was too conspicuous.

They crossed one of the bridges over the canal; the odd bullet from a German sniper whacked into the wood or went whanging off the iron. Solomon noticed that a narrow length of hessian cloth ran along one of the iron rails of the bridge. Like a skimpy towel, the space below it revealed your lower legs, while above it failed to cover your head and shoulders. This screen was intended to give people confidence, but Solomon sensibly thought a solid pile of sandbags on the enemy sides of bridges would give real rather than imaginary protection.

They scrambled up the canal's slippery bank and peered over discreetly. Solomon studied how the tree trunks grew out of the steep slope on the other side, facing the Germans. He was wearing a thick mackintosh and as they moved along the old towpath, keeping low and moving from tree to tree, he found it hot work and hard going.

The thick yellow mud sucked at his feet. Panting in a culvert dugout, Solomon peered through a slot at an enemy machine-gun post in a concrete pillbox known as 'the mushroom'. The existing willow trees in their clumps and copses were too small for his purposes, but Solomon proposed to General Percival that he make two steel-jacketed trees on his return to the UK.

Back in London, Solomon saw General Scott Moncrieff at the War Office. Scott Moncrieff recommended a firm in Holborn, Messrs Roneo, who worked with hardened steel. Solomon gave them scaled drawings and asked them to make two plywood models. What he wanted was an oval-shaped hollow steel conning tower, bolted together in two-foot sections, just wide enough for a man to climb up in order to see out. To make this OP pass as a convincing tree he needed real bark for the outside. Solomon decided to go right to the top. He would ask King George V's permission to get a decayed willow from Windsor.

On Saturday, 18 December, from his home at 18, Hyde Park Gate, Solomon hand-wrote a letter to the Keeper of the Privy Purse:

Sir,

I have just returned from G.H.Q. where I was invited to report on some matters connected with 'Invisibility in Warfare'.

The French are making what they call 'camouflage' objects to serve as artillery observation outposts and have offered to make these for the British Army, but much time will have elapsed before such things can be produced for our use by them.

The General Officers of the Second Army have impressed upon me the urgency of their need of armoured outposts – mainly imitations of existing trees – and they sent me to the front that I might study their requirements. They gladly agreed with my proposal to experiment and endeavour to provide them (with the assistance of engineers and others) with the much needed posts of vantage before the development of the spring and summer vegetation might render them ineffective.

In discussing this matter with General Sir George Scott Moncrieff at the War Office, we came to the conclusion that as trees (pollard willows in this instance) are needed, and that as secrecy in the affair is of the highest importance, that it would not be prudent to approach any private owner of such trees and that the safer course would be to ask His Majesty's permission to allow me in the first instance to study pollard willows that are on the Royal Estates and to collect bark and branches wherewith the imitations of such trees might be made to serve for the use of artillery observers in the Ypres district . . .

Believe me Sir
Yours obediently
Solomon J Solomon

Sir Frederick Ponsonby (older brother of Arthur Ponsonby, the author of *Falsehood in War-Time*) replied from the Privy Purse office at Buckingham Palace on Monday, 20 December:

His Majesty was much interested to hear that you had taken up the question of making the Artillery Observation Posts invisible. The King will be glad to give you every facility you require either at Windsor or at Sandringham.

Solomon arranged to go to Buckingham Palace the following afternoon. Ponsonby then wrote a 'Private and Confidential' letter to W. Archibald Mackellar, Head Gardener at Windsor:

Dear Mr Mackellar,

Mr Solomon J. Solomon of the Royal Academy is anxious to carry out some experiments with regard to trees. He has been commissioned by the War Office to make Observation Posts in the shape of pollarded willows and as this must be kept secret, he is anxious to have some place where he may try his experiments. The King wishes him given every facility and, if necessary, foresters or workmen placed at his disposal.

On 23 December 1915, the head gardener himself met the 10.42 train from Paddington in a dog cart and trotted Solomon along the winter riverbank. They selected an old willow which was later cut down and brought entire to the fruit conservatories at Frogmore. Working in the warm royal greenhouse with two scene painters and a theatrical prop maker, Solomon constructed a realistic tree cover for a steel core. Sections and strips of bark were sewn and glued to canvas which could be wrapped around the metal.

Meanwhile, the adjutant general at GHQ in France sent a letter on Christmas Eve requesting that Mr S. J. Solomon 'be despatched to this country at the earliest possible date, accompanied by sufficient personnel of his own selection to enable him to start work as soon as possible upon the construction of some urgently required special observation stations'.

Solomon started picking a team. They included the black-and-white draughtsman Harry Paget, an older man who had served in the Artists'

Rifles and had some useful military knowledge, shy Walter W. Russell, ARA, the ingenious and inventive Lyndsay D. Symington and young Roland Harker, who was by trade a scenery painter.

There was also 'a small man, very deaf, who staged the operas at Covent Garden', known to be 'a good organiser'. It was, in fact, Oliver Bernard, who experienced the sinking of the *Lusitania*. Bernard brought along F. W. Holmes of Leeds, who was head property man at the Drury Lane theatre and a master carpenter when 'Bunny' Bernard first met him in Manchester twenty years earlier. Solomon, Paget, Russell, Harker, Symington and Bernard all became officers on the low-status General List.

Winston Churchill gained experience both in and out of the trenches with the Grenadier Guards, sporting his newest headgear, a blue steel helmet given him by a French general on 5 December, in order to safeguard his 'valuable cranium'. In 1915 the French had not only pioneered camouflage but also protective helmets for infantry. This started because General Adrian met a *poilu* or ordinary French soldier who had survived a potentially fatal head wound because he happened to keep his metal food bowl under his forage cap. The first attempt, steel skullcaps, did not give quite enough cover from shrapnel, so more depth and narrow brims were added. French helmets (which required seventy operations to manufacture) had a central ridge because they were adapted from the existing dies to make helmets for *pompiers* (firemen) and *cuirassiers* (cavalrymen). G. M. Trevelyan, serving with the Red Cross in Italy, noticed how the 'shrapnel helmet' was gradually adopted there in the spring and summer of 1916. The first models were French, stamped 'R.F.' for *République Française*.

The British-made, green-painted 'Helmet, steel, Mark 1', known as a 'tin hat' by soldiers or a 'battle bowler' by officers, was copied in mid-1915 by its designer John L. Brodie from the 'chapel' helmet once worn by pikemen in the fourteenth and fifteenth centuries. George Coppard of the Machine Gun Corps tested the resistance of the steel helmets that littered the battlefield later in the war by hitting them with a pickaxe as hard as he could:

A good British helmet yielded only a moderate dent, but a dud would burst open down to the shaft of the pick handle ... Clearly, some cunning war contractor had been cheating and a War Office check hadn't been properly carried out. The duds were obviously of little use against shrapnel, and it is reasonable to assume that men had lost their lives wearing them.

Churchill's battalion at the front received its first 500 'Brodie' steel helmets on 24 January 1916. The Brodie was adopted by the US Army when it came into the war in 1917, and continued in American service until after Pearl Harbor.

The distinctive 'coal scuttle' German Stahlhelm weighed 2½ pounds and was probably the best-designed helmet, with two integral lugs for protective visors. German machine-gunners wore specially padded, thickly armoured versions that weighed 13¼ pounds and could resist all service ammunition. (However, George Coppard claimed German helmets were easier to hole with a pickaxe. Perhaps he hit them harder.) German helmets were also in 1917 the first to be painted in disruptive camouflage patterns of brown, green and black.

--

When Sir John French was replaced by Sir Douglas Haig as commander-in-chief on 19 December 1915, Churchill had to accept that the brigade of 5,000 men that French had promised would be reduced to a battalion of 1,000. Field Marshal Haig signed another letter from GHQ dated 31 December 1915:

I request that the necessary military status may be given to Mr S. J. Solomon RA to enable him to carry out his duties in connection with the camouflage work. I recommend that he be granted a temporary commission of the rank of Lieutenant-Colonel. I consider that this rank is commensurate with the responsibilities and importance of his duties . . .

Solomon's unique promotion was probably unmatched in WW1. Going, overnight, from a private in the United Arts Rifles to lieutenant colonel in the Royal Engineers meant hurried visits to military tailors and bootmakers. On 15 January 1916 Solomon wrote to Privy Purse Ponsonby:

The King's own trees will now I hope help and protect the men who will erect and make use of the armoured outposts, and I am indeed grateful for all the facilities that have been so graciously afforded me.

Ponsonby replied from York Cottage, Sandringham:

The King was interested to hear that you had finished the trees which you have been constructing at Windsor and feels sure that they will be a great success. His Majesty saw a confidential account of similar trees and disguises for the men which were being constructed in the French Army. It seems a very practical idea and His Majesty is glad to know you are taking the matter up.

On 18 January 1916, the historic first party of British *camoufleurs* crossed the English Channel to Boulogne. Oliver Bernard gloomily remembered the torpedoing of the *Lusitania*, but Solomon was excited. It was ten years and a week since he had become a Royal Academician, and now, in another year with a lucky six, he was a lieutenant colonel leaving 'good old England' on his way to war. Ominously, the army-commandeered London bus that met them on the quay broke down halfway to St Omer. It was a chilly night, and when they met Major General George Fowke they did not look very soldierly in their new and ill-fitting uniforms. Solomon pushed forward Paget to do the military blarney. Fowke did not know much about art or opera but he used to go to fancy dress balls at Covent Garden. 'Is Willie Clarkson still alive?' he asked Oliver Bernard. 'I used to get my fancy costumes from him.' 'Bunny' Bernard, grateful to be in the forces at all and therefore more ready to adapt to his new hosts, realised quite soon that there was a great gap between Solomon and the military.

Solomon had the artistic vision to point the way to the use of military camouflage, but he had no understanding of the human and material organisation required to achieve it. There was already a class divide among the group: Solomon tended to huddle with the artists Paget, Russell and Symington and leave out the craftsmen Bernard, Harker and Holmes. His recent dealings with upper echelons – kings, generals and so on – had exacerbated a tendency towards grandiosity. Oliver Bernard disliked 'this royal academy of camouflage, all talk of what soldiers can do, damn all to show what artists can do'. He thought Solomon was making enemies by being a know-all of theory, without the humility to find out what the military actually did and how they went about getting what they needed. Solomon dismissed all official channels of supply as 'red tape'. Trouble started when he began blithely shopping in Boulogne instead of going via Ordnance or Royal Engineer (RE) stores. Oliver Bernard's mundane skills made Solomon call him 'my business man'.

The first week they spent at Amiens, studying the camouflage and paintwork that the French had been doing for a year, combined with their use of *blindages*, or hardened steel plates, to protect observation posts. The leader of the French *camoufleurs*, Guirand de Scévola, was an impressive figure, always smartly dressed, with white kid gloves. He lent nine of his *camoufleurs* to the British, who now aimed to set up their own industry and start producing materials. The first temporary workshop was a barn found by the mayor of Poperinghe; Oliver Bernard went to Paris to get special supplies of paint and cloth. (The Director of Works, J. E. Edmonds, the future official historian, refused to pay the 468 francs Bernard spent on taxis, saying he was not authorised to use such a conveyance.)

General Fowke, meanwhile, took Solomon to Wimereux near Boulogne and showed him an abandoned feldspar factory near the golf-links, 200 yards from the cliff-edge. It stood in six acres surrounded by a ten-foot wall, on a railway branch line, with an old locomotive parked by the *atelier*. The sounds of hammering came from fifteen workmen levelling the concrete floors inside. This building would become the centre of the Special Works Park, RE, Wimereux, but it would not be ready for six weeks.

Solomon's imagination, not for the last time, got out of hand. Because the building had once been leased by Germans, he decided that the railway must have been intended to bring to the factory's raised floor a long-range giant gun which could have shelled all the shipping between Boulogne and Cap Gris Nez. When practical Oliver Bernard came to explore the building, he had no such fantasies. He mapped out where the carpenters' benches would go, the machine shop, the sawmill, the paintroom, the drying room and so on, and ordered tool chests from Army Ordnance Depot, fuel from the Army Service Corps, timber from Boulogne. He got on well with Captain Foote, in charge of RE Works at Boulogne, who was mechanically minded like him, and Lieutenant Colonel Crofton Sankey, who was amused by the fanciful 'artist officers'.

Solomon came down with a general to Ploegsteert, where Churchill was in the line, to look for a site for the observation post trees he had made in England. He tramped through muddy trenches, up 'Strand', as they had renamed it, towards 'Charing Cross' and the wood. Near 'German House' a shattered oak looked like a good possibility, because it was taller than the seven-foot-high parapet of sandbags surrounding the *bois*. Peering through a glass periscope, Solomon reckoned it was only seventy yards away from the German front trench.

Then Solomon was summoned to GHQ to meet the commander-in-chief himself, Sir Douglas Haig, at a lawyer's house in the middle of St Omer. Solomon was pleased to get a handshake rather than a salute because it felt more like a social introduction. Solomon thought Haig was a well-groomed, handsome man who was rather shy, and later painted his portrait for his alma mater, Clifton College. Haig walked over to a huge map on the wall and pointed out where they needed better observation posts. 'Is there anything I can do for you?' Haig asked solicitously. Solomon explained their art was in its infancy but he would like a field somewhere near St Omer where they could experiment on concealing guns and trenches from the air, and where staff officers could easily come and see the results.

'The whole of Flanders is at your disposal.'
'We only need a few acres nearby'.
'Well, you go to General —, and he'll let you have anything you need.'

Solomon was very full of himself when he got back, repeating to the engineer-in-chief what the commander-in-chief had said to him: 'You go to General — and get what you want.' But Fowke the future adjutant general was not pleased at all. His face went as red as a lobster and he shouted angrily: 'Everything has to go through me!'

Solomon was amazed. He simply could not understand bureaucracy. But it was how the administrative system worked. When F. E. Smith, the Attorney General of HM government, came to visit Churchill at Ploegsteert around this time, he was locked up for not having the correct military pass. (A flaming row then ensued between HMG and the office of the adjutant general, responsible for martial and military law, who finally offered a stiff apology.)

No one said sorry to Solomon. He felt further thwarted and aggrieved when his first ideas about covering the trenches leading towards camouflaged trees with mackintosh groundsheets were rejected by Fowke and Edmonds. Trenches soon filled with water and if you used mackintosh groundsheets the water pooled in them. Moreover, you could never quite match the smooth texture painted on the painted mackintosh to the natural surroundings. Solomon then tried making a cat's cradle of string, linking the groundsheet eyelets, into which small bundles of hay could be tied. This was not really satisfactory and the civilian prop-man Holmes, looking at it, said he would make something to place over the groundsheet. In a couple of hours he had woven together a square yard of netting. Eureka, thought Solomon:

I saw we were getting just what was wanted – not only for our immediate purpose, but for universal screening. Then men tied up into the meshes small bundles of long hay and this we coloured. It was all important to get back to St Omer with this sample to show Colonel Liddle, and I told my troupe that if we did nothing else, we had now justified our existence.

A correspondent in *The Times* of 3 August 1927 said that early in the war Solomon had 'extended his work to trench protection and introduced string network interwoven with branches and leaves for overhead cover'. Solomon is usually credited as the first to think of using fishing nets instead of canvas to stretch tautly over props and poles to cover guns, stores, dumps, trenches, etc. These nets, with an average size of thirty square feet, could be threaded with strips of canvas, rags of hessian, bunches of dyed raffia, and local vegetation ('plaited leafy twigs through meshes' as Seamus Heaney puts it) and then pegged at low angles to throw less shadow.

The flat-top net garnished with raffia is also claimed as a French invention. In April 1913 at Saint-Cyr, Major Anatole Kopenhague demonstrated its successful use to hide a platoon of men from a low-flying aeroplane. Unfortunately, Kopenhague's idea was formally rejected as 'lacking practical application' by the bureaucrats of the French Ministry of War (in August 1914, of all months).

Whoever invented camouflage fish netting – Holmes, Kopenhague or Solomon – nearly seven and a half million square yards of the stuff, tufted with plant products, outdid six million square yards of wire netting and a mere million square yards of painted canvas sheets to become the most used screening of WW1.

At the end of February 1916, the *blindages* that Solomon had ordered from the Holborn firm arrived. The steel was lighter than that of the French models and Solomon reckoned the oval sections could be fitted together and carried in one piece. If seven hundredweight were in a cradle hoisted by a dozen men, each man still had to heft sixty-three pounds. On 1 March he informed Lord Cavan, General Officer Commanding (GOC) 14th Army Corps in Fifth Army, that the tree was ready and he wanted to come out to the Ypres Salient.

At 3 a.m. on 5 March 1916, Solomon set out in a car with Generals

Gathorne-Hardy and Wauchope to check out the tree site. They drove towards the Yser canal with the lights out and parked behind the ruined walls of the Moulin Rouge estaminet. The duckboards of Coney Street communication trench took them into the Ypres Salient leading towards the mushroom pillbox on Pilckem Ridge. Over the canal, Solomon intended to sketch and measure a willow tree and to trace where the sap should be dug. It was muddy and cold; their breath plumed. Solomon was crossing a fallen tree when he slipped and fell into the pond below. 'That's bad!' said the general. The artist waded out laughing, thinking how his children would have enjoyed seeing their old dad in this plight. His coat had kept his body dry but his rubber boots were sloshing full and icy water had gone up his sleeves and down his neck. Sleet blew in blizzardy gusts. At long last, the chauffeur emptied his boots. Covered up in the back of the car, Solomon worried about pneumonia. He went to bed with 'quinine tabloids and a good nip of whiskey'.

The night that saw the raising of the first British observation post tree was Tuesday, 12 March 1916. Corporal Bryant bedded in the foundation of the tree over the dugout the week before. It was a steel plate, with a raised collar and boltholes around the oval opening in the middle. In the evening, fifteen men left the factory in a truck containing the tree and the lifting tackle, and Solomon and Walter Russell went by car. The truck went into Ypres, picked up the guide from a sandbagged bunker in the square and drove out past the 'DIAMANT' sign spelled in lighter bricks on the long red wall of a ruined jeweller's shop.

At a place called 'White Hart' they unloaded the tree into its cradle and the men carrying it got underneath. Other forces were moving into the Salient on their own business: a battalion in tin hats and gas capes, medieval in the moonlight. The tree-men crossed Bridge 4, went a hundred yards along the road beside Coney Street and turned into their own route. Fresh men were waiting to relieve the carriers. A flare went up and everyone froze till the swinging blue light burned itself out. German machine guns rattled in the distance. The British soldiers crossed streams on planks and struggled up the clayey canal bank with the heavy burden. At the top they tipped the cradle over and levered the weighty eight-and-a-half-foot tree upright. Solomon got into the dugout with his torch to illuminate the foundation collar's holes as the men turned the tree to align them. Greased bolts slid through into tightening nuts. Mud and grass were plastered over the plate and up the base of the tree as a man squeezed up inside (the oval measured only twenty-two inches by eighteen inches) to check the observation loopholes and their bulletproof shutters. Solomon thought it an excellent reproduction: from a few paces away you could not tell the tree was not real. Back at the barn in Poperinghe, the sappers celebrated with hot food and beer. The later official history, though, was rather sour about the first tree: 'In practice it proved of little use; it was too small to admit any but a most determined and enthusiastic man, and was too far off for good observation – faults due to inexperience.'

Solomon J. Solomon's oil painting of the event, *Our First O.P. Tree*, was shown at Burlington House in the *Art of Camouflage* exhibition that he organised in October 1919. It still hangs in the Imperial War Museum. In dim bluish moonlight, Russell and Solomon, in caps and greatcoats, look on from the right as the pollard 'tree' is tilted into place. Another Imperial War Museum painting from 1919, *Erecting a Camouflage Tree* by Leon Underwood, shows ten men, half with their shirts off but their steel helmets on, labouring to fit a curve of bark to a shiny drum under the guidance of an anxious sergeant with a wrench. This is a daylight scene. You can see the brown .303 rifle lying next to the hammer, and the muscles in their white backs.

On 22 March 1916, Guirand de Scévola hosted a dinner for the British *camoufleurs* at the little hotel in Wimereux. The Frenchmen sang songs, made witty speeches and stayed up late. It was a bittersweet occasion for Solomon, because that day he ceded command of the British camouflage section, now authorised by the War Office as Special Works Park, RE, and officially integrated into the BEF, to Major F. J. C. Wyatt, RE. The unit had a nominal establishment of ten officers and eighty-two other ranks, which was increased on 1 July 1916 to fifteen officers and 157 ORs.

The heavy metalwork for OPs and the industrial production of nets and canvases, some involving Chinese labour, remained at the parent factory at Wimereux. There were two forward Special Works Parks, as the camouflage units were called: Aire covered the northern and Amiens the southern half of the British line. Major J. P. Rhodes was in charge at Aire from 8 November 1916, and it was his idea to start employing hundreds of French women in the net factories. There were labour problems, but Walter Russell proved to be good at soothing woes. At Amiens, Captain Paget was in charge, working with the French, until December 1916. There metal lathes and woodturning machinery were available in the workshops where L. D. Symington put his hands to good use. He developed the camouflaged Symien sniper suit for the SOS schools (nearly 4,800 of them were manufactured), and made the prototypes of some 3,000 dummy papier-mâché heads. The portable observation post was another speciality (armoured or unarmoured): a cowl of chicken wire covered in plaster of Paris with a slot of fine copper-wire mesh to see through, all of which could be camouflaged to fit any parapet. At Amiens they also made 12,000 cut-out dummy 'Chinese attack' figures that could be snapped upright by electrically exploded detonators to look and sound like British troops advancing and to divert attention away from real attacks. Symington also designed a new kind of machine-gun post that was completely camouflaged from both aerial photography and direct observation. An armoured box fitted into the terrain had a simple lever inside that raised the elaborately camouflaged roof just enough to open a narrow aperture through which a machine gun, set back far enough to show no muzzle flash, could fire unseen. The Canadians used this design with great success in the fighting around Monchy-le-Preux, near Arras.

But Solomon, who had liked doing reconnaissance and showing new artist officers the ropes, became miserable as he declined into a more advisory role, detached from the practical work. In April 1916 he was in effect sacked and went home on leave. He came back in May and managed to nobble Sir Douglas Haig, busy preparing for the massive Battle of the Somme, on a visit to a camouflage factory:

SOLOMON: 'I hope my services will be retained.'
HAIG: 'We have to beat the Germans.'

Solomon could not quite fathom this Delphic utterance.

In May 1916, both Churchill and Solomon were back in England. Churchill returned to his parliamentary duties on 7 May, though he did not get back into government for another year and Solomon was summoned to talk to Colonel Ernest Swinton RE at Armament Buildings. The two men got on; Swinton was an imaginative soldier, and a fine draughtsman, who had once been the chief instructor in geo-metrical drawing at 'the Shop', the Royal Military Academy, Woolwich. He was also a good writer: *The Defence of Duffer's Drift* and *The Green Curve* turn military problems into enjoyable fiction. Swinton was Kitchener's first official war correspondent, 'Eyewitness', and had always been interested in new technology – man-lifting kites, railways, machine guns. Captain Liddell Hart, in his two-volume history of the Royal Tank Regiment and its predecessors, *The Tanks*, credits Swinton and Churchill with the invention of the tank. During the early days of the war, Swinton saw almost naked men walking towards enemy fronts bristling with barbed wire and machine guns, and this impressed on him the need for armoured protection. Solomon wrote:

Mr Winston Churchill, by a long way the best military mind among our statesmen, shared his views, and the outcome of their deliberations, with the assistance of engineers, was what is known as the tank.

The word 'tank' was coined by Swinton to maintain secrecy by referring to the armoured vehicles as straightforward riveted metal containers. The Russian for 'Handle With Care: Petrograd' was painted in large white Cyrillic letters on the side of a prototype Mark 1 tank, to deceive people that it was just a tank of oil for the Russian army.

Swinton took Solomon up to Thetford in Norfolk on 31 May to show him the secret battle-training area he had created on Lord Iveagh's pheasant-shooting estate for what was then called the Heavy Branch of the Machine Gun Corps. Solomon's job was to help hide these first hot, smelly, carbon-monoxide-poisonous tanks on their way to their objective. The artist thought that smokescreens or mist would accomplish this best, or perhaps silhouettes of perforated zinc. Mere paint was not really up to it.

Solomon went back to France in early June 1916 to find out where the tanks might be used and to study soil and vegetation colours. By the middle of June he was back in Norfolk, at the tank battle-training area. Every day he rode his pony to the new work: 'I had to take off my tunic and put on overalls, just like an ordinary house painter.' Solomon painted the first tanks and their large canvas covers in Fauvist blotches of pink and grey and green and brown; tank officer Basil Henriques described the final effect as 'a kind of jolly landscape in green against a pink sunset sky'.

Solomon was present on 21 July when the politicians, including Lloyd George, the Secretary of State for War, and Edwin Montagu, the Minister of Munitions, and the military top brass, led by the CIGS, Sir William Robertson, came up to the Elveden Explosives Area in Norfolk to watch two dozen thirty-ton rhomboid tanks driving over trenches, bashing down trees, knocking through walls and sandbag parapets protecting wooden 'enemy' machine guns. As the tanks roared noisily over terrain once familiar to Boudicca's chariots, Solomon spotted and picked up a clutch of pheasant's eggs, smooth, brown and delicate.

The tanks first crawled into action on the Somme on 15 September 1916. After a massive artillery barrage, three dozen tanks pressed home an attack. Most broke down, got stuck or were too slow, but nine of them did some damage to the enemy. Between Flers-Courcelette and Gueudecourt, where the British tanks D-5, D-6, D-16 and D-17 caused panic, German soldiers ran away.

Philip Gibbs burst out laughing when he first saw the lumbering, smoking tanks, with their unsilenced engines roaring: 'They were monstrously comical,' he wrote, 'like toads of vast size emerging from the primeval slime in the twilight of the world's dawn.'

# 7

# Guile and Guerrilla

The First World War marked the rise of the geopolitical region known as 'the Middle East'. The term was first coined in 1902 by the American theorist of naval strategy, Rear Admiral Alfred Thayer Mahan, to indicate the Arab and Persian area between the 'Near East' of the Mediterranean Levant and the 'Far East' of India and China. In WW1, Winston Churchill's broad strategic vision naturally made him an 'Easterner' among the British policy makers, not a 'Westerner' solely concerned with France and Belgium.

Another well-known 'Easterner' was T. E. Lawrence (1888–1935) who became world-famous as 'Lawrence of Arabia', leading the Arab Revolt. Thomas Edward Lawrence ('Ned' to his family) was a curious figure, described by Aubrey Herbert as 'an odd gnome, half cad – with a touch of genius'. He still excites both vilification and hero-worship. He was not really 'Lawrence' at all, but the illegitimate son of a baronet called Chapman who ran away with a governess who was herself born out of wedlock. The elusive 'T. E.' kept making up stories and changing identity: 'Ross' and 'Shaw' were names he assumed later. The consummate actor who became 'Lawrence of Arabia' was known for his camouflage of Arab robes, remarking once in a letter: 'The leopard changes his spots for stripes, since the stripes are better protection in the local landscape.'

Among the ambiguities about Lawrence are his attitudes to time and technology. He looks backwards and forwards. An archaeologist and classical scholar, romance places him in an older way of life, among tents and camels in the desert. But the real T. E. Lawrence was fascinated by modernity. His cottage at Cloud's Hill had no electricity but it did have industrially canned food and the latest wind-up gramophone for recorded music. He loved printing presses, tinkering with Rolls-Royce engines and RAF speedboats. He rode a Biblical camel but carried with him in the saddle a stripped-down 'air-Lewis' light machine gun in a canvas bucket. Lawrence was killed in 1935 riding his seventh Brough Superior motorcycle, powered by oil.

Technological change is impelled by war. Winston Churchill helped to make WWI the first war of the petroleum age. Although 1.5 million grass-eating bullocks, camels, donkeys, horses and mules laboured for the British from 1914 to 1918 (and some half a million animals perished), this was the first war in history when machines with internal combustion engines began to take the strain. The British Army started the war with fewer than 900 motor vehicles, but had over 120,000 by its end in November 1918. As First Lord of the Admiralty from 1911–15, Churchill revolutionised 'bunkering' by changing the fossilised-sunlight diet of the Royal Navy's vessels from native coal to foreign oil. Virtually all the new Royal Navy warships built in 1912, 1913 and 1914 were oil-fuelled. To guarantee those Royal Navy oil supplies, the British government spent £5 million in 1912 to gain the controlling interest in the Anglo-Persian Oil Company (later British Petroleum) which first struck black gold in the Persian Gulf in 1908. Its refineries were at Abadan, close to Basra. This shrewd investment paid for the mighty British fleet in a decade, but it gave oil – and oil-rich regions – a strategic importance they had not previously enjoyed as Britain fought Ottoman Turkey first in Iraq, then Gallipoli, and then in the region's holy lands.

The genesis of Britain's WWI foray into the Middle East lay in Ottoman Turkey's decision to join forces with Imperial Germany and to attack Russia on 31 October 1914. Germany wanted to stir up the Islamic world and when, on 14 November, the Sheikh-ul-Islam in Constantinople declared a jihad on the Triple Entente of Britain, France, Russia and their allies, this conflict became truly a world war.

The hundred-mile-long Suez Canal had been built in 1869 with French and Egyptian money, but Britain had bought its way to majority share-holding in 1875. Constitutionally, Egypt was still under the notional suzerainty of the Ottoman Empire, and was ruled by a Khedive or governor alongside an Egyptian prime minister, but it had been effectively under British administrative control since July 1882, when the Royal Navy shelled Alexandria and the British army defeated the Egyptian nationalists at Tel-el-Kebir.

Ottoman Turkey's new military alliance with Imperial Germany directly threatened the Canal. So, on 18 November 1914, Egypt was formally annexed as a British Protectorate, with a British high commissioner; the Ottoman Khedive was deposed, and a new Sultan imposed. Britain's once 'veiled protectorate' of Egypt now stood revealed in uniform, its garrisons reinforced by Imperial and Dominion troops, both Indian and Anzac, en route for the trenches of Europe and Asia Minor.

The Ottoman Turkish army tried to approach through the Sinai desert and seize the Suez Canal early in February 1915, but the attack was thwarted on the east bank. Their objective was to sever the Imperial lifeline of the canal. The German Field Marshal von der Goltz's mission in Baghdad was to clear the British and Russians out of modern-day Iraq and Iran, then invade India from Afghanistan. As Sir Walter Bullivant briefs Dick Hannay in *Greenmantle*:

'There is a dry wind blowing through the East, and the parched grasses await the spark. And the wind is blowing towards the Indian border . . . We have laughed at the Holy War, the Jehad that old von der Goltz prophesied. But I believe that stupid old man with the big spectacles was right. There is a Jehad preparing.'

Once the great game began again, the British thought that two could play at making mischief among the other fellow's natives. If the Turco-German alliance was going to foment Islamic discontent in the British Empire, then the British would cynically encourage Arab Nationalism inside the crumbling Ottoman Empire. An Arab, Sharif Hussein ibn Ali al-Hashimi, ruler of the Hijaz, the Red Sea coast province of the Arabian Peninsula, and great-grandfather of the current King of Jordan, now played the first significant hand.

When Turkey summoned the faithful in 1914 for a pan-Islamic jihad against Britain, France and Russia, Sharif Hussein declined to take part. Since the Hijaz was the holy land for Muslims and the birthplace of their faith, this mattered. Hussein was called *sharif* because he was a direct descendant of the Prophet Muhammad. Hussein, installed as the legitimate Amir of Mecca in 1908, thus began to prove an independent-minded presence inside the Ottoman Empire. An 'honourable, shrewd, obstinate and deeply pious' man according to T. E. Lawrence, but a two-faced schemer according to his enemies, he was also a new kind of pan-Arabist. He was in touch with the Syrian secret societies of urban intellectuals and army officers, al-Ahd and al-Fatat, who were beginning to think politically of nationalism, as well as with the old-style Bedouin chieftains of Arabia whose first loyalties were to family, clan and tribe. However, Hussein was based in the Hijaz, a hot desert region that could not feed itself and which drew most of its income from Islamic pilgrims on the Haj. If the maritime blockades of the Great War stopped the pilgrims coming, the Hijaz would have to become even more dependent on Ottoman Turkey. Sharif Hussein needed a powerful external ally, able to supply guns and money and keep the trade and travel routes open. Germany and Britain both fitted this bill, but Hussein's first approaches were to the British.

In April 1914, four months before the war started, Hussein's second son Abdulla came to make a private request of the British ruler of Egypt, Earl Kitchener of Khartoum. He enquired whether, if the Hijaz Arabs ever rose up against their Turkish masters, the British might possibly assist with a few little machine guns. At this stage, however, Kitchener informed Abdulla that the British government's only interest in Arabia was the protection of British Indian pilgrims going on the Haj to Mecca. But all that changed when WWI broke out.

In late September 1914 Kitchener sent a secret messenger to ask whether 'the Arabs of the Hejaz would be with us or against us'. On 31 October, Kitchener sent his salaams to Hussein's son Abdulla and also requested help against the Germans and Turks, in a telegram: 'If Arab nation assist England in this war England will guarantee that no intervention takes place in Arabia and will give Arabs every assistance against external foreign aggression.' Negotiations on what Britain might concede politically to gain Arab support continued in 1915 in a somewhat ambiguous correspondence between Hussein and the British high commissioner who had replaced Kitchener in Egypt, Lieutenant Colonel Sir Henry McMahon.

Sharif Hussein said he wanted a single independent state carved out of the Ottoman Empire, an Arab bloc that would embrace today's southern Turkey, Syria, Iraq, Kuwait, Lebanon, Jordan, Israel, Palestine, Saudi Arabia and Oman. The letters between McMahon and Sharif Hussein never quite agree on the vital topic of what territory was to be excluded from the plan. Prominent in the area of disagreement is the region of southern Syria between the river Jordan and the Mediterranean that the British called 'Palestine'. High Commissioner McMahon knew northern India better than the Middle

East; he spoke no Arabic, just signing what his advisers put before him. His Oriental Secretary, Ronald Storrs, admitted his side of the correspondence was prepared by 'a fair though not a profound Arabist' and checked by him in haste or not at all, and that Hussein's letters were in 'obscure and tortuous prose . . . tainted with Turkish idioms and syntax'. It was not a recipe for clarity.

Not all the inhabitants of the Arabian peninsula supported Hussein, the Hashimite Sharif of Mecca. At Riyadh to the east there was a rival dynasty, the House of Saud, whose descendants still rule Saudi Arabia. Their leader was the tall and fierce desert warrior, Abdul Aziz Ibn Saud, who had military strength but lacked the moral power that the control of Mecca and Medina gave to his rival Sharif Hussein.

Ibn Saud, opposed to both the Hashimite Sharif and the Ottomans, could see the usefulness of an alliance with the British. But which British? Those in London, Cairo or Delhi? British Imperialism was never as wholly coherent as its enemies imagined. Arabia came within British India's sphere of influence, and in the end Sir Percy Cox of the Indian Political Service and his new agent Harry St John Philby (the Muslim convert father of the spy Kim Philby) spent the rest of 1915 concluding a deal with Ibn Saud. Regular supplies of guns and ammunition and a £5,000-a-month subsidy kept the Saudis onside with the British, neither attacking their allies nor helping their enemies, for the rest of WW1.

Another consideration was Mesopotamia, which we now call Iraq, of strategic interest in 1914 again because of proximity to India. From Bombay it was a short sea voyage west to Basra, the entrance to the Fertile Crescent, which the Viceroy of India, Lord Hardinge, saw as a potential granary and a good place to resettle surplus Indians of the densely populated British Empire.

Hardinge was not in favour of supporting Arab revolts against their Turkish overlords because he did not want to upset Sunni Muslims in India, and he disliked native nationalists who had tried to kill him. He agreed to use Indian army troops to secure the Anglo-Persian oil installations at Abadan island on the Persian Gulf.

The 'Mespot' campaign of Indian Expeditionary Force 'D' turned into D for Disaster. Lloyd George later excoriated it as 'a gruesome story of tragedy and suffering resulting from incompetence and slovenly carelessness on the part of the responsible military authorities'. Force 'D' in Iraq did manage to secure the oil regions of Basra and the Shatt-al-Arab in the south, but then became over-confident and thought they could also take Baghdad, on the cheap. They had no planes, few heavy guns, inadequate river transport; few tents, mosquito nets, ambulances, medical supplies, blankets, clothing. The initial advances were checked at Ctesiphon in November 1915 by tougher Anatolian Turkish troops, under German direction.

The surviving Indian Expeditionary Force retreated to Kut-al-Amara where they were besieged for five months from December 1915; they sickened and were never relieved, though attempts to reach them cost 23,000 casualties. Inadequate air drops only prolonged their agony of hunger, dysentery, scurvy, cholera, malaria, heat, filth and flies.

Arabic-speaking Captain T. E. Lawrence was sent to what he called 'blunderland' with Turkish-speaking Aubrey Herbert at the end of April 1916, on a mission to bribe the Turks up to £2,000,000 to let the British forces go. The attempt failed, and Lawrence was appalled at the wastage: 'All the subject provinces of the Empire to me were not worth one dead English boy,' he wrote later, when the British casualties mounted to over 92,000 in Mesopotamia. Of the 14,000 British and Indian soldiers who finally surrendered to the Turks at Kut, over a third died as prisoners of war in the Iraqi desert or as chain-gang labourers on the German railway that was destined to link Berlin to Basra, Prussia to the Persian Gulf.

T. E. Lawrence knew about the Middle East. In 1909, he wrote a BA thesis on the influence of the Crusades on castle-building in Europe which, together with four seasons' work on Hittite archaeology at the British Museum's excavations at Carchemish, had taken him through large tracts of Turkish-administered Lebanon and Syria. His scholarly interests were used as camouflage for military intelligence work in the winter of 1913–14 when his supposedly archaeological exploration of 'the Wilderness of Zin' saw the light of day as the *Military Report on the Sinai Peninsula*, a survey commissioned by Lord Kitchener from the Royal Engineers. After the outbreak of war, Lawrence served with the General Staff Geographical Section (MO4b) Asia Sub-Section in London, and at the end of 1914, aged 26, he became the youngest member of the Department of Intelligence run by Gilbert Clayton in Cairo, whose mission was to keep an eye on the Ottoman Empire. Lawrence's varied intelligence duties in Egypt included making maps from aerial reconnaissance photographs and continually updating the Turkish Order of Battle with evidence collated from agents' gossip, travellers' notes, prisoners of war, captured documents and photographs, newspapers and radio intercepts, which he contributed to the *Handbook on the Turkish Army*, edited by Philip Graves.

Lawrence was particularly interested in those Turkish army units whose Arab officers might not be wholly loyal to the Ottoman Empire. While in Cairo, he claimed to have learned from a defector from the Turkish side at Gallipoli, Lieutenant Muhammad Sharif al Faruki, that there were other members of Syrian secret societies dedicated to Arab nationalism among the Turkish soldiers sent to reinforce the Caucasus front against Russia at the end of 1915. Robert Graves stated in *Lawrence and the Arabs* that the fall of Erzerum in eastern Turkey to Russian forces was somehow 'arranged' by Lawrence, and Lawrence himself claimed that he had 'put the Grand Duke Nicholas [of Russia] in touch with certain disaffected Arab officers in Erzerum. Did it through the War Office and our Military Attaché in Russia.' In his biography of Lawrence, B. H. Liddell Hart wrote:

In the spring of 1916 [Lawrence] had a long-range hand in a more important matter, the 'capture' of Erzerum by the Russian Caucasus Army after a curiously half-hearted defence – readers of John Buchan's subsequent novel, *Greenmantle*, may find it worthwhile to remember that fiction often has a basis of fact.

It is hard to establish what the facts of the matter were. Both Graves and Lawrence were prone to exaggeration and romancing.

--

1916 was a year of rebellions; the Irish rose at Easter in Dublin and were hanged; the Arab Revolt flared up early in June. Sharif Hussein, fearing he was about to be deposed by the Turks, gave the signal for his four sons, Ali, Abdulla, Feisal and Zeid, and thousands of Hijaz tribesmen to attack the Turkish garrisons in the west of the Arabian peninsula. Mecca, Jeddah, Rabegh and Yenbo eventually fell to the Sharif's forces, with help from British Royal Navy ships and seaplanes. But the rebels did not gain the support of all Arabs; most chose to stay loyal to the Ottoman Sultan. There were no desertions from the Turkish army, and the Sharif's rival, Ibn Saud, did not join in. Nor were the rebels able to take the holy city of Medina, which was the terminus of the Hijaz railway.

This German-engineered, solidly made, 42-inch permanent track (completed in 1908) ran along the old Derb el-Haj, the pilgrims' camel route over 800 miles south from Damascus to Medina via Deraa and Maan, traversing some 2,000 bridges and culverts that were cut and dressed from the local stone. When the railway was built, the pious might have believed that the shining rails were, as they were told, the Turkish Sultan's generous gift to the Arabs of the Caliphate, providing them with easier transport for the Haj. But the more cynical noticed that the railway stations, set about eleven miles apart, were fort-like blockhouses with rifle-slot windows and water towers that doubled as look-out posts. In fact, the pilgrim-route railway line was the Ottoman military's direct way of shifting troops and supplies down from Damascus, Aleppo and Istanbul. The Turks, wrote Lawrence,

moved an army corps to Medina by rail, and strengthened it beyond establishment with guns, cars, aeroplanes, machine guns, and quantities of horse, mule and camel transport.

In late September 1916, a Turkish expeditionary force set out from Medina to march the 250 miles down the main western road to recapture Mecca, and began pushing the Sharif's rebel Arabs back. The British feared that if the Turks succeeded in recapturing Mecca, the German message – *Gott strafe England* – would be in all the Friday prayers in every mosque around the world. Old British Empire hands like General Sir Reginald Wingate were genuinely alarmed by the dangers of jihad. Wingate, soon to become High Commissioner in Egypt after seventeen years as Governor General and Sirdar of the Sudan, had run a formidable intelligence service there since 1887, and written a case study of Islamist fundamentalism, *Mahdiism and the Egyptian Sudan*.

There seemed to be some evidence for Wingate's fears. Across the Arabian Gulf, the Turks had invaded Aden Protectorate, threatening the British naval station there. The Germans had sent Major Freiherr Othman von Stotzingen to Yemen, to set up a wireless station to communicate with German troops in German East Africa, and spread propaganda among Muslims in the Horn of Africa. Also, Lij Iyasu in Abyssinia wanted to turn that Christian state into a Muslim one. Lij Iyasu and the Muslim Somalis might well lend support to the Germans running rings round the British in Tanganyika. There were Muslims in the Sudan like the troublesome Sultan Ali Dinar of Darfur and among the rebellious Senussi in western Egypt, men whose disaffections could easily be stirred up. Further east, the Germans had crossed Persia to Kabul in order to persuade the neutral Emir of Afghanistan to raise an army to invade India where there were millions of Muslims. Weren't the 'Hindustani fanatics' of the North-West Frontier of India already known to be under the influence of Wahhabi mullahs from Arabia? The discovery by Indian Intelligence in 1915 of letters urging Jihad from a known troublemaker with contacts in Jeddah, Maulana Obaidullah Sindhi, written on yellow silk and sewn into the coat of a student called Abdul Haq, travelling from Kabul, led to the arrest of over 220 Islamists in northern India. It seemed to people like Wingate that Jihad could erupt, right across Africa and Asia, stirred by German troublemaking.

So, to prevent *Deutschland über Allah*, it was vital that Mecca should not be recaptured. The British wanted to reinvigorate the Arab Revolt against the Turks without committing Christian troops to the Muslim heartland. There were disagreements between the Foreign Office and the War Office: should soldiers be sent to save face, or was the whole thing just a minor sideshow? In the end the Arab Bureau in Cairo took charge of things. The Arab Bureau, a 'hybrid intelligence office', was set up by the director of Naval Intelligence, Admiral Reginald 'Blinker' Hall, in 1916 with Gilbert Clayton as its chief, the *Arab Bulletin* as its restricted-circulation intelligence journal and T. E. Lawrence as one of its junior members. Located in the Savoy Hotel, Cairo, the Arab Bureau was a jangling place of 'incessant bells and bustle and running to and fro' that another member, Aubrey Herbert, likened to 'an oriental railway station'. It was Hall who dispatched Gertrude Bell, the celebrated Arabist who had made six desert journeys and knew about the tribes of Syria, Mesopotamia, northern and central Arabia, to join the Arab Bureau. Hall also recruited to the Royal Navy Volunteer Reserve the calm and impressive Oxford archaeologist David Hogarth, Keeper of the Ashmolean Museum and Fellow of Magdalen College (described by Lawrence as 'Mentor to all of us'). Reginald Hall saw the strategic and naval importance of the Hijaz: if the Germans and Turks managed to set up a submarine base on the Red Sea coast, they could harry the British Empire's vital marine traffic to and from the Suez Canal.

The Arab Bureau therefore supported the Arab rebellion with crates of weaponry delivered by the Royal Navy: 54,000 rifles and 20 million rounds of ammunition in the first six months. They also sent the so-called 'cavalry of St George', British gold sovereigns, £1 sterling coins, each of which bore the mounted figure of the dragon-slayer on the obverse. The first £10,000 had been delivered to Jeddah early in June 1916 on board HMS *Dufferin*. The field treasury that doled out later instalments was a small stone and cement building in Akaba, piled up to the ceiling with dozens of squat ammunition boxes. Each box contained five canvas bags; each sealed bag held £1,000. Peake Pasha of the Egyptian Camel Corps described signing a chit and getting a special leather holster to carry the jingling swag back to his tribal mercenaries. In all, Great Britain spent some £11 million bribing Arabs in WW1.

A high-level British mission arrived from Egypt across the Red Sea in mid-October 1916 to ginger up the Arab Revolt. The party was led by Ronald Storrs, the waspish diplomat whom T. E. Lawrence called 'the most brilliant Englishman in the Near East' (but whom Aubrey Herbert dubbed 'the Monster of the Levant'). Wearing tropical whites stained scarlet up the back from sweating on a red leather seat, he had been engaged in pistol practice against bottles on the deck of the ship with Abdul Aziz el-Masri, an Arab Nationalist who had defected from the Turkish Army and was now, briefly, a General for Sharif Hussein.

Captain Lawrence, now a 28-year-old intelligence officer, had taken ten days' leave to come along for the jaunt. Storrs described him in his contemporaneous diary as 'little Lawrence my supercerebral companion'. This voyage to Jeddah by ship was the beginning of Lawrence's first visit to Arabia.

But when at last we anchored in the outer harbour, off the white town hung between the blazing sky and its reflection in the mirage which swept and rolled over the wide lagoon, then the heat of Arabia came out like a drawn sword and smote us speechless. It was mid-day; and the noon sun in the East, like moonlight, put to sleep the colours. There were only lights and shadows, the white houses and black gaps of streets: in front the pallid lustre of the haze shimmering upon the inner harbour: behind, the dazzle of league after league of featureless sand, running up to an edge of low hills, faintly suggested in the far away mist of heat.

Chapter VIII, *Seven Pillars of Wisdom*

*Seven Pillars of Wisdom*, Lawrence's account of the Arab Revolt, is subtitled 'A Triumph', but it is really an epic romance of failure. Like most aspects of Lawrence's life, it is the subject of burning debate: Robert Irwin thinks it a 'great work' best considered as a novel. Its attitude to the truth is not journalistic, but literary. Edward Said suggested it was more literature than history. Lawrence himself felt 'one craving all my life – for the power of self-expression in some imaginative form'. In the introductory chapter to *Seven Pillars* that he later suppressed, Lawrence says that the book was written as a work of suasion to inspire: 'a designed procession of Arab freedom from Mecca to Damascus . . . an Arab war waged and led by Arabs for an Arab aim in Arabia'. The book became a work of Arab propaganda, and it required the appropriate Arab hero.

Lawrence met the sons of Sharif Hussein. 'I found Abdulla too clever, Ali too clean, Zeid too cool.' On 23 October 1916, Lawrence says that he found his man. The third son, the patient and self-controlled Feisal (portrayed by Alec Guinness in David Lean's classic film about Lawrence), was the one required to regenerate the Revolt. Feisal had the fire and character to become the leader that the Arabs needed, and he also had the necessary political flexibility to support British interests. Lawrence lobbied Sir Reginald Wingate, the sirdar in Khartoum who was the GOC in Hijaz during the Arab Revolt, got approval from his line-manager in Intelligence, Gilbert Clayton in Cairo, and then went back to Arabia to play his part as Feisal's military adviser and liaison officer.

In chapter XX of *Seven Pillars of Wisdom*, Feisal suddenly asks Lawrence to start wearing Arab clothes like his own while in the desert camp, and Lawrence agrees 'at once, very gladly'. His willingness to step into costume in December 1916 distinguishes the fluid Lawrence from the more rigid Captain Shakespear who had negotiated on Britain's behalf with Ibn Saud two years earlier. Shakespear was killed in a tribal clash, wearing his Lancers' khaki uniform and pith helmet, because he insisted on staying put, taking panoramic photographs with his clockwork mechanical plate-glass camera, rather than entering the ebb and flow of Arab battle.

In August 1917, the *Arab Bulletin* published 'Twenty-Seven Articles', Lawrence's advice to other liaison officers and advisers who might work with the Bedu or Hijaz Arabs. What strikes the reader today is its wise humility. It advocates paying attention to the Arabs, fitting in judiciously, not giving orders, and never manhandling them: 'It is difficult to keep quiet when everything is being done wrong, but the less you lose your temper the greater the advantage.' The first article begins 'Go easy for the first few weeks. A bad start is difficult to atone for,' and the second says: 'Learn all you can about your Ashraf and Bedu. Get to know their families, clans and tribes, friends and enemies, wells, hills and roads. Do all this by listening and by indirect inquiry. Do not ask questions. Get to speak their dialect of Arabic, not yours.' The following nine articles are about deferential management, unobtrusively dealing with the leader: 'Wave a Sherif in front of you like a banner and hide your own mind and person.'

Articles 17–20 are specifically about clothing: 'Wear an Arab head-cloth when with a tribe. Bedu have a malignant prejudice against the hat,' is the opening. 'Disguise is not advisable,' he goes on to say. 'Except in special areas, let it be clearly known you are a British officer and a Christian.' But then Lawrence thinks that 'Arab kit' will help you to 'acquire their trust and intimacy to a degree impossible in uniform. It is, however, dangerous and difficult.' Allowances are not made, he says, nor breaches of etiquette condoned: 'You will be like an actor in a foreign theatre, playing a part day and night for months, without rest, and for an anxious stake.' It is perhaps easier to stay in British uniform: 'Also then the Turks will not hang you, when you are caught.' But, he suggests, if you can pay the psychological price of the change of clothing, the prize of success may be far greater. 'If you wear Arab things, wear the best,' he states, and 'If you wear Arab things at all, go the whole way. Leave your English friends and customs on the coast, and fall back on Arab habits entirely.'

And so *Seven Pillars of Wisdom* relates how Lawrence, dressed in the fine white silk wedding robes that later became his emblem, begins to enter the desert Arabs' harder, harsher world. The water gave him dysentery; he endured lice, fleas, ticks, heat. In chapter XXXIII, weak with illness, he was confined to a tent for ten days, dozing and dreaming about the algebra, the biology and the psychology of war. In his smelly, sweaty tent, a theory of guerrilla warfare began to cohere in his mind.

Lawrence, a classical scholar who later translated the *Odyssey*, now remembered a useful parallel from the ancient world. In book three of

Xenophon's *Anabasis*, a group of Greek mercenaries find themselves near Baghdad, a thousand miles from home, their generals and captains all treacherously murdered, surrounded by Persians thirsting for their blood. Xenophon steps forward, dressed in his finest clothes, to address the survivors. He is not a soldier like them but a leisured gentleman, a pupil of Socrates who has come along for the ride, in order to see the world and to report on it. Xenophon tells the Ten Thousand what he thinks they will have to do to get back to Hellas: first burn the wagons, the tents and all their baggage; carry only food, water and weapons, and live off the land. Thus freed to react fast, and to improvise new weapons and tactics, the Greeks fight through and survive.

Lawrence, considering his own situation late in 1916, germinated his own ideas about guerrilla warfare. This became the subject of his brilliant post-war essay 'The Evolution of a Revolt' which appeared in October 1920 in the first issue of *The Army Quarterly*, a journal founded and edited by Guy Dawnay, the intelligence officer who sent Compton Mackenzie to Lesbos. It is one of the most important things that Lawrence ever wrote. Later reworked by Liddell Hart as 'Science of Guerrilla Warfare' for the 13th edition of the *Encyclopaedia Britannica*, and incorporated into chapter XXXIII of *Seven Pillars of Wisdom*, the piece overturns conventional military thinking and analyses irregular warfare in a new way for the modern age:

Suppose we were (as we might be) an influence, an idea, a thing intangible, invulnerable, without front or back, drifting about like a gas? Armies were like plants, immobile, firm-rooted, nourished through long stems to the head. We might be a vapour, blowing where we listed.

--

In 1916, the Turks had moved their soldiers down by train to Medina and were advancing the 250 miles south towards Mecca. Lawrence soon grasped that the Arab tribesmen were not strong enough to attack the Turkish front lines head on and were quite incapable of defending fixed positions, so he attempted neither. Instead, he moved a force northwards behind the Turks to threaten the 800-mile Hijaz railway line. This threw the enemy on the defensive: the Turks recoiled and retreated to Medina, then split their force, one half to garrison and fortify the holy city, the other half to protect their supply line. Lawrence realised there was no point in taking Medina in a conventional battle of pointless bloodshed: if the Turks could be made to stay in their fort it would turn into their prison. So let the Turkish soldier languish there, consuming his own supplies and eating his transport animals, harmless and helpless without a target to shoot. 'He would own the ground he sat on, and what he could poke his rifle at.' This left 99 per cent of the Hijaz to the Arabs.

Nine-tenths of tactics are certain, and taught in books: but the irrational tenth is like the kingfisher flashing across the pool, and that is the test of generals.

'The Evolution of a Revolt' describes what is now called 'asymmetric warfare'. Fighting the Arab rebellion, Lawrence says, would be messy and slow for the Turks, 'like eating soup with a knife'. There would be no pitched battles, because the Arab irregulars – all valuable individuals, not mere units – could not afford casualties; there would be no contact, because Lawrence's Arabs would instead wage 'a war of detachment: we were to contain the enemy by the silent threat of a vast unknown desert, not disclosing ourselves till we attacked'. The Arabs would not engage Turkish troops but only attack empty stretches of the Hijaz railway line. They blew up the tracks not to destroy them permanently but to give the Turks maximum aggravation in protection and repair. Cutting the telegraph wires had an intelligence purpose: it made the Turks use the wireless more, which the British could then intercept and listen to. Waging mobile warfare with Bedouin irregulars required the best intelligence, and careful attention to the tribesmen's mood and morale.

Our cards were speed and time, not hitting power, and these gave us strategical rather than tactical strength. Range is more to strategy than force. The invention of bully-beef has modified land-war more profoundly than the invention of gunpowder.

Guerrilla sorties in the desert were more like naval operations, with camel raiding parties as self-contained as ships. Each man carried six weeks' frugal supply of food on his camel, a hundred rounds for his rifle and a pint of water to last him between wells. Ranging over a thousand miles, the tactics were 'always tip and run, not pushes, but strokes . . . We used the smallest force, in the quickest time, at the farthest place.'

The Arab irregular volunteers were contracted by honour, not discipline, and their war was conceived as one-on-one. Ideally every action was a series of single combats, in which good-quality fighters kept cool, and used speed, concealment and accuracy of fire to prevail. Lawrence reckoned that 'Irregular war is far more intellectual than a bayonet charge'.

The Armistice arrived before Lawrence could prove the idea that a war might be won without fighting battles, but he was moving that way. These simple ideas have become conventional nowadays, but then they were as revolutionary as quantum mechanics. After all, in 1916–17, the British thought it normal to lose 60,000 men in a single day on the Somme, or 142,000 in four days at Arras, or 275,000 in four months at Passchendaele. 'The Evolution of a Revolt' ends:

Granted mobility, security (in the form of denying targets to the enemy), time, and doctrine (the idea to convert every subject to friendliness), victory will rest with the insurgents . . .

That formula has had an almost incalculable historical effect around the world ever since. Communist insurgents, including the Vietnamese, learned the lesson well. The entire first printing of Robert Taber's study of *Guerrilla Warfare Theory and Practice, The War of the Flea* (1964–5), was bought up by the US military during the Vietnam War. Forty years later, there are still lessons being learned in Iraq; the sort of American military officers who read *Small Wars Journal* also possess well-thumbed copies of *Seven Pillars of Wisdom*.

In December 2006, the US Army issued its new field manual on Counter-Insurgency, written by General David Petraeus and Colonel Conrad Crane, advocating a radical change from conventional American heavy-handedness and massive firepower towards a more 'hearts-and-minds' approach. The third of its source notes cites T. E. Lawrence's 'Evolution of a Revolt'; the sixth refers to his 'Twenty-Seven Articles' from the *Arab Bulletin* of 20 August 1917, of which Article 15 reads:

Do not try to do too much with your own hands. Better that the Arabs do it tolerably than that you do it perfectly. It is their war, and you are there to help them, not to win it for them.

# 8

# The Twice-promised Land

Military deception and bluff would play a key role in the WWI Palestine campaign. In June 1917, Lawrence accompanied the warrior Sheikh Audah abu Tayi (the colourful character played by Anthony Quinn in the film *Lawrence of Arabia*) in a surprise attack on the Arabian Red Sea port of Akaba. This was defended by the Turks against attack from the sea, from which an assault by the French or British was expected, but they did not anticipate attack from the desert behind. Lawrence swept in from the hinterland with several hundred Arab irregulars in a camel charge. Having secured another port for the Royal Navy to land supplies, successful Lawrence went up to Cairo to meet for the first time the newly arrived British commander-in-chief, General Edmund Allenby (played by Jack Hawkins in the film). Moths had chewed up Captain Lawrence's khaki uniform, so he was still dressed in his Arab gear. Lawrence described him in *Seven Pillars*:

Allenby was physically large and confident . . . He sat in his chair looking at me – not straight as his custom was, but sideways, puzzled. He was newly from France, where for years he had been a tooth of the great machine grinding the enemy. He was full of Western ideas of gun power and weight – the worst training for our war – but, as a cavalryman, was already half persuaded to throw up the new school, in this different road of Asia, and accompany Dawnay and Chetwode along the worn road of manoeuvre and movement; yet he was hardly prepared for anything so odd as myself – a little bare-footed silk-skirted man offering to hobble the enemy by his preaching if given stores and arms and a fund of two hundred thousand sovereigns to convince and control his converts.

Allenby studied the map while Lawrence told him about Eastern Syria and its inhabitants. If given guns and gold, this strange little man seemed to be promising him useful diversionary help against the Turks on the right flank. 'At the end [Allenby] put up his chin and said quite directly, "Well, I will do for you what I can", and that ended it.'

Allenby's mission from Lloyd George was to conquer Palestine from Egypt, drive out the Ottoman Turks and take Jerusalem. He soon moved his headquarters from the fleshpots of Cairo up to the spartan desert front north of Rafa and, restlessly moving about among his men (which made him popular with the Australians), resupplied and reorganised his three corps for a major battle with the Turks entrenched at Gaza in Palestine where two previous British frontal attacks had failed.

Allenby's intelligence achieved more than his famed bullishness as a general. He was a cavalryman, and because he used thousands of horses and camels it is possible to think of his campaign as old-fashioned, but in fact it was also innovatory, making intelligent use of state-of-the-art technology in aviation, photography, mechanisation and wireless to deceive and outmanoeuvre the enemy. In her 2007 study, *Military Intelligence and the Arab Revolt*, Polly A. Mohs considers it 'the first modern intelligence war'.

Palestine in 1917 was the last great campaign in the annals of war where horses and camels were used strategically en masse. It was an utterly different context from trench warfare in France where artillery was everything and cavalry were useless. In Palestine, Harry Chauvel's Australian Light Horse fought thirty-six battles against the Turks in thirty months and won them all. Two of Allenby's three corps in the Egyptian Expeditionary Force were commanded by cavalrymen, including the brilliant tactician Philip Chetwode of XX Corps.

Chetwode's chief staff officer on the advance from Egypt was Brigadier Guy Dawnay, the merchant banker involved in deception in the Dardanelles. 'Dawnay was not the man to fight a straight battle' wrote T. E. Lawrence in chapter LXIX of *Seven Pillars of Wisdom*. It was Dawnay and Chetwode who persuaded Allenby that the Third Battle of Gaza should not be head on.

Broadly, Allenby's plan was like a soccer player taking a penalty: his arm and body signal towards the Turkish right at Gaza, but his foot shoots left to the Turkish left at Beersheba and curves the ball into the back of the net. It was risky: the desert route to Beersheba was suitable for horses and camels, but not for vehicles, and there would be problems supplying the animals with water. It was vital that the Turks be led to believe that any movements in that direction were only routine reconnaissance or feints; they had to think the real British attack was going to come up the coast to Gaza.

T. E. Lawrence wrote that Guy Dawnay

found an ally in his intelligence staff who advised him to go beyond negative precautions, and to give the enemy specific (and speciously wrong) information of the plans he matured. This ally was Meinertzhagen, a student of migrating birds drifted into soldiering, whose hot immoral hatred of the enemy expressed itself as readily in trickery as in violence.

Major Richard Meinertzhagen (1878–1967) was the future author of *Nicoll's Birds of Egypt*, *Birds of Arabia* and a study of avian robbery, *Pirates and Predators*. His commander-in-chief, Allenby, was also an ornithologist, and perhaps their knowledge of the aggressive world of birds led both men towards deception. Dick Meinertzhagen (the surname is German) recorded his life in seventy-six volumes of diaries full of horror and hilarity. According to him, when he first met Adolf Hitler in 1934, the Führer threw his right arm up and said 'Heil Hitler!' Slightly puzzled, Meinertzhagen put his arm up and said 'Heil Meinertzhagen!' It is a great story, except that he never met Hitler.

You get the flavour of Meinertzhagen as a schoolboy from the dialogue he says he had with Lord Salisbury during inspection on a Volunteer Corps field day at Hatfield Park in the summer of 1892:

'One of my rabbits?'
'Yes, sir.'
'And how did you kill it?'
'With a stone.'
'Well done. Are you going to eat it?'
'No. It's for our eagles at Harrow.'

According to his *Diary of a Black Sheep*, when Meinertzhagen was at his second prep school near East Grinstead he was sexually abused and sadistically flogged by a schoolmaster called Walter Radcliffe. Young Dick claimed to have felt so abandoned that evil crept into his soul. From prep-school age he determined to be the predator, never again the weaker prey. All his life rage lurked behind Meinertzhagen's armour of opinionated intransigence. It could be expressed in violent words or deeds, or suddenly masked in urbane humour and charm.

Meinertzhagen had joined the Army in India and spent 1902–6

seconded to the King's African Rifles, often up-country in Kenya where he enjoyed slaughtering both animals and humans. It was while leading a punitive expedition against the Kikuyu and the Embu people in 1904 that he managed to discover the new species of eastern giant hog which now bears his name, *Hylochoerus meinertzhageni*. Meinertzhagen finally left Kenya after facing three courts martial because of the machine-gunning of twenty-three Nandi-speaking tribesmen on 19 October 1905. They had been resisting the building of the British Mombasa–Uganda railway (dubbed the 'Lunatic Express') through their land, and their chief was due to meet Meinertzhagen to sue for peace. When they moved to shake hands, Meinertzhagen pulled out his pistol and shot Koitalel arap Samoei dead. He claimed it was self-defence against a wicked old man who was about to murder him and use his body parts for a magic broth.

He was a life-long obsessive field naturalist who used deception when hunting. In Kenya, he constructed a dummy ostrich by stretching the skin of a female ostrich over a bamboo frame. Holding the separate head and neck in his right hand and his rifle in his left, he could get as close as twenty-five yards to most game if he approached upwind. Meinertzhagen thought 'the hunting of men – war – is but a form of hunting wild animals'.

According to T. E. Lawrence, Meinertzhagen

was logical, an idealist of the deepest, and so possessed by his convictions that he was willing to harness evil to the chariot of good. He was a strategist, a geographer, and a silent laughing masterful man; who took as blithe a pleasure in deceiving his enemy (or his friend) by some unscrupulous jest, as in spattering the brains of a cornered mob of Germans one by one with an African knob-kerri.

Meinertzhagen was shocked by this portrait of himself and begged Lawrence to remove it from *Seven Pillars of Wisdom*. But his own diaries give an even worse impression. When he was at the Staff College in Quetta, Baluchistan, Meinertzhagen says that when he found his syce or stable groom mistreating his ponies, he beat the man to death with a polo mallet. He claims to have hushed the matter up with the police and got the dead man registered as a plague victim. In the East African campaign against the Germans in Tanganyika, Meinertzhagen laid dead birds and animals around a clean water-hole and signposted it POISONED so as to deny it to the enemy but keep it safe for his own use. As the British Intelligence officer, he once sent a suspected German spy 1,500 rupees and a thank-you note and made sure the Germans intercepted it, so they would shoot their own man and save Meinertzhagen the trouble. Running an effective network of agents in East Africa, Meinertzhagen discovered that the German officers' latrines were a good source of soiled documents and letters, yielding 'filthy, though accurate information'. This was the vigorously amoral soldier who played a part in British deception in Palestine.

Meinertzhagen spent some time secretly preparing the famous 'haversack ruse' of 10 October 1917 which Allenby credited with a major role in the successful attack on Gaza, and always claimed to have carried it off, in person and alone. In essence, Meinertzhagen said he rode out into the country north-west of Beersheba, deliberately tangled with a Turkish patrol and got himself shot at. He slumped in the saddle, dropping his water bottle, field glasses, rifle and, most important of all, a khaki haversack stained with his horse's fresh blood, containing personal letters, papers and £20 in notes. Then he rode away, mimicking the tactics of the lapwing, pretending to be wounded to draw predators away from its nest. But he lingered long enough to see the rifle and haversack were picked up.

The abandoned papers in the haversack looked absolutely genuine but were all forgeries. A British staff officer's notebook, Army Book 155, was filled with 'all sorts of nonsense about our plans and difficulties'. The supposed agenda for a staff conference would have told the Turks and Germans the main attack was coming at Gaza, preceded by a mere feint at Beersheba, the opposite of the truth. There was also an ardent letter from a wife announcing the birth of a son called Richard (written, Meinertzhagen first claimed, by his sister who had never had a child and was miles and weeks away in England, though he later said it was composed by a nurse at El Arish). There were also notes on a cipher which would enable the enemy 'to decipher any camouflage messages we might send later on'.

Meinertzhagen claims he backed up his stratagem by sending out anxious wireless messages about the haversack and furious divisional orders regarding proper security of papers; Turkish and German intelligence officers monitoring radio traffic took the bait to the commander of the Eighth Turkish Army, General Friedrich Kress von

Kressenstein, who swallowed it. Two British soldiers, not in the know but captured while genuinely searching for the haversack, confirmed its credibility, as did the finding of a copy of Desert Corps Orders (apparently carelessly thrown away in the wrappings of an officer's lunch) requiring the lost notebook to be returned to GHQ.

'The haversack ruse' is a terrific story. Something like it did, indeed, happen. But in his 2007 biography, *The Meinertzhagen Mystery*, Brian Garfield says that the idea for it came from another man, Lieutenant Colonel J. D. Belgrave, and that the actual perpetrator was a man called Arthur Neate, on 12 September, and not Meinertzhagen at all. If Garfield is right, Meinertzhagen swooped in to steal the credit for other men's initiative and courage.

Lieutenant Colonel Archibald Wavell was a direct participant in these Palestine events as the liaison officer between the Chief of the Imperial General Staff in London and the commander-in-chief in the field, Edmund Allenby. Later, when Wavell himself had become commander-in-chief, Middle East, he would write an admiring biography of Allenby, subtitled *A Study in Greatness*. Allenby's use of intelligence, deception and guerrilla forces in what Churchill called the 'brilliant and frugal' operations in Palestine profoundly influenced Wavell in the next World War.

Allenby's great push for Palestine began at the end of October 1917. The RFC and the RNAS had air superiority when the Royal Navy began shelling Gaza from the sea. The British in Cairo could listen in to all German aerial and airfield communications from Syria to Sinai, which meant they could send up planes to intercept any German efforts at aerial reconnaissance. Because of this, the enemy never spotted 40,000 British troops slipping eastwards on the night of 30 October. The British infantry seized the garrisoned town of Beersheba by surprise and, following a spectacular charge by the Australian Light Horse, the cavalry and camelry of the Desert Mounted Corps took the vital water wells before they could be demolished.

Then Allenby attacked Gaza on the night of 1 November. This distraction drew all the Turkish reserves westward. Deluded by false intelligence, including the haversack ruse, the Turks assumed this was the main assault of the Third Battle of Gaza. But on 6 November British mounted divisions attacked the Turkish lines from the east, from Beersheba. The Turks panicked: Gaza was abandoned the next day and Turkish troops began retreating north along the coastal plain.

Meinertzhagen later claimed that many of these soldiers were drowsy and fuddled, because of another of his tricks: thousands of cigarettes, 'heavily doped with opium', had been dropped on their lines. Meinertzhagen says he later tried one himself: 'They were indeed strong. The effect was sublime, complete abandonment, all energy gone, lovely dreams and complete inability to act or think.' There is absolutely no confirmation of this story from anyone but Meinertzhagen.

According to Meinertzhagen, his haversack ruse had completely wrong-footed the enemy, and the German General Kress von Kressenstein was subsequently relieved of his command. We are told that Allenby later wrote on Meinertzhagen's confidential report: 'This officer has been largely responsible for my successes in Palestine.' But Dick Meinertzhagen made sure the official historian knew all about his exploits, and the source of Allenby's assessment is Meinertzhagen's own edited diaries. No man was a greater burnisher of his own reputation.

In Palestine, Allenby's forces pushed on fifty miles in ten days, took the Mediterranean port of Jaffa where the Royal Navy could land supplies, then turned east into the Judean hills toward the great prize of Jerusalem. On 8 December 1917, the Turkish forces abandoned the Holy City. The mayor of Jerusalem came out in a frock coat and fez, carrying a white flag and the keys to the city, which he offered, in a moment of bathos, first to some army cooks from London, then a sergeant, then some gunnery officers, then a brigadier, until, at last, a general could be found.

So Allenby gave Lloyd George and the British people the gift they had asked him for in time for Christmas 1917. This Allied victory was both symbolic and historic: for the first time since 1244, the Christians wrested the Holy City back from the Muslims. But their plan to share it with the Jews sent shock waves through the region that still make the world tremble to this day. Lloyd George had seen that the support of international, and especially US, Jewry for the Allies was invaluable and that the Zionist movement could be used (in John Marlowe's phrase) 'as a wooden horse of Troy to introduce British control into Palestine'. He had discussed it all with his Foreign Secretary, Arthur Balfour.

Arthur Balfour's bland letter of 2 November 1917 to the Zionist federation, by way of Lord Rothschild, contained a single sentence, the famous 'Balfour Declaration', which has caused much grief. Its slippery surface is why Palestine has been dubbed 'The Twice-promised Land':

His Majesty's Government view with favour the establishment in Palestine of a National Home for the Jewish people, and will use their best endeavours to facilitate the achievement of this object, it being clearly understood that nothing shall be done which may prejudice the civil and religious rights of existing non-Jewish communities in Palestine . . .

At midday on 11 December 1917, six weeks to the day after his attack on Beersheba, General Allenby made his official and symbolic entry to Jerusalem through the Jaffa Gate, on foot. (By contrast, in 1898, the German Kaiser had arrogantly ridden in on horseback. The Catholic spin-doctor Mark Sykes hoped that the faithful of the local three great religions would appreciate this British gesture of humility towards their Holy City.) Allenby censored all references to the recent Balfour Declaration as potentially inflammatory to Arab feelings. The Press Bureau of the Department of Information (which of course was run by John Buchan) issued a D-Notice to the media on 15 November:

The attention of the Press is again drawn to the undesirability of publishing any article, paragraph or picture suggesting that the military operations against Turkey are in any sense a Holy War, a modern Crusade, or have anything whatever to do with religious questions. The British Empire is said to contain a hundred million Mohammedan subjects of the King and it is obviously mischievous to suggest that our quarrel with Turkey is one between Christianity and Islam.

British propagandists had already spread the word that the name 'Allenby' was a version of the Arabic *al-Nabi* meaning 'The Prophet', and the Haram-esh-Sherif or Temple area of the city known in the Islamic world as Al-Quds was conspicuously put under the guard of Indian Muslim soldiers from the Egyptian Expeditionary Force.

Other soldiers, English, Scottish, Irish, Welsh, Australian, New Zealand, French and Italian, were drawn up at the Jaffa Gate. It was a cold but sunny morning. General Allenby appeared flanked by the French and Italian commanders and followed by twenty of his principal staff officers and the commander of XX Corps, Sir Philip Chetwode. Among them, walking next to Colonel A. P. Wavell, was Major T. E. Lawrence, joking about his borrowed British uniform. Three weeks before, he had endured what Ronald Storrs called 'hideous man-handling' (homosexual rape and flogging) by Turkish soldiery after being captured near Deraa. At the Citadel, the proclamation of martial law was read out in seven languages to 'the inhabitants of Jerusalem the Blessed and the people dwelling in its vicinity'. The declaration said that 'every sacred building, monument, holy spot, shrine, traditional site, endowment, pious bequest, or customary place of prayer of whatsoever form of the three religions will be maintained and protected'. Then Allenby met the city's notables and religious leaders, before the military reformed their procession and went back out through the gate to where they had left their horses. This simple but impressive ceremony at the Jaffa Gate was for Lawrence 'the supreme moment of the war'.

But politics continued as ever. That afternoon, Lloyd George announced the news of the capture of Jerusalem to a cheering House of Commons; the War Cabinet was soon urging the occupation of all Palestine. The appalling battle for the ridge at Passchendaele had ended the month before with a gain of five miles of bloodsoaked mud at the cost of more than 250,000 Anzac, British and Canadian casualties. Allenby's defeats of Turkey in the Middle East, 'knocking out the props' as it was called, were seen as an escape from the horrendous stalemate on the Western Front. The 'Easterner' faction got a great boost.

Early in 1918, T. E. Lawrence was given a new role – helping the Arab Northern Army, led by Feisal, but under Allenby's command – to harry the left of the Turkish Fourth Army as they retreated northwards. Lawrence was under increasing nervous strain after his brief capture by the Turks at Deraa, when 'the citadel of my integrity had been irrevocably lost', as he put it. Now he accused himself of 'accessory deceitfulness' and 'rankling fraudulence' towards the Arabs in his 'pretence to lead the national uprising of another race, the daily posturing in alien dress, preaching in alien speech'. He wanted to be free of it. In fact he wanted to be tinkering with Rolls-Royce armoured cars among men of his own sort. But Allenby needed him to help take Damascus and if possible Aleppo.

There was no escape for me. I must take up again my mantle of fraud in the East . . . It might be fraud or it might be farce: no one should say that I could not play it.

117

Allenby gave Lawrence more money, 2,000 camels, armoured cars and aircraft to push forward attacks by Feisal's army against the Amman–Deraa–Damascus sector of the Turkish line, in order to make them reinforce east of the river Jordan. The final phase of the Palestine campaign in September 1918 called for more British trickery. This time they feinted right, up the Jordan valley from Jericho, but really broke through on the left, straight up the coastal plain from Jaffa. Lawrence records, in chapter XCVIII of *Seven Pillars of Wisdom*:

After the Meinertzhagen success, deceptions, which for the ordinary general were just witty hors d'oeuvres before battle, became for Allenby a main point of strategy. Bartholomew would accordingly erect (near Jericho) all condemned tents in Egypt; would transfer veterinary hospitals and sick-lines there; would put dummy camps, dummy horses and dummy troops wherever there was plausible room; would throw more bridges across the river; would collect and open against enemy country all captured guns; and on the right days would ensure the movement of non-combatant bodies along the dusty roads, to give the impression of eleventh hour concentrations for an assault.

The deception was wholly successful: three whole divisions moved west to the coast by night, hiding in orange groves by day, without being seen. Soldiers doubled up in the tents, daylight movement was prohibited, and horses could only be watered at fixed hours when the RAF was up in full force to deter enemy spotter planes. Dummy cavalry horses were left behind in the Jordan valley, made of wood and canvas, stuffed with straw. Mules pulled wattle hurdles to raise dust at the time these 'horses' were meant to be trotting to water. An Indian havildar or sergeant deserted and told the Turks the true facts, that a great attack was coming up the coast, but the new German commander Liman von Sanders dismissed him as a plant – another haversack ruse – because any evidence to confirm what he said had been skilfully hidden.

The Arab role in all this was to harass and disrupt communications and transport in the east so as to convince the Turks that their left at Deraa was under major attack. As Liddell Hart put it, 'Lawrence wove a web of feints and fictions to persuade the Turkish command that Allenby's attack was coming east towards Amman instead of north to Galilee.' Lawrence, in his own words 'the godless fraud inspiring an alien nationality', was now feeling increasingly guilty about his role.

He spent his thirtieth birthday agonising over his inadequacies. In the Sykes–Picot agreement of May 1916, Britain and France had already broadly divided up their spheres of influence in the Levant. Lawrence knew that this 'old-style division of Turkey between England, France and Russia' took no account of Arab nationalism; he had warned Feisal about it privately, but also convinced him that the only way out was 'to help the British so much' that they would be shamed into granting a decent peace. Lawrence says he 'begged' Feisal 'not to trust in our promises, like his father, but in his own strong performance'. Nevertheless, Lawrence came to fear that he was duping ignorant people in a glorified swindle:

Yet I cannot put down my acquiescence in the Arab fraud to weakness of character or native hypocrisy: though of course I must have had some tendency, some aptitude for deceit, or I would not have deceived men so well, and persisted two years in bringing to success a deceit which others had framed and set afoot. I had had no concern with the Arab Revolt in the beginning. In the end I was responsible for its being an embarrassment to the inventors. Where exactly in the interim my guilt passed from accessory to principal, upon what headings I should be condemned, were not for me to say.

In September 1918, a mixed British and Arab force gathered at Azrak, east of Amman, under Lawrence and his immediate superior, Colonel Joyce. Joyce's fighting force blended the new with the old, and communicated in English, French, Arabic and Hindustani. It comprised two aircraft, five Hijaz Armoured Car Company vehicles with their tenders, a couple of ten-pounder guns on Talbot cars, four French mountain guns, a score of Indian machine-gunners, and hundreds of Bedouin Arab horsemen and camel riders. They proceeded to attack the railway line north and south of Deraa, blowing up several kilometres by placing under the iron sleepers thirty-ounce gun-cotton charges which bent and warped the steel track into 'tulips' beyond repair.

To cut off all Palestine, as well as the Hijaz, by destroying the railroad from Damascus, Constantinople and Germany, Joyce's men took Mezerib station, west of Deraa, illuminating their evening meal by burning the Turkish trains and petrol tankers. They snipped the telegraph wires, slowly, with ceremony. 'It was pleasant to imagine Liman von Sanders' fresh curse, in Nazareth, as each severed wire tanged back from the clippers.' Lawrence blew up what he claimed was his seventy-ninth railway bridge at Nisib, lighting a thirty-second fuse on 800 pounds of gun cotton in one go. But not everything was going their way. They were under Turco-German attack from the air, and Lawrence had been bombed on camel-back, in a car, on foot. Now he wanted air support.

When an aeroplane brought news that Allenby's advance was working well, Lawrence flew back in it. The gulf of the Jordan valley and the Dead Sea cut off direct communication between the Arab army in the east and Allenby's forces in Palestine. But the aeroplane abolished geography, and Lawrence flew to Headquarters at Bir Salem, near Ramleh, to see his commander-in-chief.

In a cool, airy, whitewashed house, proofed against flies, Lawrence was shown Allenby's plans for three Imperial thrusts over the Jordan: the New Zealanders to Amman, the Indians to Deraa, and the Australians to Kuneitra. All would converge on Damascus, with Lawrence's Arabs assisting on the right flank. Lawrence explained his air problems, and Allenby summoned the RAF. Lawrence admired 'the perfection of this man who could use infantry and cavalry, artillery and Air Force, Navy and armoured cars, deceptions and irregulars, each in its best fashion!' They planned for a bomber, loaded with petrol and stores, and two Bristol fighters to be sent over to Lawrence.

The huge Handley-Page bomber, which could carry a ton of supplies, impressed Lawrence's Arabs. 'Indeed and at last they have sent us THE aeroplane, of which these things were foals.' 'These things' were the few much smaller fighter biplanes which had occasionally assisted the Arab Revolt since November 1916, the first air support for a guerrilla force in history.

The Arab guerrillas had done their duty to Allenby, and with increased Arab attacks and bombings, the Turkish Fourth Army was slowly collapsing; the Arabs had earned their gold sovereigns and could stand down and go back to their flocks and herds. But Lawrence wanted to push on, to Damascus.

I was very jealous for the Arab honour, in whose service I would go forward at all costs. They had joined the war to win freedom, and the recovery of their old capital by force of their own arms was the sign they would best understand.

In *Seven Pillars of Wisdom*, Lawrence tells how the Arabs waded through blood to reach Damascus, which they were the first to enter, and where they successfully restored order from 1 October 1918. Wavell says this was 'not the whole truth', but it was a central tenet of the Arab propaganda myth that Lawrence needed to create.

Allenby, the commander-in-chief, 'gigantic and red and merry', turned up in his grey Rolls-Royce at the Victoria Hotel, Damascus. According to Lawrence, he approved 'in ten words' all Lawrence had done, confirmed his appointments, and took over the hospital and the railway. Then Feisal, 'large-eyed, colourless and worn', arrived by special train from Deraa, smiling through the tears which the welcome of his people squeezed from him. Allenby and Feisal met for the first time:

They were a striking contrast – the burly confident Englishman, accustomed to command and to dominate by sheer force of personality, and the slight ascetic Arab with his princely bearing, to whom the arts of the politician were more natural than the vigour of a soldier. Both were men of fine quality, and appreciated and trusted each other.

Wavell, Allenby: A Study in Greatness

But then Allenby started imposing the terms of the Sykes–Picot agreement about French and British spheres of influence. Feisal must deal with a French liaison officer, and not attempt any control of Lebanon, even though Feisal's country Syria needed a Mediterranean port. The French dislike of the Hashimites began to emerge. Feisal, arguing for self-rule, denied knowledge of any Anglo-French deal and Lawrence – who had, in fact, betrayed the existence of the Sykes–Picot agreement to him – claimed ignorance of it too. What the Arabs should have earned by their own courage and endurance in the campaign was now trumped by imperial realpolitik. Lawrence backed away in disgust from the great deception he had been part of, inducing people to fight for what would never be given. Wavell described Lawrence as 'overstrained in mind and body'. He asked Allenby's permission to go, and left Damascus on 4 October 1918. The last word in *Seven Pillars of Wisdom* is 'sorry'.

---

T. E. Lawrence has had many detractors, but also powerful friends. Chief among his admirers was Winston Churchill, who refers to Colonel Lawrence's 'astonishing personality' in *The World Crisis*, and wrote an essay for the anthology *T. E. Lawrence by His Friends*: 'I deem him one of the greatest beings alive in our time . . . one of Nature's greatest princes.' Churchill also singled Lawrence out for admiration akin to hero worship in *Great Contemporaries*.

Churchill's own first experience of real war had been seeing 'the Spaniards out-guerrilla-ed in their turn' by the rebels in Cuba in 1895. He read *Seven Pillars of Wisdom* as the story of one individual directing 'audacious, desperate, romantic assaults' against a narrow steel railway track running through blistering deserts, the Achilles tendon which if severed would bring down Turkey, then Germany. Churchill identified with Lawrence as 'someone strangely enfranchised, untamed, untrammelled by convention, moving independently of the ordinary currents of human action'. Churchill's later encouragement of the Special Operations Executive, of guerrilla and partisan armies, of commandos and Special Forces and their raiding tactics – 'butcher and bolt' – owed an enormous amount to the example of Colonel Lawrence.

In the spring of 1921, Winston Churchill took over the Colonial Office. The Middle East 'presented a most melancholy and alarming picture' of turmoil and turbulence. There was rebellion in Iraq, Egypt was in ferment, there was tension between Arabs and Jews in Palestine and disgruntled desert chiefs were rousing the Bedouin beyond the Jordan. Churchill formed a new department to deal with the area and invited T. E. Lawrence to join. He proved an admirable civil servant.

In March 1921, at the Hotel Semiramis in Cairo, Colonial Secretary Churchill gathered the top British civil and military administrators of the region (nicknaming them 'the forty thieves') all together for a ten-day conference. Churchill and Lawrence then effectively redrew the map. They split the British-controlled territory west of Iraq in two, along the line of the river Jordan. The 23 per cent of the land west of Jordan, already under a Jewish High Commissioner, Sir Herbert Samuel, was to become the 'national home for the Jewish people' promised in the Balfour Declaration. The 77 per cent of the dry territory east of the river, now named Trans-Jordan or Transjordania, was for the Arabs, and was to be ruled by Sharif Hussein's son, the Hashimite Emir Abdulla. His brother Feisal, who had been ejected from Damascus by the French in July 1920, now received his consolation prize, the Kingdom of Iraq, a place he had never visited.

Lawrence was pleased to seem a kingmaker in Jordan and Iraq and to reward the Hashimite Sherifians, but Churchill was anxious to save some of the 'ruinous expense' of imperial over-reach: garrisoning Mesopotamia or Iraq with 40,000 troops and suppressing the insurgency there in 1920 had cost Britain £33 million. He wanted to maintain the security of the new country on the cheap by withdrawing the soldiers and just using the RAF, then being nurtured on a shoestring by Hugh Trenchard.

When Churchill was in his previous job, Secretary of State for War and Air, he had told the House of Commons on 15 December 1919: 'The first duty of the Royal Air Force is to garrison the British Empire', and the quiet success of a swift air campaign in Somaliland, in January/February 1920, convinced him that 'air control' was the way of the future. Half a dozen RAF planes, supporting 500 Camel Corps and a battalion of Kings African Rifles on the ground, apparently managed to smash an Islamist rebellion by Mohammed Abdullah Hassan, known as 'the Mad Mullah of Somaliland', in three weeks, and impose a peace that lasted for the next twenty years. It all cost only £77,000, so the Air Ministry made the claim that colonial policing by aeroplane, independent air action involving a little judicious bombing or mustard gassing of rebellious tribes, was the most economical use of the iron fist under the velvet camouflage of independence. After Feisal was crowned King of Iraq on 23 August 1921 he was protected from his enemies by eight RAF squadrons as well as some armoured cars and gunboats. By 1923 they had paid their way by saving Mosul and its oilfields both from Turkish invaders and Arab rebels.

Another fan of T. E. Lawrence was the foremost military critic in Britain between the wars, Captain Basil Liddell Hart, who wrote a biographical study of him, '*T. E. Lawrence*': *in Arabia and After*, published in 1934. Liddell Hart's famous 'indirect approach', his 1920s rethinking of infantry tactics and strategy to avoid the butchery he saw as a company commander at the Somme, drew to an extent on Lawrence who, Liddell Hart said, 'fore-shadowed what I believe will be the trend of the future – a super-guerrilla kind of warfare'.

Lawrence aroused love and hatred in equal measures. David Cannadine dismisses Lawrence as a 'homosexual egomaniac'. Sir John Keegan sees the encouraging of guerrillas by developed nations, in which Lawrence played a pivotal role, as the tragic irresponsibility which unleashed modern terrorism. Yet John Buchan said in his auto-biography 'I could have followed Lawrence over the edge of the world' and called Lawrence 'the only man of genius I have ever known'.

John Buchan and Lawrence first met in 1920, although Buchan had heard about Lawrence from mutual friends like D. G. Hogarth and Aubrey Herbert. They had much in common: both were small, energetic, tougher than they looked, classically trained, with similar tastes in literature and the same benign vision of the future of the British Empire (a voluntary association without racial prejudice). They also shared a lifelong interest in unconventional warfare. 'The science of war had always been one of my hobbies,' Buchan wrote.

Buchan changed Lawrence's life when, as director of information in 1917, he sent the American journalist and film-maker Lowell Thomas out east to cover Allenby's campaign. There he was introduced to Lawrence by Ronald Storrs, military governor of Jerusalem, with the words, 'I want you to meet Colonel Lawrence, the Uncrowned King of Arabia.' The cinematic travelogue and lecture that Thomas put together after the war, 'With Allenby in Palestine and Lawrence in Arabia', packed out the Royal Opera House, the Albert Hall, the Philharmonic Hall and the Queen's Hall in London from August 1919, and then toured the world for four years, culminating in a book, *With Lawrence in Arabia*, which launched the Lawrence legend and made him a kind of matinee idol.

It was John Buchan who suggested to Liddell Hart that he put Lawrence's essay 'The Evolution of a Revolt' into the *Encyclopaedia Britannica* as its entry 'Guerrilla'. John Buchan (according to his son William's *Memoir*) delighted in Lawrence's rare and secretive visits to his Oxfordshire home, and incorporated Lawrence into later incarnations of his fictional hero Sandy Arbuthnot. In Buchan's 1929 novel *The Courts of the Morning* (which reworks Joseph Conrad's *Nostromo*), Sandy Arbuthnot leads a horseback guerrilla uprising in a mineral-rich South American republic called Olifa, and blows up a railway, just as Lawrence did. For the increasingly desk-bound Buchan, Lawrence represented a last link to the world of adventure.

# 9

# A Dazzle of Zebras

Back in England, at the end of 1916, long-haired, white-moustached David Lloyd George had stepped up to lead what he later called 'the bloodstained stagger to Victory', the long last phase of the war in which camouflage, deception and propaganda played a vital role. On 6 December 1916 King George V asked David Lloyd George to become Prime Minister, and the 'Welsh wizard' set about forming a national government, drawing his administration from the Conservative, Labour and Liberal parties, with a war cabinet reduced to five. In the same month, Solomon J. Solomon set up a camouflage school in Hyde Park.

Lloyd George had more drive and initiative and a greater sense of urgency than his predecessor Asquith. A cartoon of him in *Punch*, entitled 'The New Conductor', showed the new premier as a vigorous figure in evening dress, baton upraised for the 1917 overture. It was a mammoth task. The country was two years into a military effort that was draining the exchequer (the war cost £5.7 million a day) and straining national resources. The land battles slaughtered soldiers and the air raids scared the citizens of London, but it was the war at sea that was doing the most economic damage as U-boats attacked the cargo ships that supplied the British Isles. By the end of 1916, Britain had lost a fifth of its merchant fleet.

The Admiralty seemed paralysed in the face of the submarines, telling the Government in November 1916: 'No conclusive answer has as yet been found to this form of warfare . . . We must for the present be content with palliation.' The only defences against submarines were underwater steel nets and not very reliable mines; U-boats could only be attacked when on the surface, by ramming or shooting. The new weapons that would eventually make a difference – hydrophones for detection and depth charges for destruction – took time to research and develop.

Churchill had feared from the beginning that enemy submarines could destroy British sea power and win the war. German U-boat attacks had diminished after the bad publicity they gained by sinking the *Lusitania* in 1915, and the Kaiser had curbed the renewal of torpedoings. But in 1917, with deadlock in the trenches and block-aded Germany reduced to a diet of potatoes, his Imperial Majesty was desperate: 'I order the unrestricted submarine campaign to begin on 1st February with the utmost energy.'

Ironically, the power of the U-boat weapon would actually ensure that Germany lost the war. The isolationist United States of America only entered the fray after Imperial Germany began its strategy of indis-criminate submarine attacks on all ships, neutral or Allied, military or merchant, hospital or passenger, within huge zones of blockade. Two days after the Kaiser announced unrestricted submarine warfare, the USA cut off diplomatic relations with Germany. But this was not enough. Britain needed American manpower and industrial muscle actively on its side in the war. German submarines tipped US opinion in favour of the Allies, but a piece of deception clinched it.

--

The famous coup by the British Naval Intelligence Department that helped bring America into the war was not strictly naval, but diplomatic. The director of naval intelligence was still Admiral Reginald Hall, that 'demonic Mr Punch in uniform' as Barbara Tuchman described him. Hall controlled Room 40, Old Building (OB40) at the Admiralty, the heart of British signals intelligence. Here were employed 800 wireless operators and around 80 cryptographers and clerks, who intercepted some 15,000 German secret communications in WW1. John Buchan's 1927 short story 'The Loathly Opposite' describes wartime cipher work done by a unit of disparate amateurs very like those of OB40.

'Gentlemen do not read each other's mail,' said US Secretary of State Henry L. Stimson piously in 1929 when he closed down Herbert O. Yardley's cipher bureau. The British – and Reginald Hall in particular – were less scrupulous about enemy communications in wartime. On 17 January 1917, OB40 illicitly intercepted, on American territory, a diplomatic cable message from the German Foreign Secretary, Arthur Zimmermann, to Von Eckhardt, the German Minister in Mexico. Two of OB40's cryptographers, Nigel de Gray and the Reverend William

Montgomery (a scholarly expert on St Augustine of Hippo) cracked the code and were astonished by what they read. Zimmermann's cable said that if the USA came into the war on the Allied side, then Germany would propose an alliance with revolutionary Mexico and help the Mexicans reconquer lost territory in Texas, New Mexico and Arizona.

Hall had to play a clever game when he passed on this ticking bomb to the American Government in late February 1917. Hall's goal was to get the USA to join the combatants, so he had to convince the Americans that the cable threatening to foment revolutionary war from Mexico was genuine, without letting slip that the intercept had been made in violation of US neutrality. Moreover, Hall could not allow the Germans to suspect that their codes had been broken.

To camouflage his real source, the telegraph cable to the USA he was still tapping, Hall ensured that Edward Thurston, the British minister in Mexico, obtained a copy of the Zimmermann telegram in the form it had been received at the Western Union office in Mexico City. On 22 February, when Hall showed the American embassy in London the telegram dated 19 January, he could more or less honestly say that it had been obtained in Mexico and cracked in London.

The deciphered telegram shot to the US Secretary of State and then on to President Woodrow Wilson, who exclaimed 'Good Lord!' several times as he read it on 27 February. When published all over the front pages of the US press on 1 March 1917, 'the Zimmermann Note' caused a ruckus. Senator Stone of Mississippi and other isolationists suspected a trick by devious Brits trying to hornswoggle the USA into the war. The press magnate William Randolph Hearst (on whom Orson Welles based *Citizen Kane*) instructed his newspaper editors to treat it as 'in all probability a fake and a forgery'. There were over eight million German-Americans in the USA, and they remembered previous anti-German propaganda campaigns by British agents. But on 3 March, when Zimmermann himself naïvely admitted to an American reporter in Berlin that he could not deny having written the note, the floodgates of righteous indignation opened. The idea of Mexican revolutionaries like Pancho Villa and Emiliano Zapata being aided by Prussians to storm across the Rio Grande was too much. Pro-Germanism was swept away and the USA was inexorably impelled towards war. Later that month, 26,000 more US sailors were enlisted. Churchill said, 'A new Titan long sunk in doubt . . . now arose and began ponderously to arm.'

The United States was taking up arms just at the time Britain and France's ailing ally, Russia, was letting them fall. On 15 March 1917, amid widespread strikes and the eruption of 'soviets' or workers' councils, Tsar Nicholas II of Russia was forced to abdicate. A provisional government of liberals and moderate socialists was formed under Kerensky. They were, at least, parliamentarians, and the USA was the first government to recognise them, on 22 March.

President Wilson finally spoke up for the Allies, including Russia, on 2 April 1917. He declared that 'the world must be made safe for democracy' and called the Imperial German Government 'a natural foe of liberty'. The US Congress formally declared war on 6 April 1917, pledging 'all the resources of the country'. In terms of resources, the USA produced more steel and more automobiles than any nation on earth, but its army of 5,000 officers and 123,000 men was not then very much bigger than the original BEF of 1914. The draft began in May 1917 and soon the USA had ten million fit young men under arms, being equipped and trained to go overseas. American soldiers first fired at the German enemy in late October 1917.

The all-out German U-boat threat to the vital food, fuel, and industrial supply lines of the British Isles, France and Italy required the urgent development of new methods of protecting ships from submarines. But the Royal Navy did not think naturally in terms of predators and prey. Incredibly, Lloyd George and Maurice Hankey had to struggle to make the hidebound Admiralty accept even the simple idea that grouping merchant ships into convoys gave security of numbers, and allowed them to be shepherded safely by destroyers carrying depth-charges. But after the convoy system began in July and August 1917, the losses from U-boats began to fall.

It was almost as hard to convince the naval establishment that camouflaging ships would confuse submarines. British Royal Navy battleships and cruisers had followed the style of the German and French fleets from 1903 in being painted a neutral blue-grey to blend with the sea and the sky, although most destroyers and flotilla leaders stayed black. Ships that did close support work for terrestrial forces (as in the Dardanelles) began getting mottled paintwork from 1915. But the Germans' intensifying use of submarines and torpedoes in 1917 called for something more daring.

The Scottish Professor of Zoology Sir John Graham Kerr was

among the first in 1914 to propose something like the painter Abbott Thayer's 'countershading', concealing objects by reversing the natural positions of light and shade. Kerr suggested that Royal Navy warships should use white paint as well as the standard grey. But the Admiralty did not run with his biologically based idea, nor with Abbott Thayer's notion of painting submarines blue 'like high swimming open sea fish'. During the 1917 submarine crisis, another painter came up with a dramatic new idea.

Norman Wilkinson was a 38-year-old professional marine artist who later became President of the Royal Institute of Painters in Water Colour. Living near Portsmouth, he was a yacht racer from an early age, and was encouraged to break into commercial art by Arthur Conan Doyle, then working as a doctor in Southsea, long before he became famous as the creator of Sherlock Holmes. Wilkinson's painting of Plymouth Harbour, commissioned by the chairman of Harland and Wolff, hung over the mantelpiece of the smoking-room on the *Titanic*, the biggest passenger ship that yard had ever built. Wilkinson worked consistently for *The Illustrated London News* from 1901–15 (the heyday of the black and white illustrator) and claimed to be 'the father and mother of the "artistic" poster on English railway stations'.

Wilkinson joined the Royal Navy Volunteer Reserve (RNVR) and, as we have seen, painted in the Dardanelles. In 1917 he was posted to Devonport in the English Channel, and with the rank of lieutenant in the RNVR, was in command of an eighty-three-foot motor launch which swept for mines and patrolled off Portland Bill with two depth charges ready for enemy submarines. Wilkinson knew from the Dardanelles how alarming and effective submarines could be; now U-boats were sinking sixty vessels a week. On Channel patrol, he watched scores of troop- and supply-ships sailing across to France. Painted black and starkly silhouetted, he saw they were ideal targets for a U-boat commander's periscope sight.

Wilkinson also happened to be a lifelong dry-fly fisherman (his oils, drypoints and etchings of angling scenes are well known). Fishing requires tactics and camouflage; trout have to be persuaded, cautiously and intelligently, to rise to a deceptive fly. 'The good fisherman', Arthur Ransome observed, 'is always engaged in the active exercise of his imagination. He is the fish he catches.' In a chilly railway carriage, travelling back to Plymouth from a weekend's trout fishing in Devonshire in the spring of 1917, Wilkinson had a sudden vision. If it was impossible to paint a ship so that no submarine could spot her, 'the extreme opposite was the answer – in other words, to paint her . . . in such a way as to break up her form and thus confuse a submarine officer as to the course on which she was heading'. He arrived at Plymouth consumed with excitement, went straight to the Royal Naval Barracks and asked if he could see the commander. Wilkinson made a rough draft of a camouflaged ship, marked port and starboard with odd shapes in green, mauve and white, went to see the Flag Captain of HM Dockyard, Charles Thorpe, got him excited too, and drafted a letter to the Admiralty Board of Inventions and Research, dated 27 April 1917:

The proposal is to paint a ship with large patches of strong colour in a carefully thought-out pattern and colour scheme, which will so distort the form of the vessel that the chances of successful aim by attacking Submarines will be greatly decreased . . . The idea is not to render the ship in any degree invisible, as this is virtually impossible, but to largely distort the external shape by means of violent colour contrasts.

The director of naval equipment, Captain Clement Greatorex, picked up the idea, and gave it the name 'dazzle painting'. At the end of May, a small store ship, HMS *Industry*, was then test-painted according to Wilkinson's designs, and coastal stations and other ships were ordered to report what they saw of her.

Wilkinson was informed that there was no room to do this at the Admiralty in London, so he would have to find other premises to develop his proposals. Walking along Piccadilly, he bumped into an old friend, the sculptor Derwent Wood, RA, outside the Royal Academy at Burlington House, where Solomon J. Solomon used to drill with the United Arts Rifles in their white jerseys. Wood suggested using the Royal Academy schools, and by the middle of June 1917 Wilkinson had managed to get the use of four studios for his 'Dazzle Section'.

Wilkinson also outflanked the cautious Admiralty by selling his idea to the vigorous Glaswegian shipowner Sir Joseph Maclay, newly appointed by Lloyd George as the controller general of merchant shipping. J. P. Maclay saw the benefit of camouflaging Merchant Navy ships to protect them from submarines. Going behind the Admiralty's back caused 'a ding-dong row', but because Wilkinson was not a regular naval officer he got away with it, and 'Dazzle' was transferred from Royal Navy to merchant shipping.

As Norman Wilkinson had realised on the Plymouth train, a ship with smoke unravelling from its funnels, moving against a changing sky and sea or sharply outlined on a horizon, was hard to hide. But, by using the 'razzle-dazzle' geometry of bold stripes, curves and zigzags in black, white, blue and green to break up the structural outline of the hull, Wilkinson hoped he could disrupt the low-down periscope view from a U-boat. The distorting of perspective might make its commander doubt the target vessel's course, speed and distance in the same way as hunting lions miss the outline of an individual zebra, their vision confused by the flickering herd. The point of dazzle painting was deception. A camouflage officer once explained to a merchant skipper who objected to the vivid painting of his vessel:

Dear Sir, – The object of camouflage is not, as you suggest, to turn your ship into an imitation of a West African parrot, a rainbow in a naval pantomime, or a gay woman. The object of camouflage is rather to give the impression that your head is where your stern is.

Wilkinson assembled his team of naval *camoufleurs* in the Royal Academy Schools at Burlington House. The three modellers put together a series of one-foot-long, flat-bottomed models of merchant ships; then one of the five RNVR lieutenants designed a dazzle scheme which was painted on in washes by one of eleven young women with art-school training. (One of these 'lady clerks', Eva Mackenzie, later married Wilkinson.)

The model ships could be revolved on a turntable in front of different sky backgrounds and viewed through a periscope set about ten feet away in order to judge the most distorting effects of slopes, curves and stripes. The designs all had to be different so U-boat captains could not get used to them, but their aim was always to make onlookers uncertain of the whereabouts of the bow, stern and bridge of the ship. Lines and stripes had to be carried round and over the ship, including funnels and lifeboats, so that it was deceptive from all angles. When Wilkinson was satisfied, the colour layout was copied on to a 1 foot: $\frac{1}{16}$-inch white paper chart showing both port and starboard side of the design, and then dispatched to the port where the real ship was lying. There, the ten dock officers, who were usually artists in RNVR uniform, supervised the painting of the stripes on to the vessel, using black, white, blue and green as the principal colours, either in primary form or mixed to various tones. When the British Mercantile Marine began jazzing up the fleet in October 1917, the young Vorticist artist Edward Alexander Wadsworth was in charge of repainting in the dockyards of Bristol and Liverpool. Heavy engineering met the avant-garde; the results are still astonishing.

Judging by the positive reports from sea-going skippers from August 1917 on, dazzle painting worked:

September 25th, 9.55 a.m. sighted HMS *Ebro*, in the Sound of Mull on the port bow, end on.

She appeared to alter course to port immediately after and seemed to continue to do so, whereas, in reality, she was altering her course to starboard.

I should think confusion would be caused in aiming gun or torpedo.

I was so sure that she was trying to cross my bows that I was on the point of stopping my engines and going full speed astern to avoid a collision, when I discovered that she was altering course to starboard. After passing the vessel it was almost impossible to say how she was steering.

In October 1917 the Admiralty ordered the repainting of all merchant vessels and armed merchant ships and a number of cruisers, destroyers and minelayers on convoy duty. At the end of that same month, King George V came to visit the Dazzle Section at Burlington House, intrigued to learn how something could be camouflaged by being made more visible, not less.

Norman Wilkinson, newly promoted to commander, did his party trick in the room where they tested the designs. He invited the bearded sailor king (a noted yachtsman with various exotic tattoos from his years in the Royal Navy) to act the part of a submarine captain. The king had to look through the shielded periscope at the latest painted model ship on the turntable, and then estimate what course it was steering by placing an unpainted model at the correct position on a compass card to his right. The painted ship was heading ESE; the king reckoned it was going S by W. Incredulous, he walked round to study the little ship, and congratulated Wilkinson. 'I have been a professional sailor for many years and I would not have believed I could have been so deceived in my estimate.'

By that time Solomon J. Solomon's new camouflage school in central London's Hyde Park, between the Bayswater Road and the eighteenth-century powder magazine, just north of the Serpentine, had been running for almost a year. It was just a few minutes' walk across Kensington Gardens from Solomon's London home and studio at Hyde Park Gate. It was doubly convenient because, as he was not being paid for military work, Solomon needed to keep painting to earn money. An increasing number of grieving families in the officer class were commissioning posthumous portraits of husbands, sons and brothers killed in the war.

The British Army School of Camouflage was run by regular Royal Engineer officers, led by Major, later Lieutenant Colonel, John P. Rhodes, but Solomon was retained as honorary technical adviser. Created to experiment with new ideas, to instruct artist-officers in techniques of concealment, and to run courses that familiarised officers and NCOs with the basic principles of camouflage, the school also placed itself well, politically. Like the experimental training-area that Solomon had wanted near Haig's GHQ in France, it became a handy and safe showplace for top brass, politicos and the press to view aspects of trench warfare, and helped to market the idea of deception and the term 'camouflage'.

The institution gained the seal of royal approval when the King and Queen came to visit on 8 March 1917. King George V's note in his daily diary is one of the first recorded usages of the French loanword 'camouflage' in English: 'May & I went to Hyde Park close to powder magazine where we saw a demonstration of the use of camouflage in warfare (which is concealment) most interesting . . .'

Solomon also went to Scotland to advise on camouflage in the Firths of Forth and Tay, and to Hull after it had been bombed by zeppelins. He went up in balloons and aeroplanes to see how potential targets looked from the air and how they might be made to look like something else. As aerial bombing increased, with night and day raids by heavy aircraft like the Gotha bomber, so did the need for large-scale strategic camouflage, hiding key landmarks that enemy pilots would navigate by.

Meanwhile, France, which had a head start on Solomon and the British *camoufleurs*, was leading the way again in *défense contre aeronefs* or airships, fitting painted covers to disguise lakes and canals and the confluences of the rivers Seine, Marne and Oise. Paris installed arrays of smoke generators (*engins fumigènes*) to pump out a fog of obscuring cloud. This was part of anti-aircraft defences that included rings of anti-aircraft guns and hundreds of barrage balloons attached to two-kilometre-high steel cables which would damage any planes that flew into them. Spotters north-east of the capital telephoned a twelve-minute warning of enemy bombers, and wailing air-raid sirens sent hundreds of thousands of Parisians down into thousands of air-raid shelters and dozens of metro stations.

By 1918, the French were trying large-scale visual deception, *camouflage par faux-objectifs*. Giant models of the Gare de l'Est railway station, together with fake boulevards and avenues made of wood and canvas, were set up in fields north-west of the city, with strings of lights that stayed on when Paris blacked out its street lights. But the British Royal Engineers remained sceptical of these kinds of *objectifs simulés* as antidotes to air raids. When enthusiastic amateurs wrote suggesting 'the erection of a replica of London at some little distance in the country, meanwhile covering the real London with imitation fields', the ideas were (as a witty letter to *The Times* by Colonel J. P. Rhodes pointed out) 'received with reverence', but 'reluctantly discarded as unsuited to this imperfect world'. However, these ideas would be picked up later in WW2.

--

Meanwhile, the *camoufleur* Oliver Bernard was having a different kind of war in France. Bernard stayed in the field because he was determined to show the bastards who called him 'a cock sparrow' that a little man could prove himself a proper soldier, and also that he was not a stuck-up 'artist-officer' like Solomon. Bernard liked the clarity of army field service manuals, and learned from them, so that when he was asked to take rifle inspection on parade one morning, he knew just what to do. He understood that good discipline must be consistent and authority certain. Bernard had bollocked bad workmen when he was in the theatrical world, and now his soldiers had to accept that dirty kit, lost 'pull-throughs' for cleaning rifles and unshaven chins would not pass muster. His need to fit in was far greater than Solomon's. Oliver Bernard was an orphan who now found a place to belong; his autobiography, *Cock Sparrow*, is dedicated to the Corps of Royal Engineers.

Bernard's baptism of fire with Second Army in the Ypres Salient came in early May 1916 after he was appointed the erecting officer of the second, third and fourth camouflage trees at Burnt Farm, Belle Alliance and Hill Top Farm respectively. He was determined that all his OP trees would be better designed and placed than Solomon's had been, and that he would not lose face by showing fear. Bernard described his Wimereux-manufactured 'Oh Pips' as:

hollow imitations of pollard willow trees, consisting of bullet-proof steel cylinders composed of elliptical sections, assembled and cased in outer jackets or blindage of thin sheet iron; the blindage being framed, contoured and hammered, finally dressed to reproduce the external appearance of existing trees which were so replaced to accommodate observers.

The sun was setting as the *camoufleurs* cut through the springtime woods towards the canal barges at Essex Farm. There were stray shells bursting, splintering trees and blowing reeking holes in the ground. As the sky darkened, the violence became almost pretty: shrapnel shells burst orange in purple patches of smoke. Oliver Bernard noticed that his companion and reconnaissance adviser, the *camoufleur* André Mare, was sweating profusely. His own new Brodie 'tin hat' was heavy and uncomfortable, so he complimented the Frenchman on the design of his lighter shrapnel helmet. Mare shook his head gloomily, '*Non, non, pas bon pour les petits morceaux, votre chapeau est le meilleur.*' ('No, no, no good for little bits [of shrapnel], your hat is better'.)

It took two nights' quiet work to erect that first tree. After their moment of triumph Bernard and Mare were challenged by a British sentry not in the know, and taken along trenches to a battalion HQ in a rough-hewn dugout and questioned by candlelight. Only a telephone call to heavy artillery HQ at Vlamertinghe confirmed that they were not spies. On the third night their party of forty-odd sappers and gunners carrying nearly a ton of equipment for their second tree was shelled heavily, and the guide lost his way. Bernard got very angry. After swearing blue blazes and threatening to shoot anyone who left the kit or the trench, he stormed off with a stolid lance corporal called Kearvel who claimed to know how they could get to the line of pollard willows at Belle Alliance.

The two men clambered over the parapet and stumbled eastward into a cratered moonscape fitfully illuminated by star shells, taking turns to fall into holes. Bernard was hard of hearing but even he could not miss the machine guns chattering like magpies and the deep baying of the big guns. On he went through an old communication trench, with Kearvel behind him. The trench deepened; turning to look back in the shadowy flicker of a fading star shell, Bernard glimpsed not his lance corporal but two figures with coal-scuttle German helmets. In momentary darkness he scrambled out of the trench and lay flat behind the enemy parapet. He carried a Browning .45 Colt automatic in his officer's leather holster. Across his wrist, Bernard says, he shot the first man through his gas mask, and then his puzzled companion, who toppled sideways. The third man, an officer, took two bullets before a fresh shell blast sent Bernard jumping and stumbling back towards his own lines. He ran into Kearvel, who had located the tree site, and then they found the work party, with André Mare sitting sheltering beneath sandbags half sheared through by machine-gun bullets. Bernard says he never enjoyed a cigarette more. They moved the gear to the site, put up a protective breastwork and dug a sap for the following night, when the tree would actually be erected. Bending over a toolbox, the gallant Corporal Kearvel was shot clean through the buttock to much ribaldry from his mates.

There were worse incidents. On his first job with the First Canadian Division's heavy artillery in June 1916, Bernard stepped into the entrails of a sentry who had been blown apart while he was crawling around in no-man's-land near Maple Copse trying to find a shattered fir tree that would accommodate an OP periscope. The smell of the sentry's blood was 'surprising'. The same month found Oliver Bernard at German House in Bois de Ploegsteert, sawing down a stout oak that Solomon had spotted months before as suitable for an 'Oh Pip'. They dug a sap twenty feet out from the frontline trench; when they were ready to substitute their fake tree, which had to be three feet higher than the original oak in order to give them a better view, they also had to raise an entire parapet of sandbags in one night to look commensurate with it.

Oliver Bernard worked with the 1st Canadian Division until October, often in company with Major Norton, DSO, a tall Survey Officer, Royal Artillery (SORA). Bernard thought of them as 'the big and little wizards' of the Ypres Salient, crawling on reconnaissance missions, helping to erect snipers' hides, periscopes, and fake tree observation posts in sites from Boesinghe to Arras.

On 4 July 1916, the fourth day of the Battle of the Somme further south, Bernard made his first examination of the remains of a shattered windmill, Verbranden Molen, near Krustaat, which had a vertical oak beam sticking up like a fingerpost from its rubble of ruins. Bernard had discovered that the Ross company of London (who made the best spyglasses for the Lovat Scouts) also manufactured excellent periscopes ten feet six inches long. He reckoned that if they could dig a concealed observer's cabin under the mill, sink a further eight-foot well down which they could lower the periscope for lens cleaning, then hide a couple of periscopes in the upright beam, they could establish a panoramic view of the German batteries hidden behind Wijtschate (or 'Whitesheet') hill.

Delicate jobs were not easy. The special works sappers might have to work for sixteen nights in a conspicuous, elevated position only a few hundred yards from the German lines. They had to try not to be seen or heard as they gouged and hacked two channels in the tough oak, deep enough to accommodate ten-foot lengths of periscope in bulletproof casing, which was then sealed up flush with hammered sheet iron treated to look like weathered wood. Every time a star shell or Verey light went up in the darkness they froze. There was no talking. And yet the rest of the British Army remained amazingly noisy. Lorries came roaring up to the trench tramway to the skeleton mill and then dumped engineering supplies with the kind of din that drew down enemy gunfire. Oliver Bernard sometimes reflected bitterly that there were three kinds of military clients for his camouflage: the very few who believed in making things difficult for the enemy; a greater number who believed in making things difficult for everybody but the enemy; and a great lump of rigid intractables who thought that any form of concealment was somehow a breach of King's Regulations. These idiots were facing German enemies who did not hesitate to copy all camouflage ideas from the French and British.

But Lieutenant Bernard came from the theatre, where the show must always go on. Early one summer morning in 1916 the job at the windmill was done. On the campbed in his hut, a few hours later, Bernard was woken from the sleep of the just by the general and his brigade major. 'The mill is finished,' they said. He agreed it had been completed at 2.30 a.m. 'No, the mill is *finished*,' they said. 'The Boche finished it.' After the German artillery blitz, all Bernard could do was collect the object glasses and eyepieces of the bent periscopes, and curse the fortunes of war.

The little wizard got around the Salient unscathed, until a place called Vormazalee. There, on 4 August 1916, the day after his brother Bruce was killed with the New Zealanders, a machine-gun bullet hit Oliver Bernard just below the left kneecap. After a spell in hospital at Wimereux, he was shipped back to England. 'I'm not from the Somme,' he said to the lady who pushed a basket of fruit into his ambulance at Charing Cross Station. Major Rhodes got Bernard into the efficient Clock House hospital on Chelsea Embankment.

One day, Bernard hirpled out on crutches to see Solomon J. Solomon at Hyde Park Gate. At the front door he hesitated, wondering whether the artist would appreciate his old 'business man' dropping in, but Solomon was genial and friendly, inviting Bernard into his studio and settling him down to paint his portrait. Beneath the affable conversation, Bernard could sense that Solomon was hurt and disappointed by what had happened to him in the army. The cynical and worldly Bernard had always wondered what motivated Solomon. Was his energy fuelled by ambition or greed? Now, as Bernard talked to him in his studio, Solomon 'revealed a man with the heart of an irrepressible child, gifted, generous, spoiled, unaccustomed to hard knocks and opposition which are the common enemy of all pioneers'. In Bernard's opinion, although Solomon was not fitted to run a military unit and his political tactics were unwise, he had been the first person to grasp the potential of the new French idea and to press it energetically on the authorities. Solomon would have received more recognition had he been prepared to advise rather than dictate.

---

In France, deceptive camouflage work continued as the two sides shelled each other over no-man's-land. Early in 1917, Oliver Bernard limped back to his post, sporting a wound stripe next to his Military Cross. He dug Oh Pip observation posts into chalk near Vimy Ridge, and behind brick walls by Mount Kemmel. A few days before Easter in April 1917, aged just 36, Bernard became the camouflage officer of IX Corps in Sir Herbert Plumer's Second Army. This force was preparing and training for a massive assault on the Messines-Wijtschate Ridge overlooking the Ypres Salient from the south. Sappers and gunners paved the way for the infantry: British, Australian and Canadian tunnellers were secretly digging through the clay to place twenty-one giant mines under the German positions, listening through microphones to the ordinary sounds of enemy life that would soon be cut short. At a place the British called the Bluff, the tunnellers had bravely dug underneath waterfilled craters. Now the *camoufleurs* assisted by poking up disguised periscopes only seventy yards from the German lines. Meanwhile, over 2,200 howitzers and big guns were assembled, and coordinated with 400 heavy mortars and 700 heavy machine guns. The Second Army also had air superiority: eight tethered observation balloons were backed up by II Brigade Royal Flying Corps whose 300 aircraft were already attacking airfields, railways and German reserve camps as well as patrolling and photographing the enemy lines. An enormous scale model of Messines Ridge and its defences, the size of two croquet lawns, was constructed in detail from RFC reconnaissance evidence. Officers studied the model from scaffolding built up around it.

When Bernard was summoned to tell his corps commander, Lieutenant General Sir Alexander Hamilton Gordon, about camouflage, he took some aerial photographs with him. Gordon's lugubrious disposition had earned him the ironic nickname 'Sunny Jim' but Bernard really did feel 'as if a ray of sunshine had unexpectedly penetrated the unhappiest depths of his weary but persevering soul' as he at last was able to explain what he thought was going wrong to somebody who could do something about it. Camouflage had to be a forethought, not an afterthought, said Bernard; built-in from the beginning. If the enemy spotted your first diggings for an ammunition dump or an artillery emplacement, all later attempts to hide it would only advertise its importance the more. The wrong sort of camouflage was worse than none, Bernard added, showing Hamilton the aerial photographs of gun-pits being prepared in the open and then covered with extraordinary wigwams, square tents of hessian and light green canvas with sloping sides that were as conspicuous as block houses. 'What damn fools we all are!' exclaimed the corps commander. (Those sites were left in place as excellent decoys when they quietly relocated the guns.)

The subsequent assault on Messines or Mesen Ridge was perhaps the greatest British success so far in the stalemated, deadlocked war. At

3.10 am on 7 June 1917, nineteen of the twenty-one mines buried by the tunnellers under the German lines went off in a rolling sequence that lasted an appalling twenty-eight seconds. Great pillars of flame reached up to the sky and then collapsed in dirt and debris and smoke. Philip Gibbs said it was 'as though the fires of hell had risen'. A million pounds of explosive produced a shockwave that could be heard and felt on the other side of the English Channel by the sleepless lying awake in London. Huge craters, 250 feet across, punctured the blasted landscape. Perhaps 8,000 German soldiers perished immediately in their shattered bunkers and trenches. Walking behind a massive creeping barrage of artillery, mortars and machine guns, 80,000 British and Anzac infantry moved forward, took the entire ridge and moved down the other side. More than 7,000 dazed Germans surrendered. Philip Gibbs noticed the camouflage sacking on the helmets of the Germans as well as their complete ignorance of how much Germany was hated. After four days' fighting, half of the 25,000 British killed and wounded were Anzacs.

After June 1917, when the Anzacs took over the Canadian sector of the Ypres Salient, camouflage stepped up from retail to wholesale. The First Australian Division, who had fought at Gallipoli, asked Special Works if all visible roads from Poperinghe to Ypres could be screened from German observation balloons. They did not want to lose more men, equipment and transport through visual carelessness. From this date onwards, there is photographic evidence of banners of hessian, ranked in arches across roads, forming overlapping layers against a distant observer. Production of road screening materials shot up to 25,000 square yards in June and July, from nil in May. By Christmas 1917, 112,000 square yards were flapping in the wind.

Only very slowly was the BEF beginning to understand that the key idea of camouflage was deception, not just concealment. But new ideas were slowly getting through. Bernard was moved up to the coast near Dunkirk to help camouflage naval guns in sand for a major coastal attack in July 1917 which in the event was thwarted by a massive German pre-emptive strike, using mustard gas. Part of Bernard's duties involved disguising an RFC airfield. He managed to persuade an RFC officer not to build any new huts, but instead to occupy existing farm buildings, neither altering the grounds nor making new tracks. 'Damn good idea, and better than any camouflage,' said the squadron commander. 'Not at all, that is camouflage,' replied Bernard. When Plumer was sent to northern Italy in November 1917 to prop up the flagging Italian Allies against the Austrians, Bernard went along as camouflage officer.

Of course, the enemy was using camouflage, too. Bernard wrote of his 'magnificently trained and perfectly equipped opponents who designed the most scientific means for protecting and concealing themselves in and behind their own lines throughout the western front'. In April 1917, after the Germans withdrew from Adinfer Wood to the Hindenburg Line, Captain J. C. Dunn recorded how they made use of the whole wood:

On its front, hidden in the beech hedge, are machine-gun emplacements of concrete and armour-plate, like large letter-boxes. Within it are gun-emplacements and shelters built of large boles, planted over with ferns and grasses for concealment; smaller shelters are woven cleverly of branches, some growing and some partly or wholly cut. Its trees are erect and unbroken. Moss and ivy, violets, bluebells, anemones and wild strawberry carpet it. The relics of its occupation are unobtrusive . . .

By 1917, the Special Works Parks in France were not only using their French female workers to produce screening and netting but also artist-ically creating a wide range of realistic hollow dummies as hides: trees, walls, dead horses, human corpses. In July 1917, when King George V and the Prince of Wales came up from Cassel to visit the Special Works Park on the Wormhout–Wylder road, the *Daily Mail* reported: 'The King saw all the latest Protean tricks for concealing or, as we all say now, for 'camouflaging' guns, snipers, observers.' *The Times* special corres-pondent also followed that ten-day tour of the Western Front:

On Friday, July 6, the King drove first to the home of the high priests of the great mysteries of camouflage, a magician's palace in a Belgian farm, where nothing is what it seems to be. It is a bewildering place, which, of course, cannot be described in detail – a land on the other side of the looking-glass, where bushes are men and things dissolve when you look at them and the earth collapses, where visions are about and you walk among snares and pitfalls . . . It is the grown-up home of make-believe. Here the King was received by the chief magicians, who showed him their black arts and made him privy to all their secrets.

# IO

# Lying for Lloyd George

Lloyd George was a Liberal, but his manoeuvrings brought down the last Liberal government that Britain would ever have. After the press helped to get rid of Asquith and to bring him to power in December 1916, Lloyd George rewarded the great press barons by giving them jobs in government and changing the whole publicity machine. Under this Prime Minister, propaganda became mass-market.

A crucial figure in presenting the right stories and managing public perceptions was that genius of British newspapers, Lord Northcliffe, who was born Alfred Harmsworth in 1865 in Chapelizod, Dublin. He was self-educated, working as a freelance journalist to support his mother, brothers and sisters after his barrister father declined into alcoholism. He learned from George Newnes's publication *Tit-Bits* in the 1880s that the newly literate classes wanted information made accessible and entertaining. Emotionally impulsive himself, Harmsworth had a knack for understanding people's crazes and curiosities, and so excelled at popular journalism. 'Smiling pictures make people smile,' he said. 'People like to read about profiteering. Most of them would like to be profiteers if they had the chance.' He started with *Bicycling News*, and his stable of popular magazines, including *Comic Cuts* and *Marvel*, was selling a million copies a week by 1892.

The first daily newspaper Harmsworth acquired was the ailing *London Evening News* in 1894. He turned around its fortunes by changing the format, simplifying the reporting (stressing human interest), making the subbing and headlines snappier, and adding a woman's column. On 4 May 1896, he launched an entirely new paper, the *Daily Mail*, priced at only half a penny. The first issue sold nearly 400,000 copies, almost as many as all the penny papers combined. Skilful use of Britain's railway network pushed its distribution right across the country, and made the *Daily Mail* the first truly national mass-market newspaper, more attractive than the stodgy broadsheets. Although Lord Salisbury snobbishly complained that the new rag was 'run by office boys for other office boys', it had bright short paragraphs that made things simple and clear.

Patriotism and imperialism sold papers, and Harmsworth harnessed propaganda to profit. The *Daily Mail*'s jingoistic coverage of the Boer War brought daily sales to nearly a million, the highest circulation in the world. Throughout the early years of the twentieth century, Harmsworth beat the drum in the *Daily Mail* for a bigger navy, a larger army, and stronger defences, playing up invasion scares and the menace of foreigners. In 1903, Harmsworth launched the tabloid *Daily Mirror*, aimed at the 'New Woman', with an all-female staff, and in 1908 acquired *The Times*, the newspaper of the British establishment, which he modernised, attracting more advertising. He paid his hacks well and encouraged the infant National Union of Journalists. In March 1914, he dropped the news-stand price of *The Times* from threepence to a penny and tripled its circulation to nearly 150,000.

By WW1, Harmsworth had his peerage, and the new Lord Northcliffe was eager to play the part of tribune of the people, challenging governments and vested interests. Now he threw himself into the Allied cause, splashing German atrocities over his pages, fighting censorship, championing the common soldier, yelling for more recruits and better munitions, applauding conscription. His papers trumpeted Kitchener as saviour in 1914, then blamed him for the shells crisis in May 1915. He persecuted Lord Haldane for being a Germanophile largely because the two men had a pre-war quarrel about the future of air power, and hounded Churchill over the debacle of the Dardanelles. Northcliffe, wrote Churchill, 'wielded power without official responsibility, enjoyed secret knowledge without the general view, and disturbed the fortunes of national leaders without being willing to bear their burdens'. By the end of 1916, Northcliffe had become fed up with the wily and idle Asquith (who hated and distrusted him in return) and helped to elevate Lloyd George. The increasingly megalomaniac press lord was triumphant when Asquith's government fell. A full page of the *Daily Mail* on Saturday, 9 December 1916, was headlined 'The Passing of the Failures'.

Lloyd George would have liked Churchill to join his administration but the kingmaker Lord Northcliffe made it clear via the *Daily Mail* and *The Times* that Churchill still carried the black spot. Moreover, an inquiry into the Dardanelles and Churchill's role in the adventure was still *sub judice*. Churchill felt betrayed when he was not at once brought back into government, but six months later, when Northcliffe was away in North America, Lloyd George asked Churchill to come back, first as chairman of the Air Board, and then as the vital Minister of Munitions.

On 12 June 1917, Lord Northcliffe was having breakfast at the Hotel Gotham, New York City, when an energetic young Canadian called Campbell Stuart came to see him. Northcliffe had just become chairman of the British War Mission to the United States, and faced an enormous task. The UK had already spent £3.71 billion on the war so far, and needed food from the USA as well as immediate loans of $200 million a month. Campbell Stuart was appointed military secretary to the British War Mission, and soon became the attaché, secretary and fixer for the press baron in his crusade to persuade America to commit itself to winning the war against Germany. Stuart was with him in Kansas City on 25 October 1917 when Northcliffe met the cream of the newspapermen of the Middle West, 'in which every shade of opinion was represented', and later told how:

Northcliffe talked to these men with extraordinary frankness about their isolationist tendencies, their provincialism, their ignorance, and so on, as I doubt any other Englishman at that time could have done, and his words had an enormous effect.

Sir Campbell Stuart *Opportunity Knocks Once* (1952)

Lloyd George had offered Lord Northcliffe the directorship of a proposed new Department of Information to coordinate propaganda, but he turned it down. This is how John Buchan, after a busy war writing his *History of the War* and other books and also working for General Haig's chief of intelligence, Brigadier John Charteris, got the top job in February 1917.

Buchan's new Department of Information brought together foreign propaganda and war publicity, but they were still scattered in different places. The department was still a provisional organisation, a 'mushroom ministry' always in danger of being wolfed by bigger, historic centres of government power. 'The only real war was in Whitehall,' wrote the novelist Arnold Bennett, then employed in propaganda work. 'The war in Flanders and France was merely a game, a sort of bloody football.' Charles Masterman continued to run the Production section from Wellington House, which was responsible for books and pamphlets, as well as photographs and paintings. Its Pictorial Propaganda Committee selected the first 'Official' war artists, including Augustus John, Muirhead Bone, Wyndham Lewis, C. R. W. Nevinson and William Orpen.

The Press and Cinema section was based in the Lord Chancellor's Office in the House of Lords. Buchan wanted a well-informed press using true stories, and he encouraged the film-makers towards authenticity. This section also dealt with cables and wireless, and Buchan brought in the chief executive of Reuters news agency, Roderick Jones, as an unpaid part-time adviser. (After Baron Herbert de Reuter committed suicide in April 1915, leaving Reuters news agency financially weakened, Jones had done a deal with the British government. Using a £550,000 loan arranged by Herbert Asquith's brother-in-law, Jones bought out Reuters, became its chief shareholder and ensured that its wartime news-gathering was presented 'through British eyes' and that its worldwide distribution network was available to the British government.)

The Intelligence section of the Department of Information in Victoria Street replaced the old Neutral Press Committee, and was meant to get good news out of various branches of government as swiftly as possible. The Administrative section where Buchan himself sat was based in the Foreign Office. Buchan had to keep in touch with the King at Buckingham Palace and to report to the Prime Minister in Downing Street, though Lloyd George preferred to hobnob with other press cronies.

The war hit John Buchan personally very hard. Herbert Asquith's son Raymond was the first of his Oxford friends to be killed, then Bron Lucas, an airman with the RFC, was shot down in November 1916. The worst blow fell on Easter Monday, 9 April 1917, when Buchan lost his best friend, Tommy Nelson, his partner in the publishing house, and his youngest brother, Alastair. Both were killed in France, half a mile from each other, in the Battle of Arras.

--

Soon after becoming Prime Minister in December 1916, Lloyd George had told a suspicious Labour and Socialist deputation: 'I hate war; I abominate it. I sometimes think "Am I dreaming? Is it a nightmare? It cannot be fact." But . . . once you are in it you have to go grimly through it, otherwise the causes which hang upon a successful issue will all perish . . .' By the end of the war, the polemical Irish Socialist, dramatist and pamphleteer, George Bernard Shaw, shared Lloyd George's view that war is hell, but you still have to win it. Shaw was 60 when he went off to France in 1917 to visit the trenches and write up his conclusions in a daily newspaper. Major General George Macdonogh, the director of Military Intelligence who ran the propaganda unit M17b, had asked Philip Gibbs to recommend a writer to visit the Front 'who might produce something good about the life and heroism of our men'. Gibbs replied, half in jest, 'What about Bernard Shaw?' The response was laughter. 'Good heavens, what an idea!'

Winston Churchill called Bernard Shaw a 'double-headed chameleon', and described him as a 'bright, nimble, fierce, and comprehending being'. In the cold, snow-bright January of 1917 Shaw came out to the trenches, 'with his beard blowing in the wind of France and Flanders'. He had tackled the idea of it in drama – *Major Barbara*'s arms-dealing millionaire Andrew Undershaft calls himself 'a profiteer in mutilation and murder' – but now the playwright was within the force field of real war. Undershaft also said, 'The more destructive war becomes the more fascinating we find it,' and Shaw repeatedly uses the word 'fascination' to describe war's hypnotic appeal. Shaw wrote that great war correspondents 'like Philip Gibbs, finely sensitive to the miseries of the troops' were 'fascinated by the spirit which drives men to endure and defy so much outrageous mischief and danger'.

Philip Gibbs was with Shaw at places like Ypres, Vimy and Arras, and enjoyed his wit. A general at luncheon once asked Shaw when he thought the war would be over. Shaw replied, 'Well, General, we're all anxious for an early and dishonourable peace.' The general did not laugh, but his junior staff officers did. Philip Gibbs also records Shaw recommending 'parallel lines of thought' about the war. 'One of them is that it is a complete degradation of all that we mean by civilisation. The other is, my dear Gibbs, that we've got to beat the Boche.'

Philip Gibbs, John Buchan and the other early war correspondents had been guided in their visits to the front by Hesketh Prichard, but by 1917 visitors were being escorted by the white-haired Captain Charles

Montague, formerly a leader writer and theatre critic on the *Manchester Guardian*. Montague had dyed his silver hair yellow in order to get into Kitchener's 'New Armies' as a private soldier, at the age of 47, and rose to sergeant through merit before accepting a · commission in Intelligence. Montague kept cheerful at war: 'I have found a hobby in bomb-throwing,' he wrote home, 'which unites the joy of bowling googlies and playing with fireworks.' Ethically, war was incompatible with Montague's Christianity, and he had been passionately opposed to the Boer War, but he saw WWI as a necessary vileness to be got over as quickly and efficiently as possible, without 'slacking and shirking and boozing'. C. E. Montague described how the New Armies lost their illusions in *Disenchantment* (1922), one of the great books of WWI.

Impervious to fear and elated by shelling, Montague escorted his journalistic charges as close to danger as possible. In January 1917 he took George Bernard Shaw around the devastated, shell-pocked landscape for a week. Shaw was in khaki camouflage uniform like everyone else. But Flanders was white with snow in January 1917, and a Romanian general in a dove-coloured cape was invisible while Shaw's khaki was so glaringly conspicuous that the playwright felt he might as well have worn a medieval herald's bright tabard. The three pieces that Shaw wrote about his experiences in France appeared in the *Daily Chronicle* in March 1917, collectively entitled 'Joy-Riding at the Front'. Gibbs says that the title and tone deeply offended people at home, who thought it heartless and mocking. But ninety years on, it seems as fresh as paint. 'Joy-Riding at the Front' is full of ironies: gunners industriously shell trenches that the enemy has already abandoned, and there are inept demonstrations of fire, flame and gas that make friendly French villagers cough and senior staff officers scuttle away. Shaw's second article, 'The Technique of War', analyses the imprecision of artillery and aerial bombardment, but also aims to reassure anxious readers at home that most projectiles miss their loved ones at the front, and he turns the extravagant wastefulness of war – 'It burns the house to roast the pig' – to propaganda effect: 'Therefore, my tax-payer, resign yourself to this: that we may fight bravely, fight hard, fight long, fight cunningly, fight recklessly, fight in a hundred and fifty ways, but we cannot fight cheaply.'

In war, Shaw says, keeping a cool head is better than seeing red:

'Hatred is one of the things you can do better at home. And you generally stay at home to do it.' This idea may have come out of his talks with C. E. Montague, who observed in *Disenchantment*: 'Hell hath no fury like a non-combatant.' Serving soldiers understood that the morality of war was different from the morality of peace, 'just as the morality of an interview with a tiger in the jungle is distinct from the morality of an interview with a missionary.' Shaw was not a pacifist, and he saw that people went to fight out of solidarity, not selfishness: 'It is not that you must defend yourself or perish: many a man would be too proud to fight on those terms. You must defend your neighbour or betray him: that is what gets you . . .'

George Orwell said that for the 'enlightened' of his generation, '1914–18 was written off as a meaningless slaughter', and some writers like Wilfred Owen tend to pity the soldiers of WW1 as passive victims, 'those who die as cattle'. Shaw, however, saw it as his patriotic duty to report more encouraging news in 'Joy-Riding at The Front'. For all his clear sight, he was not above putting a favourable gloss on things: 'Men torn from civil life of the most prosperous and comfortable kind, and engaged in the most perilous service . . . say without affectation that they have never been so happy . . .'

--

'The Duty of Lying', an interesting chapter in C. E. Montague's *Disenchantment*, begins:

To fool the other side has always been fair in a game. Every fencer or boxer may feint . . . In cricket a bowler is justly valued the more for masking his action.

In war your licence to lead the other fellow astray is yet more ample . . . For war, though it may be good sport to some men, is not a mere sport . . . A good spy will lie to the last, and in war a prisoner may lie like a saint and hero . . . Even the Wooden Horse of the Greeks has long ceased to raise moral questions . . . Ruses of war and war lies are as ancient as war itself, and as respectable.

Montague saw the press as a perfect weapon for deception. Enemy intelligence read everything in the newspapers:

. . . worrying out what it means and which of the things that it seems to let out are the traps and which are the real . . . priceless slips made in unwariness.

Here is a game, Montague suggests, to exercise the rat-like cunning of the intelligence officer: sniffing out real crumbs of information from poisoned bait.

Montague, in peacetime a *Guardian* journalist, would have known many of the secrets of the Western Front. He suggests that the use of 'camouflage stories' in the press was never fully exploited by either side, but what little he reveals of the practice is intriguing. A popular science journal he does not name, late in the war, gave 'a recklessly full description' of the 'listening sets' used by the British to eavesdrop on German telephone calls in the field. This article was actually 'a camouflage', planted by GHQ as 'the last thrust in a long duel'.

The Germans had been listening to British field telephone conversations from the very beginning of trench warfare. In early February 1915, the day after the Life Guards had replaced a French regiment near Ypres, secretly, at night, with all their identifying badges removed, Captain Stewart Menzies, future head of the Secret Intelligence Service (SIS), was astonished to get a message in a bottle lobbed over the wire from an Alsace regiment in the German trenches opposite welcoming 'the English cavalry' to their section of the line.

Observation Posts were linked by telephone lines to gun batteries further back, but these forward lines were often leaky. The British did not realise until July 1916 how good the Germans were at intercepting traffic on British field telephones. *Fernsprech Truppen* ('Telephone Troops') tapped Allied calls either by directly attaching a cable to the line, or by earth induction from any lines that were not wholly metallic, picking up the electrical signals as they went through the ground (this could be done from up to 3.3 km away), then amplifying and monitoring them on Moritz listening sets. Thus careless British talk cost more lives than sinister 'foreign spies'. Idle trench chatter helped German intelligence to build up the complete British order of battle, and often let them know when and where attacks were coming so they could prepare their machine guns and artillery. Montague points out that the British did not grasp the extent of what was going on until the Battle of the Somme:

When the war opened the Germans had a good apparatus for telephonic eavesdropping. We had, as usual, nothing to speak of. The most distinctly traceable result was the annihilation of our first attack at Ovillers, near

Albert, early in July 1916. At the instant fixed for the attack our front at the spot was smothered under a bombardment which left us with no men to make it. A few days after, when we took Ovillers, we found the piece of paper on which the man with the German 'listening set' had put down, word for word, our orders for the first assault. Then we got to work. We drew our own telephones back, and we perfected our own 'listening sets' till the enemy drew back his, further and further, giving up more and more ease and rapidity of communication in order to be safe.

--

In truth, pettifogging British staff bureaucracy meant that nearly two years were wasted shuffling files between different (and jealous) departments before the signals problem could finally be brought under control. It was only late in the war that the British started using other media to mislead the Germans. Hence an apparently indiscreet article in a rather out-of-the-way wireless journal in which, according to Montague, 'the reach of our electric ears was, to say the least of it, not understated. Few people in England might notice the article. The enemy could be trusted to do so.'

Montague also writes about the deception plan that accompanied the expanded British Fifth Army's attack on Pilckem Ridge, north-east of Ypres, on 31 July 1917, which turned into the notorious Battle of Passchendaele by the time it petered out exhausted in November. Under the overall command of Field Marshal Douglas Haig, the initial push tried to employ the sort of deceptions successfully used by Plumer's Second Army, taking Messines in June, and by the Canadians, capturing Vimy Ridge in April. Canadian soldiers had surprised the Germans by a brilliant *coup de théâtre*, emerging in the middle of no-man's-land from tunnels dug through the chalk. They had also disassembled, transported and reassembled a church steeple so the Germans wasted shells bombarding the wrong place. As had happened before the Messines assault, replicas of the ground and models of the defences were now used for training near Ypres while the big guns were got into place. The British did an elaborate feint, a 'Chinese attack', much further south at Lens, to make the Germans think the push was coming there rather than further north:

Due circumstantial evidence was provided. There were audible signs that a great concentration of British guns were cautiously registering, west of Lens. A little scuffle on that part of the front elicited from our side an amazing bombardment – apparently loosed in a moment of panic. I fancy a British Staff Officer's body – to judge by the brassard and tabs – may have floated down the Scarpe into the German lines. Interpreted with German thoroughness, the maps and papers upon it might easily betray the fact that Lens was the objective.*

Then an apparently indiscreet general in 'one small edition of one London paper' blabbed that the British push was aimed at Lens and a supposedly outraged MP asked a question in the House of Commons about greater control of the press. Montague believed that deception worked: 'The Germans kept their guns in force at Lens, and their counter barrage east of Ypres was so much the lighter, and our losses so much the less.' But Robin Prior and Trevor Wilson in their authoritative study of the battle, *Passchendaele: the Untold Story*, disagree. Their assessment is brutal: 'Haig's deception plan, in sum, seemed to have the capacity only to deceive Sir Douglas himself.'

Other kinds of tricks were going on behind British lines. The enemy dead yielded much useful information: their diaries, documents, maps, letters, pay books, photographs, postcards, identity discs and shoulder straps, even the markings on their weapons and kit, could all help intelligence fill out the 'order of battle', the identifying of enemy units and formations. But you could learn even more from living prisoners of war. Hundreds of thousands of Germans were captured in the Great War, and many gave away much more vital knowledge than their name, rank and number. This was not always achieved through formal interrogation. The sympathetic approach often worked well: a chair, a cigarette and a friendly chat with someone who took no notes seemed harmless enough. The boastful could be drawn out with appreciation, the quiet ones coaxed to unburden themselves. Away from the interrogation rooms, the holding cages and cells were also wired for sound with concealed microphones and fluent German-speakers eavesdropping via listening-sets. And there were stool pigeons:

* This anecdotal aside, lightly shielded by 'I fancy', 'may have', 'might' evidently hints at an actual event. The Scarpe is the canalised river that ran west from British-held Arras into German-held territory. It sounds like a trial run for one of the most famous of WW2 deceptions, operation mincemeat, 'The Man Who Never Was', a quarter of a century later, in spring 1943.

A 'pigeon' was a renegade German or an Englishman speaking perfect German, dressed up in German uniform and introduced into an assembly of prisoners in order to 'direct their conversation into the proper channel'. The 'pigeon' would proceed to talk of forthcoming operations, or of losses, or of food and discipline or of anything else upon which he had been primed beforehand by the British Intelligence staff.

Once, in 1915, a wounded German officer was captured at Ypres and taken to hospital in Poperinghe . . . he refused to open his lips. Subterfuge was thereupon resorted to. A British officer, posing as a wounded German officer, was carried into the cot adjacent to that occupied by the genuine German. The camouflaged officer had his head shaved in the approved Teuton style and his arm and leg all bandaged up and in splints. And so the two were left next to one another through the night. The real German moaned; the camouflaged German followed suit. The real German asked, '*Sind Sie Deutscher?*' The camouflaged German replied: '*Yawohl. Bin auch offizier.*' The camouflaged German didn't encourage conversation; he was morose and taciturn . . .

These stories come from *The Secret Corps: A Tale of 'Intelligence' On All Fronts*, written by Captain Ferdinand Tuohy, former head of the GHQ wireless service. In chapter 6, 'The Brain War', Tuohy berated the infuriating slowness of the intelligence system in grasping the importance of deception:

To begin with, one concentrated almost entirely on finding out what the enemy was doing. In the next phase, one took measures to prevent the enemy finding out what you were doing. Finally, one saw to it that the enemy was thoroughly well deceived and hoodwinked into making false deductions. This final development of 'Intelligence' will rule supreme in any future war . . .

Tuohy thinks this initial slowness was because the British General Staff gave up all ideas of tactical surprise after the failed attack of Neuve Chapelle in spring 1915. He says plans for deception operations were still being rejected in 1916. But by the middle of 1918, deception was standard practice at GHQ in France.

Memoranda from Lieutenant General Herbert Lawrence, Chief of the General Staff, sent out in August and September 1918 show that deception could only begin to succeed against the enemy when the British got a complete grip on their wireless security (by, for example, changing call signs daily) and began sending the fake chatter and 'dummy traffic' that they wanted the Germans to hear; when they started going through all the motions of physical camouflage in unnecessary places; and when they supplied false information to their own troops as well as the enemy, so anyone captured would, quite sincerely, reveal consistent chaff.

The most advanced example of this kind of thinking comes from the final three weeks of WW1, in a memorandum on 'Security' by Colonel Richard Meinertzhagen for Brigadier General, General Staff, GHQ, Intelligence, dated 23 October 1918. The usual counter-espionage sense of 'security' is preventing the enemy finding out about you. This entails keeping documents safe, stopping leakages from wireless and telephone, and instructing all ranks how to behave when captured (i.e., after giving the required name, rank and number, repeating 'I cannot say' to questions.) It also entails warning your own men about some of the devious methods employed by the enemy to extract information.

Meinertzhagen, however, had his own novel definition of security. He thought the aim of security went well beyond 'preventing the enemy divining our real plan', and was actually 'feeding him with sufficient material to induce him to believe he is in possession of our real plan'. In his parlance, 'security' means 'deception', as in 'Security feeds the enemy with material served up in as acceptable form as possible.' And when he writes that 'badly prepared camouflage will have the same effect on a trained Intelligence Officer [at enemy HQ] as badly served food – it will be refused or if accepted will not be digested', the word 'camouflage' now means 'disinformation'.

He saw the kind of physical camouflage done by Royal Engineers at Special Works Parks as 'negative camouflage, i.e., camouflage designed to conceal information, inanimate objects or troops from the enemy'. In the Meinertzhagen view 'positive camouflage' is whatever is 'designed to carry false information to the enemy', as well as 'to our own troops and indeed to the world in general'. Meinertzhagen thought it essential that 'positive camouflage' – deception – 'be controlled by the one brain' at GHQ and then filtered down through Armies, Corps, Divisions, Brigades etc. He summarised the various forms of 'positive camouflage' under the following headings:

(a) Construction of aerodromes, hospitals, hutments, railway sidings, gun positions, dumps, etc. As far as resources permit, camouflage of this description should be real, but when dummy work has to be introduced, it must be designed to deceive our own troops as much as the enemy.

The R.A.F. should be constantly asked to report on all such work.

(b) The spreading of false news by various means.

(c) Wireless, Power Buzzer and telephone camouflage. Both dummy messages and degrees of activity have in the past proved useful camouflage weapons.

(d) Troops and transport movement, attitude and dispositions of troops of all arms, railway activity and great aerial or A/A [anti-aircraft] gun display: Artillery registration.

(e) Preparation of maps and documents designed to deceive the enemy and deliberately allowed to fall into the enemy's hands or enemy agent's hands.

Not for another five decades would the full extent of Meinertzhagen's commitment to deception, both in private and in public, emerge.

Like Bernard Shaw, C. E. Montague understood that the morality of war was not the morality of peace, 'so you may stainlessly carry deception to lengths which in peace would get you blackballed at a club and cut by your friends'. He felt, however, that the British, in their amateurish gentlemanly way, were still half-hearted about using the press to deceive. Montague fantasised what whole-hearted use of this weapon would be like:

If we really went the whole serpent the first day of any new war would see a wide, opaque veil of false news drawn over the whole face of our country. Authority playing on all the keys, white and black, of the Press as upon one piano, would give the listening enemy the queerest of Ariel's tunes to follow. All that we did, all that we thought, would be bafflingly falsified . . . The whole sky would be darkened with flights of strategic and tactical lies so dense that the enemy would fight in a veritable 'fog of war' darker than London's own November brews . . .

Montague thought that during WW1 'the art of Propaganda was little more than born'. He wondered what would be the long-term effects of another war in which propaganda had really come of age, 'and the State . . . used the Press, as camouflaging material, for all it was worth'. He reckoned 'a large staff department of Press Camouflage' would be needed: 'The most disreputable of successful journalists and "publicity experts" would naturally man the upper grades.' Argument and reason would be replaced by emotion and 'the practice of colouring news, of ordering reporters to take care that they see only such facts as tell in one way'. The moral downside of it was that the untruthful journalist, the 'expert in fiction', having gained high distinction by his 'fertility in falsehoods for consumption by an enemy', would continue to thrive after the war was won, in 'that new lie-infested and infected world of peace'.

On 10 February 1918, Buchan's Department of Information became the Ministry of Information. The energetic Canadian Max Aitken, Lord Beaverbrook, proprietor of the *Daily Express*, became Minister of Information, and had a seat in Lloyd George's cabinet. Max was clear about his new role:

The Ministry of Information is the Ministry of publicity abroad. Its business is to study popular opinion abroad and influence it through all possible channels, of which the chief is the overseas press. Its object is to state the British case to the world.

Of course, this was done differently from official diplomacy. For the press baron, 'The Ministry of Information represents the democratic and the popular side of the Foreign Policy', so he used journalists and writers in the work. Rudyard Kipling was often asked for personal advice by Beaverbrook, although he was never officially employed in a propaganda capacity as he was felt to be too fierce and vengeful a 'Hun-hater' for public consumption. Buchan rose to Director of Intelligence. Of supplying information to neutrals, he minuted:

The department must work to a large extent secretly, and as far as possible through unofficial channels. Camouflage of the right kind is a vital necessity. It can advertise its wares, but it dare not advertise the vendor.

Beaverbrook's biggest coup was landing Lord Northcliffe as the Director of Propaganda in Enemy Countries, reporting directly to the Prime Minister and the War Cabinet. Northcliffe's organisation came to be known as 'Crewe House' after the Marquess of Crewe placed his splendid residence of that name in Curzon Street, London, at their disposal.

The man who organised, recruited and ran Crewe House was Sir Campbell Stuart. In February 1918 he was charged with putting together a team to produce and distribute propaganda to the Central

Powers. His managing committee included the editor of the *Daily Chronicle*, the foreign editor of *The Times*, the managing director of Reuters news agency and the celebrated novelist H. G. Wells. What was even more impressive in Whitehall terms were the links Crewe House made with other government departments to ensure smooth delivery. These included a healthy line of credit with the Treasury; full co-operation from HM Stationery Office who printed millions of leaflets in myriad foreign languages; ample use of the Ministry of Information wireless service; and full-time dedicated go-betweens to the War Office and the Air Ministry, the Political Intelligence Department of the Foreign Office, the director of Naval Intelligence, and the director of Military Intelligence.

Collaboration with the secret world was necessary for distributing the millions of cartoons, leaflets and pamphlets that Crewe House produced. At first RFC planes were used, but they did not have a satisfactory means of scattering the sheets, and after two leaflet-disseminating planes were shot down and their pilots given long terms in prison for spreading seditious messages, the War Office changed tack. Military Intelligence (MI7) had regularly been using large hydrogen balloons to get agents and crates of carrier pigeons into enemy territory at night, and now they used thousands of smaller balloons to deliver paper eastwards on the wind. Some 2,000 hydrogen balloons of specially 'doped' paper, about twenty feet in circumference, were produced every week. Each could carry up to 1,000 leaflets, which ones depending on that night's wind direction:

The leaflets were sewn onto a slow-burning fabric fuse, which was ignited before launching. As the fuse burnt, the leaflets fell off one by one, thus serving as ballast. For the first hour or so the fuse carried leaflets designed for German troops, then some hours of leaflets for friendly civilians and, finally, leaflets for German civilians. On every suitable night millions of these leaflets were despatched by teams strung along our front line.

*The Inner Circle: The Memoirs of Ivone Kirkpatrick*

Late in the war, many British propaganda leaflets went by internal post all over Austria, Bavaria and Germany, thus avoiding the strict censorship of foreign mail. This happened in two ways. First, they were smuggled in bulk via the book trade, which was not closely supervised, especially if the volumes had the covers of German classics.

Second, they were carried over the border from neutral Holland by *Gastarbeiter* and sent via the normal post inside enemy territory to neutrals, potential sympathisers, the intelligentsia and the newspapers, using counterfeit postage stamps engraved and printed by Waterlow of Watford for one of the British secret services.

Production of propaganda was split into three enemy areas: Austro-Hungary, Bulgaria, and Germany. Northcliffe was persuaded by H. Wickham Steed, the foreign editor of *The Times*, that the dual monarchy of Austria and Hungary was the weakest link in the chain of the Central Powers, and therefore the first place to start hammering. This was because the sprawling Habsburg Empire contained many peoples and nationalities who were potentially pro-Ally. 'There are thus in Austria-Hungary, as a whole, some 31,000,000 anti-Germans, and some 21,000,000 pro-Germans', wrote Northcliffe. 'The pro-German minority rules the anti-German majority.'

On 24 February 1918, Northcliffe asked Lord Balfour, the Foreign Secretary, for a clarification of inter-Allied political attitudes towards the Habsburg Emperor's dynasty and the ethnic minorities he ruled. Clearly, policy had to precede propaganda: he needed a clear line to follow. Balfour agreed four days later that 'a propaganda which aids the struggle of the nationalities, now subject to Austrian Germans or to Magyar Hungarians, towards freedom and self-determination, must be right'. It was the continuation of the divide-and-rule strategy formerly directed against the Ottoman Empire.

So, in early April 1918, H. Wickham Steed and another member of Campbell Stuart's team, the academic Dr R. W. Seton-Watson, were in Italy, attending the three-day Congress of Oppressed Habsburg Nationalities in Rome where Italians, Poles, Czechs, Slovaks, Serbs, Croats, Slovenes and Rumanians found common cause in 'the right of peoples to decide their own fate'. Six months before, the Italian front at Caporetto had buckled under a surprise attack by the Austrians. A hundred thousand Italians had been taken prisoner and 700 guns lost in the retreat. Now the tide was turning. Steed and Seton-Watson set up a polyglot printing press at Reggio Emilia. It published a weekly news journal and patriotic and religious leaflets in six languages which were fired across the trenches by mortar, rocket and rifle-grenade, dropped by aeroplane, and even thrown by contact patrols of ardent deserters who volunteered for the task. Troops of doubtful loyalty were assailed across no-man's-land by loudspeaker propaganda and gramophone records playing Czecho-Slovak and southern Slav songs. Deserters began coming across, carrying the leaflets, singly or in groups. Some Czech troops mutinied. The Austro-Hungarian military authorities were further alarmed when the Italian fightback started from June.

From May to July 1918 the director of propaganda literature against the Germans was H. G. Wells, the imaginative and largely self-educated author of many successful books, including *The History of Mr Polly* and the WWI novel *Mr Britling Sees it Through* in which a grieving father seems to find God. When he joined Crewe House, Wells agreed that policy had to be clarified before any positive propaganda could begin, and he urged 'a clear and full statement of the war aims of the Allies'. Typically, he wanted an ideal vision of the future, something people could believe in after the war. 'The thought of the world crystallises now about a phrase "The League of Free Nations".' He wanted to hold out a beacon of hope, the dream of perpetual international peace. Wells believed in this ideal but thought the British government was cynical about it. 'We were in fact decoys. Just as T. E. Lawrence of the "Seven Pillars" was used all unawares as a decoy for the Arabs.'

Though they came from his department, Wells did not personally write the texts of the British-produced German-language propaganda leaflets that showered down on Germany in such quantities (over ten million in the last three months of the war). These aimed to inspire fear rather than hope, and they were demoralising because they were true: reminders of shortages and social problems, maps and diagrams of lost ground and military defeats against inexorable Allied success, graphic depictions of growing American strength, name lists of dead and captured German U-boat commanders, pictures of happy smiling Germans who had given up and were not being tortured, the surrender of Bulgaria in September.

Losers are more susceptible to propaganda than winners, because more fearful and anxious. Nor was it only leaflets working on their minds. Many German newspapers translated interesting pieces from the neutral press – Dutch, Scandinavian and Swiss – and such papers were skilfully bombarded with 'camouflaged articles' from Crewe House that, without banging a drum, showed the social, economic, commercial and scientific conditions in Allied countries in a glowing light. German readers could make their own comparisons with drabness and depression at home. So the gnawing discontents and sapping of the will continued.

A German trench newspaper appeared, *Heer und Heimat*, with a picture of the Kaiser between two oak-leaf clusters, and a subtitle *Der Ruf zur Einigkeit*, 'The Cry for Unity', which featured a front page cartoon showing the German political parties at home fighting each other rather than the enemy. This paper looked and seemed thoroughly German, but it too was produced by Crewe House. General von Hutier was perhaps right to warn his troops against British 'ruses, trickery and other underhand methods.'

The German awareness of British methods – triumphantly boasted about afterwards by people like Sir Campbell Stuart – had far-reaching consequences. Both Ludendorff in his memoirs and Adolf Hitler in *Mein Kampf* believed that the British propaganda campaign had corroded the German will to resist; that the German armed forces were never really defeated at the front, but only stabbed in the back; and that devious foreigners were responsible, not decent Germans. Self-pity and self-deception would be stirred into the toxic resentments of the nascent German National Socialist Workers' Party.

# II

# Deceivers Deceived

As the war dragged on towards the end, Solomon J. Solomon brooded in London. In February 1918 he bought a radiographer, a kind of magic lantern that projected a flat picture on to a screen, for use with his camouflage students. Its combination of mirrors and powerful electric lights enabled Solomon to study enlarged and illuminated aerial reconnaissance photographs in considerable detail. He became obsessed with three of them in particular, taken in autumn 1917, because his painter's eye had spotted curious anomalies. Some of the shadows seemed to be wrong in relation to the sun. As he pored over the pictures Solomon began to suspect that this was a landscape deliberately designed to fool the camera. New tricks were being used to see through camouflage as it became more sophisticated; colour-blind spotters, for example, sent up in aeroplanes, had been successful in picking out artificial greens from natural ones. But Solomon became convinced that German camouflage was still pulling the wool over Allied eyes.

Solomon eventually laid out his ideas in *Strategic Camouflage* (1920), a book whose conspiratorial tone is set from the opening epigraph:

When a man would commit a crime in a room overlooked from another the first thing he does is to pull down the blind; and if he is using a light, he closes the shutters too.

War is a crime, and this war was, and henceforth every other war will be overlooked, and the first thing the participants need to do is to devise and prepare their blinds.

The Germans did not neglect this precaution.

Germany was a technically advanced country. Of the first hundred Nobel Prizes for Science, Germany won thirty-three to Britain's eighteen, and Germans set the technological pace of twentieth-century warfare. Solomon insisted that the Germans had surpassed the Allies in their camouflage of men and equipment.

'Camouflage and the interpretation of aerial photography were war babies, and are still in their infancy,' he wrote in *Strategic Camouflage*, convinced that no official reader of photographs was as well equipped as a painter to do the job (early photographic interpreters included a diplomat and a stockbroker). Like Abbott Thayer, Solomon overrated the artist's expertise. He was unimpressed by the carefully illustrated official manual, *Notes on the Interpretation of Aeroplane Photographs*, and remained convinced that the British had often been 'fooled by the devices of the enemy'.

He argued that the Germans had managed to construct, in back areas five to nine miles behind their lines, low hangars the size of whole fields to conceal thousands of men by day. These barely sloping structures blended with the hedges and roads of Flanders and hoodwinked both aircraft spotters and photographic interpreters that all they were seeing was a patchwork of cultivated fields. Solomon thought the Germans understood exactly what would show up in the Allied aerial photographs. 'Keep everything low' was one of the camouflage instructions found on a German prisoner. Bridges, for example, stuck out as clear targets for bombs or artillery. But a dark-painted bridge, dropped three or four inches below the surface of the river it spanned, was still navigable and far less conspicuous. Solomon thought German *camoufleurs* had managed to cover roads over completely with wire and canvas, so that a hasty spotter would see only a route empty of traffic. This was the equivalent of a conjuror's false bottom, Solomon reckoned. Enemy transport could continue to flow underneath painted buckram and mosquito-netting.

Solomon claimed to spot errors in German 'skiagraphy' or painting of shadows. When a house cast no shadow on the ground, he was convinced that the Germans had erected an imitation field alongside it, coming up to its eaves. He pointed out houses that were too small, tree shadows that did not move with the sun, pipes pretending to be paths. *Strategic Camouflage* becomes dizzying as Solomon asks us to scrutinise muddy blow-ups of black-and-white photographs as well as his own impressionistic colour paintings of the wobbly shapes, odd shadows and dim contrasts in certain photographs.

Solomon also made coloured drawings, sketches and models to prove his point. In March 1918, he began pestering his superiors in the

Royal Engineers, men at the Air Ministry, the General Staff, the Prime Minister's office, and fellows at the Athenaeum Club like Sir Martin Conway, who said, 'This is the most important find since the beginning of the war,' and went off to tell General Hugh Trenchard, chief of the RFC and future father of the RAF. In a hurried survey under poor light, 'Boom' Trenchard peered at Solomon's evidence and saw only fields. But the matter should be looked into, he said, and fresh photographs procured. 'Are you prepared to go to France?' Trenchard asked Solomon.

But it seemed to be nobody's business to send him. Solomon became depressed. Back home, he stared again at the photos taken of St Pierre Capelle: the worked fields now looked astonishingly like undulating hangars, the haystacks were fake, the tree shadows all wrong. The whole area seemed to him to show a vast hidden camp for reserves, barely eight kilometres east of Nieuport, on the only rising ground in a marshy district north of Ypres.

When Trenchard sent his aide, Major John Moore-Brabazon – Britain's first-ever certified pilot, the inventor of the aerial camera and Churchill's future minister of aircraft production in WW2 – to see Solomon's evidence in his studio, the painter failed to persuade this sceptical aviator. Solomon managed to get to see the new CIGS, Sir Henry Wilson, who did not say much, but according to Solomon, 'seemed to think there was something in it'.

--

Hesketh Prichard, the king of the snipers, was called by people on his own side 'The Professional Assassin'. It was said in an admiring way, but he paid a high psychological price for the title. Although it was his job to kill Germans, the hours his grey eyes spent studying the enemy through a Ross glass were also extended exercises in empathy. Another man, seen with his unshaven face and scruffy cap magnified twenty times, cannot long remain the bestial baby-eating 'Hun' of propaganda. Prichard observed the enemy's all-too-human habits and bodily needs, trying to survive in the squalor of the trenches on the other side of the barbed wire, and saw *mon semblable, mon frère*. His sensitivity was what made him such a good hunter and sportsman. It is true that after he heard on 3 October 1915 that his great friend Alfred Gathorne-Hardy had been killed with many of his men at Loos, ten yards from the German wire, he wrote: 'If it is any satisfaction, I shot a German between the eyes at 5 o'clock today'; and that after Nurse Cavell was executed later that month he said: 'It makes me so glad when I shoot a German, and especially an officer.' But for the most part, Prichard was not motivated by revenge. He knew the Germans were both brave, and human. His work was plainly necessary, but it began to trouble him more and more that it was murderous.

The writer H. M. Tomlinson in his 1930 memoir *All Our Yesterdays* remembered how Prichard went sick for some weeks after putting a bullet through the head of a particular German sniper who had been a deadly nuisance. This same incident is written up in Prichard's own *Sniping in France* as a short story, 'Wilibald the Hun', rather than as reportage. Perhaps psychically troubling material needed to be disguised as fiction. Hesketh Prichard began getting splitting head-aches from the eye strain of spotting and shooting all day. Then he developed what was called 'trench fever' from the drains leaking excrement. But he battled on with his work, inventing new devices, developing the courses and tactics of camouflage and deception, saving thousands of lives by helping to take a few others. But the King of the Snipers was slowly sickening and wasting away from his mysterious illness: he endured fourteen operations before dying in June 1922, aged 44.

H. M. Tomlinson wrote an obituary appreciation of Hesketh Prichard for the Liberal paper *Nation and Athenaeum*. He saw the man who first guided him at the front as 'a gentleman': a privileged person from a leisured caste whose code was honour and service. The war degraded what was noble in Hesketh's philosophy, and its delicate notions were ruined by 'the senseless waste of our own men'. Tomlinson said he was shocked by this 'drainage of good life', but also by

the chicanery, the meanness, the stupidity, the intrigues, and the callousness of those of whom he wished to think well . . .

What subtle infection of his body occurred through this terrible disturbance to his settled habits of thought I do not know. But one could see that he was mortally wounded. The Press called it 'blood-poisoning'. I suppose that term will do as well as any other.

---

The first day of spring, 21 March 1918, saw the opening phase of a massive German offensive designed to break out of the trenches and to drive the British back to the French coast. After the Russian Revolution of October 1917, the Bolsheviks had reneged on Russia's alliance with Britain and France and sought a separate peace with Germany at Brest-Litovsk. With the Russians out of the picture and the Americans not yet arrived in Europe, this seemed to the Germans the right time to strike a decisive blow, so they quietly moved a million more men and 3,000 guns to the western front.

The Allies knew a big attack was coming soon but they had no exact idea of its scale and location. In *The Secret Corps*, Ferdinand Tuohy says that the German General Erich von Ludendorff had ordered every department and branch not to convey any information whatsoever to the enemy, and that Ludendorff appointed special security officers to police the concealment:

The Germans hid and even distorted their signal traffic to put us off the scent; faked road and rail activity elsewhere than in the projected area of operations; even built dummy dumps and hospitals and battery positions and aeroplane hangars at certain parts of the front so that our observers should photograph them.

Major movements of men and stores were done at night or cloaked in camouflage. The Germans had created a phantom army in front of the French sector, using signals deception and dummy wireless traffic. Field Marshal Haig's intelligence at GHQ was poor because Brigadier John Charteris told him only what he wanted to hear.

*Die Kaiserschlacht*, the Kaiser's Battle, began with the greatest artillery barrage of the war: 3.2 million shells in one day. The damp weather helped camouflage the Germans: their *feldgrau* uniforms were ghostly as they advanced over the misty fields. The bloodiest day of the war yielded 80,000 casualties on both sides.

The second phase of the German attack, code-named 'Georgette', was launched near Armentières on 9 April 1918. Four German divisions under General von Arnim struck at a weak point where the dispirited 2nd Portuguese division held the line. German soldiers flooded through the gap and Armentières, Ploegsteert and Messines were all abandoned by the Allies on 10 April. What had taken four horrendous months to gain at Passchendaele was lost in a few days. The British front was crumbling on Thursday, 11 April 1918 when Haig issued an order to hold 'to the last man'. The Germans seized a wilderness of mud, what Churchill called 'the Dead Sea fruits' of the Battle of the Somme, before they stopped, exhausted.

There were major repercussions after the offensive. First, the Allies agreed to a unified military command under the French General Ferdinand Foch, just as, in WW2, all would agree to work under the American General Eisenhower. Second, gloomy news of the big German attack with British retreats and heavy casualties caused inevitable political fallout at home. Who would carry the can for the losses and setbacks? Lloyd George shook up his government. Lord Milner replaced Lord Derby as Secretary of State for War on 18 April and his place in the War Cabinet was taken by Austen Chamberlain. The reverberations reached as far away as Palestine, where Allenby was stripped of troops he was going to use in his final push against the Turks.

Amid the recrimination and blame, Solomon J. Solomon suddenly did not seem quite such a mad obsessive in London clubland and society. The Germans had made a surprise attack *en masse*; perhaps they *had* been camouflaged in the way that Solomon suggested. The painter now cranked up his campaign. He drafted a letter to the Prime Minister and took it round for his neighbour, Sir Edward Carson, to check. Carson wrote a forceful letter to Lord Milner pointing out the importance of Solomon's discovery that German camouflage was helping to hide troops in the landscape.

A week or so later, Solomon presented his case to Lloyd George and the Secretary of State for War while lunching at Sir Alfred Mond's house. 'There is no doubt about it,' affirmed the Prime Minister. Solomon says, 'He saw at once that the March surprise had been effected in some such way by the Germans.' Lloyd George then passed Solomon on to General J. C. Smuts and Lord Rothermere, President of the Air Council. Solomon also tried to see Major General George Macdonogh, the director of Military Intelligence, to find out if any information on camouflage had been gleaned from enemy prisoners.

In May 1918, the famous American portrait painter John Singer Sargent, RA was back in London, about to be commissioned as a war artist. Solomon saw a chance to warn the American Army. According to Solomon,

'[Sargent] came to my studio, and went most carefully through the photographs and was quite satisfied that my reading of them was correct . . . the camouflage was so clever that only an artist could make an initial logical analysis of the puzzle pictures they presented to the airmen.'

In early July, Solomon heard from Professor Mantoux that General Foch was interested in his discoveries and wanted to see him at French GHQ at Versailles. Solomon underwent a blizzard of red tape, officials and bumf before starting a muddled, hopeless journey to France. Three trains, a hotel for the night and a car finally got Solomon to the British Mission at Château Breau. General Weygand arranged for Solomon to drive on to Chalons to meet the chief French expert reader of photographs, Captain De Bissy.

At Chalons, a German big gun began firing every ten minutes and everyone trooped down into the cellars. The Fête de la République, 14 July, was on a Sunday so the Germans enlivened the party with a big attack. Solomon's weekend ran into his usual problem: a grudging acceptance by the French of some points, but rejection of his idea that the Allies had been fooled for three years. Their attitude boiled down to 'the French are cleverer than the Boche, so how could they do anything better?'

Tired from lack of sleep, and having failed to persuade Captain De Bissy, Solomon began the long rail journeys back to Paris. Dirty and bedraggled, he got a twelve-franc room at the Hotel Terminus. His luggage had been lost; he had to buy a barber-shop shave and a clean collar. While waiting around, he wrote more letters and went to see people who were often out. Lying in bed on Thursday night, Solomon heard an air-raid warning, the sirens 'like a spider's web of sound throughout the city', but nothing happened. The Ministère de la Guerre sent him on to the Section de Camouflage et Inventions at 23, rue de l'Université.

The next Sunday Lieutenant Colonel Solomon J. Solomon, the first British *camoufleur*, had lunch at Chantilly with Captain Guirand de Scévola, the first French *camoufleur*. In the sunny garden afterwards, Solomon explained to de Scévola and his staff the photographs and models that he had brought with him. There is an over-insistent note in Solomon's report:

None of them dissented . . . All saw quite clearly every point, and expressed astonishment that no attempt had been made to test the validity of my findings . . . De Scévola wishes me to join the Conference of Allied Camoufleurs that they hope will soon take place.

Solomon saw more generals and their aides as he tried to get permission to go to the Belgian front to prove his theory. He wanted planes to bomb the area, and another aircraft to photograph the results. In early August he was at Rouen where Walter Russell took him to a Special Works factory at Marowne. Solomon thought that Royal Engineers camouflage was now run by staff who knew nothing. At Abbéville he breakfasted with Lyndsay Symington and asked him what he thought of the camouflage being made in France; Symington said that in his opinion 75 per cent of the stuff made was utterly wasted, through lack of knowledge and too much standardisation of the product.

From Boulogne, Solomon returned to the Special Works Park at Wimereux, where he had begun two years before. He was shown around by Major F. J. C Wyatt, who had supplanted him. Solomon was appalled by the monotony of the production: standard flat-top camouflage nets of one familiar pattern without regard to terrain. These factories were boring – without artistry, inspiration or grasp of the function of camouflage. Wyatt must have found him the most galling of visitors.

At the Officers' Rest Club in Boulogne, Solomon wrote letters to officials and badgered other guests with his models and photos. He spent the morning on the beach, and slept in the afternoon. Tapping his feet to Boulogne bandstand music, he sketched disguised aerodromes. He telephoned the British Mission: there was no news of the summons to Belgium he was imminently expecting.

Like a prophet, Solomon railed at the kings in command. The man had a burning vision that no one could see. How could people who did not even understand the meaning of strategic camouflage recognise it right before their very eyes? Back at Wimereux he received a letter from his wife that persuaded him to go home. Paget drove Solomon to Boulogne and endured photo analysis for an intense few minutes before the departure of the 2.30 ferry for Folkestone.

By 20 August Solomon was sitting in Hampstead Hospital with his photographs of St Pierre Capelle and Sparappelhoek, sharing them with his kite-balloonist nephew Joseph Hubert Solomon who had obtained the originals while serving in Belgium. Solomon's reading of the photos was 'a revelation' to the young observer. His uncle explained that the reason he had never seen any traffic along a two-kilometre stretch of the St Pierre Capelle–Nieuport road was because it was camouflaged. Solomon then wrote another letter to Lord

Milner, the Secretary of State for War, urging him to direct some artillery fire on this road – this despite the fact that Milner had already told Solomon that in his opinion the absence of traffic was because it was travelling only at night, and moreover there was no reason why the Germans should make vast hangars the size of fields – think of the labour and the cost!

Soon after, the commandant of the Camouflage School in Hyde Park, John Rhodes, brought a Captain Lejeune to Solomon's studio to examine his evidence. Solomon thought Lejeune was 'perhaps the most intelligent of all the readers and while at the studio he could find no crab in my reading. It was the first time that any official had thoroughly gone into the matter and he left, as I thought, impressed, but Rhodes afterwards told me he was still not convinced.'

Now Solomon began feeling paranoid. 'I was smiled at as a troublesome lunatic.' He felt that 'English gentlemen' were acting with 'sinister promptings', and that there were people 'behind the scenes making insinuations'. There is no hint from Solomon that any of this is anti-Semitic.

At first, however, the commandant of the Camouflage School certainly tried to give Solomon J. Solomon a fair crack of the whip. On 5 September 1918, John Rhodes wrote a letter to GHQ in Great Britain, enclosing a map reference, an enlarged photograph and an attached transparency marking the key features, and headed 'Suspected use of artificial area Camouflage by the Germans'.

An unnamed Brigadier from MI3 at the War Office replied for GHQ Intelligence on 20 September 1918.

A number of letters from Lt. Col. Solomon J. Solomon, on the subject of possible German camouflage in certain areas, have been referred to the General Staff for consideration during the last few months.

Lt. Col. Solomon's arguments have been most carefully considered and it is possible to state definitely from the examination of a large number of Photographs of the areas in question, taken under different conditions of light and at different times of year, that his conclusions are not borne out by facts . . .

In these circumstances, it is not considered that it would be justifiable to ask for a special bombing raid round ST PIERRE CAPELLE suspected by Lt. Col. Solomon, more particularly in view of the fact that this area has been bombed already.

In September 1918, John Rhodes was in France, visiting Special Works

Parks. This time he discussed Solomon's theories about Sparappelhoek and St Pierre Capelle with GHQ in France. Just before going on eighteen days' leave, he wrote to Solomon:

I saw two recent stereoscopic photographs which I am afraid do not support the idea of raised country.

In the circumstances I do not think it of much use to apply to them again to bomb the area or to make any other tests. I, personally, was quite satisfied by what they showed me that the area referred to was not raised in any way.

Back from leave, however, Rhodes wrote to Wyatt, Officer Commanding the Camouflage Park at Wimereux, regarding 'Suspected use of artificial area Camouflage by the Germans':

Now that the line is sufficiently advanced I shall be very glad if you will inform me definitely, and as early as you can, whether the areas alleged by Lieut. Colonel Solomon J. Solomon to have been covered by the Germans with Camouflage in order to form a concealed staging area, were or were not so covered . . . I am very anxious to get this report as early as I can.

MI3 wrote again to Rhodes on 27 October:

The localities of SPARAPPELHOEK and ST. PIERRE CAPPELLE, suspected by Lt.-Col. S. J. Solomon, R.E. as being camouflaged camps, have now fallen into our hands and have been carefully examined. There is no trace at either of any of the work suspected by Lt.-Colonel Solomon . . .

A copy of this was also sent to Solomon who wrote back on 5 November defending himself to the General Staff. Dropping the names of Marshal Foch, General Weygand, the Secretary of State for War and the Prime Minister, he now shifted tack, pointing to new evidence further south

. . . that the valley between Bullecourt and Creisille is largely covered with exactly the type of camouflage I had described . . . The arcaded roads found at Quéant, since fully described, the Bertha emplacement falling into our hands, where the dummy gun-setting had been bombed and that of the real gun hard by untouched, and the fact that none of the German strategic Camouflage methods had been 'read' in spite of my repeated descriptions of them, since early in March, are sufficient evidence that neither in the British or the French Armies, are there men capable of interpreting a photograph of this order, with anything like scientific accuracy or thoroughness.

When the Armistice came, Winston Churchill was in the Hotel Metropole on Northumberland Avenue, waiting for Big Ben to chime eleven, ending fifty-two months of war. The war ended for Philip Gibbs where it had begun for the British Army in 1914, at Mons in Belgium. It seemed a miraculous coincidence. At 11 o'clock on 11 November 1918 the batteries stopped firing. 'No more men were to be killed, no more to be mangled, no more to be blinded.' As the sun went down into a peaceful night, Philip Gibbs felt the fires of hell had been put out. He heard people talking happily, voices singing, bands playing.

The end of the war saw the appearance of one of its greatest parodies, Charlie Chaplin's film *Shoulder Arms*, released in October 1918. The strangest sequence is the six minutes that Charlie Chaplin spends camouflaged as a tree, running around in California's eucalypt-filled version of no-man's-land, knocking out big Germans with his sticking-out branched arms. Historically, this marked a change in the public consciousness of camouflage. What had been an official secret a few years before was now an open joke. Light-hearted Charlie Chaplin was getting laughs while Solomon Solomon was becoming more and more intense.

On 13 November 1918, two days after the Armistice that ended 'the war to end wars', an experienced British camouflage officer arrived to survey the area of Belgium indicated by Solomon as heavily camouflaged. Taking along a photographer, this Royal Engineer camouflage officer visited a series of Flemish villages: Thourout, Ostend, Sparappelhoek, Middlekerke, Slype, St Pierre Capelle, Leke, Beerst, Vladsloo, Dixmude, Essen, Zarren, Staden and Roulers. This is the moment when Solomon's theories would either be vindicated or vanquished, because only an expert *camoufleur* had the power to pass judgement. In fact, the man who had been selected to make the judgement was Solomon's old protégé and rival, Oliver Bernard.

Bernard found some roadside brushwood screening, 'wooden frames roughly 5' x 5' strung together, with brushwood applied vertically about 7' high', used to mask junctions and crossroads under observation, but these were not German tunnels or 'arched road covers'. Bernard also pointed out:

ST PIERRE CAPELLE was under continuous observation of R.N.S.G, at RAMSCA PELLE, from early 1915 until the recent operations. An attempt to artificially raise road levels to sufficient height for traffic would be conspicuous in this extremely flat locality, especially to careful observers long familiar with every landmark.

Oliver Bernard could find no officer who had entered the area soon after evacuation, nor were any photographs taken: 'there appeared to be nothing worth recording in evidence of or against the question raised in this matter.' Signing his damning report of 14/11/18, 'O. P. Bernard, Major RE, Army Camouflage Officer, 2nd Army', the scenic artist finally squashed the portrait painter, and the practical 'business man' had the last word:

Debris of existing or demolished Camouflage and structural work on roads and country side were carefully sought without a particle of evidence being found to denote either area or road camouflage.

But Solomon was unable to give up now. He made his own way to St Pierre Capelle which was largely destroyed, and to the house with no shadow. Nailed on a green board under the eaves he found a scrap of tarred brown paper which he was sure must have made up the surface of the sloping fake field. He found more paper in the rubble, some grey felt, a hardened bag of plaster of Paris, some squares of canvas over wire, but he was convinced the Germans had tried to destroy the most incriminating evidence.

It was enough for Solomon to keep his campaign going. John Rhodes became more and exasperated by what he came to call 'the wisdom of Solomon'. Everyone had taken a great deal of trouble to consider his claims, but there were limits. 'I consider his letter', Rhodes had written to GHQ on 10 November 1918, 'besides being quite unjustified by the facts, to be a gross and unwarrantable reflection on the integrity, intelligence and capacity of the officers to whom he refers. I recommend that he ceases to act as honorary adviser to the Camouflage School.'

When Solomon's book *Strategic Camouflage* was published in 1920, John Rhodes's anonymous and devastating critique appeared in the right-wing *Morning Post* under the headline 'Camouflage Gone Mad'.

As the Armistice of 11 November 1918 approached, war propaganda turned into political jockeying. Northcliffe, increasingly grandiose, wrote his own 'Peace Propaganda Policy' of unconditional surrender and demanded a seat at the Paris Peace Conference, while setting on his newspaper, the *Daily Mail*, to attack Lord Milner at the War Office.

On 7 November, the Ulsterman and QC Sir Edward Carson had stood up in the House of Commons. The man who once destroyed Oscar Wilde in cross-examination now turned his forensic attention on the newspaper tycoon:

> It is almost high treason to say a word against Lord Northcliffe. I know his power and that he does not hesitate to exercise it . . . It seems to me nothing but indecent that the gentleman engaged in foreign propaganda on behalf of His Majesty's Government should make part of his propaganda an attack on the Secretary of State for War in the Government under which he purports to serve . . . to drive him out of his office. For what? In order that Lord Northcliffe may [himself] get into the War Cabinet, so that he may be present at the Peace Conference . . . The whole thing is a disgrace to public life in England and a disgrace to journalism.

Lord Northcliffe, rebuffed by Lloyd George, resigned on 12 November, the first full day of peace. His megalomania grew worse and tipped into paranoia. Poisoned by an infection in his teeth which affected his brain and then his heart, the greatest genius the British press has ever known went mad and his last weeks in the summer of 1922 were spent raving in a hutch on the roof of the Duke of Devonshire's house in Carlton Gardens.

As soon as WWI ended, propaganda became a dirty word. Crewe House was shut down and cleared by Sir Campbell Stuart by 31 December 1918, and the government hurried to wash its hands of its own publicity machine. Lord Beaverbrook had resigned in October, with no one replacing him, and his Ministry of Information was the very first wartime Ministry to be completely closed down, also by the end of 1918.

The 'liquidator' was John Buchan. He bequeathed the Art and Photography sections of the Ministry to the Imperial War Museum which would be established by Act of Parliament in 1920. As the nation rejoiced in victory, Buchan turned his attention to peacetime, and began lobbying for the release of 1,500 conscientious objectors who were still in prison.

Unlike so many others, John Buchan had not been damaged or

deranged by deception. At Christmas 1918, perhaps he remembered his hero Richard Hannay in *Greenmantle*, a disguised fugitive in Germany, spending Christmas 1915 in an enemy household:

That night I realised the crazy folly of war. When I saw the splintered shell of Ypres and heard hideous tales of German doings I used to want to see the whole land of the Boche given up to fire and sword. I thought we could never end the war properly without giving the Huns some of their own medicine. But that woodcutter's cottage cured me of such nightmares. I was for punishing the guilty and letting the innocent go free. It was our business to thank God and keep our hands clean from the ugly blunders to which Germany's madness had driven her. What good would it do Christian folk to burn poor little huts like this and leave children's bodies by the wayside? To be able to laugh and to be merciful are the only things that make man better than the beasts.

# PART II

# 12

# Wizards of WW2

The twin avatars of British strategic deception and black propaganda in WW2, Dudley Clarke and Sefton Delmer, were both men pulled between two worlds. Dudley Clarke was an artistic type, inventive and theatrical, who had to find an outlet for his creative ingenuity within the rigidities of the British army. Sefton Delmer was brought up in Germany. During the First World War he was the sole British pupil in his Berlin school; when his family moved to Britain, he became the only German-accented boy in a wartime English public school. Both men's lives were shaped by WW1.

Lieutenant Colonel Dudley Clarke, RA, the man who was to become the *éminence grise* of WW2 strategic deception, seemed a conventional enough colonel, with his left-parted hair brushed back from the widow's peak, and his courteous manner. He liked to appear in rooms, or disappear from them, silently, and his pale oval face, with quick glances from under drooping eyelids, gave him the disquieting look of a sardonic butler. 'Sphinx-like' was how someone described the ivory mask quality of a man who became, in the words of his biographer David Mure, 'the Compleat Military Jeeves', solving his masters' problems.

He was not eccentric, but he was original, and the distinction mattered in the conformist world of the British Army. He never married, did not like children, and was conventional in his prejudices and his romantic conservatism: he later wrote a WW2 history of an elite regiment, the 11th Hussars, which won more battle honours than any other cavalry or tank regiment. 'Good old Dudley' was socially affable but quietly calculating. He said he always wanted to be 'one of those in the inner circle, watching the wheels go round at the hub of the British Empire at some great moment of history', and he camouflaged his hard work in getting there as luck, hiding his ambition under amusing, self-deprecating stories of accidents and mistakes.

Clarke could charm senior officers brilliantly, but he also got things done. His intelligence was allied to an ingenious imagination and a photographic memory. He did his best work at night, and in public places always sat with his back to a wall. You would not notice him in a crowd and he was never famous, yet Field Marshal Harold Alexander believed that he did as much to win the war as any other single officer. He ended up as Brigadier Dudley Clarke, CB, CBE, the greatest British deceiver of WW2, a special kind of secret servant.

Born Dudley Wrangel Clarke in Johannesburg on 27 April 1899, (with a caul over his head), he was the eldest son of a Yorkshireman who had gone out to South Africa to make his fortune and came back to a permanent job in a gold-mining finance company in Pall Mall. In 1912, he began his three years at Charterhouse public school, described by one old boy, Osbert Lancaster, as 'an extensive concentration camp in Early English Gothic'. It was not far from the military establishment at Aldershot where Dudley fell in love with the gorgeous full dress uniforms, glittering marching bands and jingling cavalry that had also appealed to the young Winston Churchill. Charterhouse was close to the flying base at Farnborough where the schoolboy made friends with the air mechanics of the Royal Engineers Balloon Section and the newly formed Royal Flying Corps.

The outbreak of the Great War in August 1914 found 15-year-old Dudley Clarke already in uniform, at his school's Officer Training Corps (OTC) camp in Staffordshire. Of the six boys in his tent, two would be killed in the war, one would lose a leg, and their sergeant instructor would die at Gallipoli. Dudley's father Ernest Clarke would be knighted for his voluntary war work organising motor ambulances for the Red Cross (having started out with eight, he ended up with 4,000.) Dudley's younger brother Tom, then aged 7, had a letter published in the *Daily Telegraph* in October 1914 describing how he had made little flags with the colours of the Allies and sold them to raise five shillings for 'the poor Belgian refugees'. The proud father showed this around at a Red Cross committee meeting and soon the Red Cross were selling little flags and shaking collecting tins. In his autobiography, *This is where I came in*, Tom apologised for inflicting flag days on the world.

Unlike Dudley, Tom Clarke was a born civilian who dropped his rifle on parade in the OTC and slipped off from school to go racing at

Sandown Park; his favourite uncle was a chairman of the Magic Circle who could conjure half-crowns from ears. After a varied career, Dudley's brother T. E. B. Clarke took up screen-writing for Michael Balcon at Ealing Films and ended up as the Academy Award–winning writer of such great British films as *The Lavender Hill Mob, Passport to Pimlico, The Blue Lamp* and *The Titfield Thunderbolt*.

Young Dudley shared his brother's creative imagination, which fuelled the inner man as he stamped about on parade grounds, curried in stables, box-wallahed in barracks. In February 1916, Dudley joined the Royal Horse Artillery and a month after his seventeenth birthday he was in the Royal Military Academy at Woolwich, being taught to ride by tough sergeants. Commissioned second lieutenant in November 1916, he was too young to go to France with the British Expeditionary Force and applied to join the Royal Flying Corps.

On Monday, 5 November 1917, Dudley Clarke arrived at the School of Military Aeronautics in Reading with a huge valise with no handles that he called 'the green elephant'. He was 18 years old, lively and full of himself, and keeping a diary. This shows that Dudley Clarke was already a precise man, exact about train times. But there is nothing about the war in the diary; instead he likes the glamour of showbiz. He is becoming a 'stage-door Johnny', waiting outside for the Gaiety chorus and blown away by a Theatre Royal revival of a famous melodrama called *The Whip* which simulated the Derby with a real horse race on a revolving travelator. The first thing he noticed in Reading, apart from his thrilling first encounter with girls in uniforms astride motorcycles, 'very dashing numbers', is the Hippodrome. That night he headed straight for the Reading Palace where he spent two shillings on the last seat in the house and watched Sharp's Tromboneers bring the house down with 'Yaka Hula Hickey Dula' and Perry and his Pert Pianiste attempting to twinkle, though her front teeth were missing.

One night, unable to sample the joys of Reading for lack of cash, Dudley (who once edited a home newspaper called *Knutty Knews* and enjoyed his 'brain waves'), amused himself in his room by 'raising an apparatus composed of a bootlace, a lanyard and some straps off my valise, by which I am enabled to turn out the light without getting out of bed'. He showed the same ingenuity when he was posted to Egypt for seven months in 1918, setting up 'The Problem Club', a series of enjoyable challenges for other young airmen in Egypt who, like him, were saving money of an evening by not going out.

The first problem was to produce the most original article obtained in the most original manner. Horne took a hair from each member without their knowing. Clarke produced a piece of the mess chimney after climbing a roof.

Clarke first flew solo in Egypt, in early July 1918. The flying hours of the training squadron at Suez were 4.30 a.m. to 10.30 a.m., and thereafter Clarke acquired his lifelong taste for swimming and sunbathing. He chewed Chiclet gum and read seven-penny novels in his tent, wearing a white cricket shirt and Charterhouse football shorts with no stockings, his chair a Nestlé's Milk crate and his desk made of Haig & Haig whisky boxes. He was still an aesthete, though; his tablecloth of pale blue silk matched his muslin curtains and bedspread, and he enjoyed others admiring his room and the hookah he had haggled for. After Clarke got his wings at No. 5 Fighting School, Heliopolis, he thought it 'great fun' to go out and strafe Egyptian camels in the desert – much later he regretted such arrogance – and at the Armistice he watched the Cairo celebrations turn into a drunken riot of arson and looting.

There were two strands to Dudley Clarke's military career between the wars. One was active. In the absence of major conflict, ambitious soldiers like Clarke or the young Churchill who wanted to shine had to find interesting scrapes to get into. In 1920 Dudley Clarke was stationed in Mesopotamia, learning polo and pigsticking, when the four-month Iraqi uprising occurred. Clarke evacuated Europeans and cash boxes down the Tigris on a steamer, repelling potential boarders with small arms. In September 1922, he found himself on leave in Turkey, caught up in the Chanak crisis which ended Lloyd George's political career. As British occupying troops resisted the Turkish nationalist leader Mustafa Kemal's threat to take back Constantinople and the Dardanelles by force, Clarke's job was to feed scraps of false information into the ear of his landlord, a Kemalist spy. In late 1925, again on leave, Clarke went to Morocco to cover the French and Spanish suppression of Abdel Krim's Riff rebellion for the *Morning Post*. (He found a publisher and joined the Society of Authors intending to write a book but, unlike Churchill, did not complete it.) In 1930 he joined the Transjordan Frontier Force and swaggered about black-booted in *kalpak, kurtah* and cummerbund; he learned to ride a camel, chased Ikhwan marauders, and sat with the founder of the Arab Legion in Jordan, John Bagot Glubb, over coffee in the desert. However, Glubb Pasha, in his dark, four-button suit, stiff collar, tie and fedora, seemed to Clarke disappointingly like a character from H. G. Wells, with 'none of the flamboyant fancy-dress favoured by Lawrence'.

The second strand of his early career was recreational. While stationed at the Royal Arsenal, Woolwich, in 1923, Dudley revived the pre-war Royal Artillery Officers' Dramatic Club. When General White asked him to take charge of the Royal Artillery display for the 1925 Royal Tournament at Olympia, Dudley came up with a grand pageant demonstrating 200 years of firepower and its transport, from Minden to the Marne. He talked to the circus proprietor Bertram Mills and costumier Willie Clarkson and set about hiring two elephants, two camels, sixteen oxen, eight Sikhs and fourteen of the biggest Nigerians he could find from the hundreds of black men who came to the audition. Dudley had a twelve-foot stock whip with a thirty-foot thong made at Swaine and Adeney, and used it to create a noisy and colourful half-hour show that ran twice daily for thirty performances, employing 37 guns, 300 animals and 680 men.

At the Staff College in Camberley in 1933–4, Dudley Clarke wrote and directed two Christmas pantomimes, *Alice in Blunderland* and *Al Din and a Wonderful Ramp,* which played to appreciative full houses. In 1935, however, Clarke (by then Captain D. W. Clarke) put on another spectacle whose climax was so realistic that much of the audience fled. The Aden Silver Jubilee Display was performed to honour King George V at Holkat Bay in Aden, capital city of Yemen, on 6 May 1935. This was a time of growing tension in the region, because Mussolini was marshalling his Italian fascist forces to attack Haile Selassie's Ethiopia on the other side of the Red Sea. As a finale, the combined services presented 'Invasion!' Dudley had roped in all the armed forces to simulate an attack. It started in darkness after the sounding of the Last Post. HMS *Penzance* and No. 8 Bomber Squadron RAF appeared as a hostile warship and aircraft, suddenly lit up by the searchlights of the Aden anti-aircraft section, shelling and strafing to cover enemy troops landing on the beach. The point of the show was meant to be that the valiant Aden Armoured Car section, the Aden Protectorate Levies and the Aden Armed Police would drive the invaders back into the sea. But when the ship and the planes first appeared out of the darkness, someone yelled in Arabic 'The Italians are here!' and the spectators took to their heels.

In February 1936, Clarke got the job he wanted: brigade major in Palestine. He bought a white two-seater 1929 Delage and shipped it to Port Said. In the Directorate of Military Operations he read a secret aide-memoire about Britain's military weaknesses. The tide was turning, though. On 3 March 1936 Stanley Baldwin's government issued a White Paper on Re-Armament, and the Army Estimates, published two days later, showed the fourth successive annual rise in defence spending, to £49 million from the low-point of £36 million in 1932.

In Gibraltar, Dudley Clarke visited his friend from the Staff College, fluent German-speaker Kenneth Strong (later Eisenhower's intelligence officer). After 'interesting and revealing meetings at GHQ' in Cairo, Clarke drove his car from Egypt through the Mitla Pass to Jerusalem. In the twenty years since Allenby had walked through the Jaffa Gate, tensions in the Holy Land had been increasing. Britain now ruled the country under a Mandate from the League of Nations, but increased Jewish immigration, as advocated by the Balfour Declaration, was making the Arab inhabitants feel threatened, even though their own birth rates were high and they were in the majority. Anti-Semitism and anti-Jewish laws in Nazi Germany augmented Jewish immigration to Palestine.

In April 1936, the month the lid blew, Tony Simonds arrived in Jerusalem from Egypt with the simple order to 'institute Military Intelligence'. In his autobiography, *Pieces of War*, Simonds says that intelligence was then abysmally low on the list of military priorities. Nor were the British in Palestine equipped for the outbreak of strikes, sabotage and Arab guerrilla warfare that ensued. The British armed forces consisted of two infantry brigades at Jerusalem and Haifa, commanded by Colonel Jack Evetts with only two staff officers, one of whom was Dudley Clarke. There was just one squadron of RAF planes and two troops of RAF armoured cars at Ramleh, as well as 500 British, 300 Jewish and 1,000 Arab Palestine police (who could not be trusted). They had to deal with a new kind of uprising in which hundreds got killed and over a thousand were wounded.

As senior operations staff officer, Clarke got the RAF and army staffs working closely together, as they would do much later in Combined Operations. He sent an RAF wireless tender out with army convoys, so if they were attacked they could flash a signal and call in close air support to bomb or machine-gun the assailants. Clarke also saw that the regular army was too blunt an instrument to deal with the guerrillas, who had local support and could only be fought locally, by small units who would need good clear intelligence, which Simonds got from both Jewish sources and loyal Arabs.

In September 1936, Lieutenant General John Dill, former director of Military Operations and Intelligence, was appointed to supreme command in Palestine. An additional division of British troops – 17,000 men – was sent out to help quell what the Colonial Office called a 'campaign of violence' with which 'the Arab leaders are attempting to influence the policy of His Majesty's Government'. Dudley Clarke became Dill's chief of staff at his HQ in the King David Hotel in Jerusalem, and began working late through the night, with his back to the wall and a gun to hand. Guerrilla war sharpened his wits; he fitted a clip for an automatic pistol to the steering column of his Delage and always reversed the car into parking bays to get away quickly.

In September 1937, Clarke got a new master. John Dill was replaced as military commander of British troops by a man who knew Palestine well, Major General Archibald Wavell, the biographer of Allenby and friend of T. E. Lawrence, a craggy taciturn figure who never talked when he had nothing to say. Some found Wavell's silences excruciating, but Clarke coped. Their first dialogue was on a drive from Jerusalem to Haifa:

'When did you join?'
'1916, sir.'
(An hour later) 'I meant when did you join this Headquarters?'

Clarke wrote,

I soon learned to respect these silences, and even to understand them, while I somehow came to realise that the General understood *me*. From this strange relationship I gradually became imbued with an abiding affection for the man himself.

Unconventional soldiering ran in Wavell's family. His grandfather was a mercenary who fought with the Spanish in the Peninsular War and then against them in Chile and Mexico. Having seen the stupidity of many of the tactics in WW1, Wavell liked the unorthodox thinking of people like J. F. C. Fuller, T. E. Lawrence and B. H. Liddell Hart, and wanted infantry who were 'quick-footed' and 'quick-minded'. For fourteen seasons, from 1926–39, with only a brief gap for his time in Palestine, Wavell trained British soldiers for war in annual field manoeuvres far from the barrack-room square, trying to simulate some of the real conditions of battle, including muddle, chaos and surprise. Bernard Fergusson, Wavell's first ADC, tells how 'the Chief' encouraged daring guerrilla tactics in his exercises and used two different versions of the 'haversack ruse'. In the first, he fooled a fellow brigade commander by planting a marked map covered with bogus dispositions. In the second, some years later, the same officer was commanding a rival division in manoeuvres. This time Wavell had a map made that showed his dispositions completely correctly, and gave it to a cavalryman with instructions to blunder into captivity and then pretend to destroy the map. Wavell was hoping that the officer would say to himself: 'Old Archie's . . . forgotten that he played this one on me once before. The one thing that we can be sure of is that not one single disposition shown on this map is genuine.' Everything went according to plan; the officer lapped it up and took a horrible beating.

Soon after Wavell arrived in Palestine, the District Commissioner for Galilee was murdered by 'Arab terrorists' in Nazareth. Wavell cracked down hard on the Arab Higher Committee, and several Arab leaders were deported to the Seychelles. The Grand Mufti of Jerusalem, Haji Amin el Husseini, chief of the extremists as well as the leader of the religious community, hid inside the temple area in the centre of the Old City, finally escaping to join the Nazis in Berlin on the ancient principle that 'my enemy's enemy is my friend'. The anti-imperial struggle always had international dimensions: the Mufti left behind on his desk the Arabic translation of an IRA handbook about fighting the British.

One day, Wavell was en route to visit a military post when his car was flagged down by a British officer from his Intelligence staff. A 'dark, fiery and eager' captain called Orde Wingate, having ambushed the commander on the highway, abruptly laid out his plan for dealing with the armed Arab gangs: armed Jewish gangs or Special Night Squads, trained, organised and led by British officers. Wavell gave Wingate the go-ahead to fight fire with fire. Dudley Clarke, meanwhile,

wrote a long, thoughtful appreciation, 'Military Lessons Learned from the Arab Rebellion', which was circulated in the War Office.

In his introduction to Dudley Clarke's first book, *Seven Assignments*, Wavell wrote:

When I commanded in Palestine in 1937–8, I had on my staff two officers in whom I recognised an original, unorthodox outlook on soldiering . . . One was Orde Wingate, the second was Dudley Clarke.

When WW2 got under way, and Wavell became commander-in-chief, Middle East, he encouraged the development of special forces and secret fraud by picking Orde Wingate for guerrilla war in Ethiopia and Dudley Clarke for strategic deception.

Sefton Delmer was the physical opposite of Dudley Clarke, a 'huge, breezy and bearded' rogue well known to most of the secret circles of wartime London as Mr 'Seldom Defter'. Denis Sefton Delmer, always 'Tom' to his family and friends, was born to Australian parents on 24 May 1904, in Berlin, where his father Frederick Sefton Delmer was a 'Herr Professor' of English at Berlin University. Although his Australian parentage gave him British nationality, the boy grew up speaking German at home and as a young man retained a slight German accent. 'Tom' did not start speaking English until he was five, when his mother Mabel Hook took him on an eighteen-month trip to Australia while Frederick Delmer completed *English Literature: from Beowulf to Bernard Shaw* (1911), which became the standard textbook for students in Germany.

From 1914–16, as the only English boy in a German school, the Friedrichs Werdersche Gymnasium, during 'an orgy of war-hysteria', young Tom Delmer sometimes had to defend himself with his fists. But many Germans were kind when his father was interned for refusing to become a naturalised German. The landlord lowered the rent, former students repaid old loans, the dentist would not charge, and the school did not expel him as an enemy alien. Half a lifetime later, Tom paid back one kindness. Revisiting his old neighbourhood in the wrecked and ravaged Berlin of summer 1945, Delmer found himself back in his old schoolyard, now a hospital for lung patients, crunching over the gravel in his large British Army uniform and calling out the name of 'Herr Harry Deglau'. When a frightened man in a bath chair responded, Delmer was able to hand over a carton of Players' cigarettes and a block of precious chocolate to Harry Deglau, the boy who had saved him from a beating-up thirty years before when young Tom cheered the news of the Australian battleship *Sydney* sinking the German cruiser *Emden*.

Delmer's father was found not guilty of spying and let out of gaol in the spring of 1915. The family endured the starvation winter of 1916–17 and were finally allowed to leave Germany for Holland on 23 May 1917. Tom ate five rich ice creams in the refreshment room at Oldenzaal station and was promptly sick. Soon after his thirteenth birthday he redeemed himself by telling an intelligence officer at the British Consulate in Amsterdam just how many empty trains he had counted travelling east and how many full ones travelling west, confirming that Imperial Germany was switching forces to the Western Front – his first but not his last contact with the British secret services.

The refugee boy with odd clothes adjusted to new life (rowing, not cricket) at St Paul's School in Hammersmith in a London being bombed by Gotha aircraft from his old home in Germany. His father survived a quizzing about his loyalty by Admiral Reginald Hall of Naval Intelligence himself and went on to write articles for Lord Northcliffe's *Daily Mail* and *The Times*. Delmer did not see Germany again until after WW1, when he was nearly 17. He stepped off the Warsaw Express in Cologne to meet his father who was by then working in the Inter-Allied Commission of Control, monitoring Germany's compliance with the Versailles Treaty's demands for disarmament.

Young Tom was charmed by the Weimar Republic in the spring of 1921. Drinking wine in the sunshine beside the Rhine, watching wandering Hansels and Gretels, healthy youths with knapsacks on their backs and bosomy girls in flowered dirndl dresses, singing the songs that he knew from school in the warm, blossom-scented air, he knew that the Germans would never ever want to make war again. 'All was peace, all was beauty.' *Nie wieder Krieg.* No more war. He would describe the decay of this hope in his 1972 book *Weimar Germany*, a survey of the Weimar Republic from 1918 to 1933.

Delmer read German at Lincoln College, Oxford, just missing a first, but he spent many of his 1920s vacations in Germany, where he saw his happy adolescent dream shrivel. The subtitle of his book is

*Democracy on Trial*, and from its opening page, in which the Social Democrat leader Friedrich Ebert makes a secret pact with the officer corps of the *Reichswehr*, the German Army, two days before the Armistice of 1918, Delmer argues that 'Germany's democracy was . . . a democracy with a hole in its heart'. The penultimate illustration of his book is a John Heartfield collage showing three insects on an oak tree: Ebert the caterpillar, Hindenburg the chrysalis and Hitler the butterfly. Delmer thought German democracy was merely a bit of temporary camouflage to trick the Allies. 'It would vanish when they vanished.' He also explored the cynical military pact between Germany and Russia. From 1924, Germans built weapons on Russian soil, and trained military personnel there; in exchange, the Russians got access to technical skills, new patents and manufacturing processes.

Dishonesty and self-delusion stalk the pages of *Weimar Germany*, crackpots, charlatans and Spartacist revolutionaries in a land of rampant inflation where just to buy a cabbage, your money was not counted but weighed. When inflation ended on 20 November 1923, one US dollar was worth 4,200,000,000,000 marks. But Delmer thought inflation was partly a matter of self-interest. Because the war debt was computed in marks, inflation freed Germany from its reparations; it was also something that could be blamed on the wicked Versailles Treaty, democracy and the Jews. Delmer describes how, after the mark stabilised in 1924, there was a five-year 'golden' period of art and culture until the Wall Street Crash, but he also sees the work of Weimar artists like George Grosz and Otto Dix as documentary evidence of a rancid stew of resentment and thuggery.

After Sefton Delmer left Oxford in the summer of 1927, he was living with his parents in Berlin and preparing for the Foreign Office exams to become a British diplomat when the course of his life changed. His father was by now the 'stringer' for various British and American newspapers, but had gone off on holiday leaving his son to hold the fort. When Tom got a tip-off from the porter of the Adlon hotel that Lord Beaverbrook had arrived, he immediately rang the owner of the *Daily Express* and offered his services. 'Come and see me,' growled Churchill's friend Max in his rasping voice.

Lord Beaverbrook came accompanied by the novelist Arnold Bennett (who wrote a novel around Beaverbrook as a WW1 propagandist, *Lord Raingo*) and some aristocrats travelling in connection with a film-project that never got made, cigar-chomping Valentine Castlerosse, Mrs Venetia Montagu and blonde, blue-eyed Lady Diana Cooper. Tom Delmer was slim, dark and tall – he stood six feet one in his socks – and made himself useful answering the telephone and chattering about the erotic and exotic Berlin nightlife. Later Castlerosse toured the transvestite night clubs, and Bennett asked to see some of the nudist and homosexual magazines young Delmer had mentioned. After Delmer's third trip to buy stacks of porn in Potsdamerplatz, he began to get strange looks from the woman at the news-stand. Delmer showed Lord Beaverbrook around Berlin and told him that the Germans were secretly rearming. When asked what he wanted to be, young Tom replied, 'I want to be a newspaperman, sir.' He rewrote one of his stories at Lord Beaverbrook's dictation, and Beaverbrook's secretary phoned it through to the *Daily Express* with a message from the owner: 'Tell the editor that I advise him to put it on the front page.' And so 'D. Sefton Delmer' got his first byline and joined the *Daily Express*, for whom, over the next thirty years, he would become a legendary foreign correspondent. Within a year, aged only 24, he was back in Germany running the paper's new Berlin bureau.

Berlin in the twenties was fun for a young reporter: a fount of good stories, violence and vice mixing 'in a ferment of ultramodernism and get-rich-quick hysteria'. Delmer's sense of humour did not always go down well. When he turned up at a fancy-dress ball in a child's pickelhaube helmet, a toy sword and a popgun with a cork on a piece of string plugging the barrel, he was almost lynched for insulting the German army. And when he presented both weapons 'in aid of your next war' he was forcibly ejected. But usually his fluent demotic German and enjoyment of wine, women and song made him liked everywhere, in high society, among the politicians, the rich, the bohemian, in bars and night clubs. Delmer distributed his visiting card to all the petrol-station men in Berlin, and paid them for tip-offs. The efficient German telephone system was the key to his scoops. He had phones in every room and could also put his calls through to whichever nightclub he was visiting so that he could dash off at short notice to the riot or murder scene.

Sefton Delmer witnessed first-hand the rise to power of the National Socialist German Workers' Party (NSDAP). When he first saw Adolf Hitler speaking, in February 1929, in a small hall, Delmer walked out.

The man was obviously just another Weimar crank, denouncing oranges as foreign fruit and urging Germans to eat German food. But a year later, the NSDAP or Nazi party had expanded from twelve seats in the Reichstag to 107. When Delmer went again to see Hitler speak, he sat as close as he could, fascinated by the man's hypnotic staring blue eyes, popping out of his head as he worked himself into a fury. As Hitler shrieked and gestured, sweat poured off him and the dye from his cheap blue serge suit stained his wet collar a dirty purple colour. Delmer looked around at the comfortable middle-class audience: Hitler 'stirred them into a state of aggressive exultation. It was frightening.' When Delmer neither sang nor gave the Nazi salute at the end, the man next to him wanted to knock him down: 'Just you wait till after. We'll show you. We'll teach you.'

At the end of April 1931 Sefton Delmer started getting to know leading Nazis personally. A forger introduced him to Major Ernst Röhm, the beefy soldier with the shrapnel-scarred face who was the new chief of staff of the Nazi party's paramilitary wing, the thuggish brown-shirted Storm Division or *Sturmabteilung*. Röhm and Hitler went way back to the earliest days in Munich, when Röhm had helped raised the money for the Nazi paper *Völkischer Beobachter* and filled the first stormtrooper ranks with men from his Bavarian Freikorps. Over a lavish lunch at the *Daily Express*'s expense, Röhm explained that he was removing the rowdier elements from the SA and turning them into an orderly citizen force to protect against Bolshevism, but Delmer knew that as a journalist he was the object of a charm offensive designed to reassure both foreign opinion and the German generals who really ran things. When Röhm said he was inspecting a parade of all the Berlin stormtroopers that night, Delmer asked if could come along. 'Of course you can, my dear fellow,' said Röhm genially, but told Delmer he must attend in the guise of one of his plainclothes ADCs.

There were 3,500 stormtroopers in the Sports Palace, most of them in brown shirts and breeches. A low-browed blond sadist with cherry-red lips was barking orders. This was Edmund Heines, a convicted murderer and one of Röhm's favourites. Delmer shook the assassin's hand ('a reporter will shake hands with the devil himself on a story'). Röhm made a speech telling the stormtroopers their time would come, and invited Delmer to meet Hitler in Munich.

In May 1931, Sefton Delmer was the first Englishman to visit the NSDAP or Nazi HQ in Munich. *Das Braune Haus* was closely guarded by the black-capped *Schutz-Staffeln* or SS, Hitler's Protection Squad, and its decor was like that of a grandiose railway company. Delmer met Röhm again and spotted the bald, dumpy anti-Semite Julius Streicher, editor of the racist newspaper *Der Stürmer*, wolfing white sausage and sauerkraut in the canteen. In the statistics department, with its rows of files and Hollerith punch-card machines which could tell you instantly how the party was doing anywhere, Delmer was surprised to see Austria and Czech Sudetenland already treated as part of Germany, six years before the *Anschluss*, seven before Munich. His guide showed him a vast map of Germany and Austria studded with pins: 'Each of these pins represents a unit of 100 stormtroopers.' Even the Catholic or Communist areas had pins in them. In the last month, 38,500 people had joined the Nazi party, and the *Sturmabteilung* was expanding rapidly too.

Röhm showed Delmer into the Führer's large room with tall windows leading on to a balcony overlooking the street. Hitler was in the corner talking to a bushy-browed man with a strangely simian face, Rudolf Hess, but got up and strode forward in his double-breasted blue suit, clicked his patent leather heels and saluted with a half-bent arm. When Röhm made the introductions Hitler shook hands mechanically, without smiling, and said in a guttural voice '*Sehr angenehm*' which Delmer noted as 'the German equivalent of "Pleased to meet you" . . . used roughly by the same class of people in Germany who say "Pleased to meet you" in England'. Delmer found Hitler rather ordinary with his little moustache, his unhealthy skin, his too carefully arranged brown hair. 'He reminded me of the many ex-soldier travelling-salesmen I had met in railway carriages on my journeys across Germany. He talked like one too.' But none of those bagmen talked with quite 'the passion, the volubility and the concentration' that Delmer saw now in Adolf Hitler.

The Führer was soon shouting, denouncing France for persecuting the Germans. Delmer got him on to England, and he was 'off like a bomb'. Hitler claimed to want to cooperate with Britain and Italy in a three-way Axis to checkmate the Poles and the French, raving about Nordic blood and a joint mission for the world; he wanted debt reparations cancelled and 'a free hand in the east'. 'Our people must be allowed to exploit the resources now being wasted by Bolshevik mismanagement.' Delmer reckoned that Hitler would have liked Britain to hold the pass in the west while he exterminated Soviet Russia and marched his country with giant strides towards Fascism. Their conversation was interrupted by his Imperial Highness Prince August Wilhelm, the Kaiser's chinless, knock-kneed second son, bursting theatrically into the room carrying a sheet of paper with the casualty figures of stormtroopers fighting the Marxists in the last four months: 2,400 killed and wounded. *'Mein Führer,'* he exclaimed, *'das ist Bürgerkrieg!'* ('This is civil war!'). When Hitler made the introductions, Queen Victoria's Nazi great-grandson lisped in English, 'I am enchanted to make your acquaintance, Mr Delmer.'

In the spring of 1932, there were presidential elections in Germany. In the first ballot on 13 March, Field Marshal von Hindenburg and the Social Democrats won 49.6 per cent of the vote, Hitler's Nazis won 30 per cent, Thälmann's Communists 13 per cent, and the lack of an absolute majority meant a run-off on 10 April 1932. Sefton Delmer was invited to join Hitler on his campaign tour, flying around Germany.

Adolf Hitler knew how to exploit the mystique of the air. Two years later, the opening sequence of Leni Riefenstahl's Nazi propaganda film *Triumph des Willens* would linger over the godlike approach of Hitler's aeroplane through the clouds towards Nuremberg. Delmer's place on the air tour was part of a media blitz and had been fixed as an exclusive by Hitler's PR man, Ernst 'Putzi' Hanfstängl. Hitler was prepared to have Delmer on board because he spoke German like a German, with no interpreter needed.

Adolf Hitler's Flying Circus left Tempelhof airfield on the drizzly morning of 5 April. Club-footed, dwarfish Joseph Goebbels turned up in a brown and beige Mercedes convertible, wearing a white trench-coat and a snap-brim hat, seen off by laughing Magda Goebbels in a luxurious black Persian lamb hat and coat that showed off her blonde hair and blue eyes. Hitler arrived with two black Mercedes full of brutal and camp SS bodyguards, led by Sepp Dietrich. The plane smelt of rubber and petrol and as the fawning courtiers finished their ministrations, Delmer watched Hitler slumped in his seat, cotton wool stuffed in his ears against the noise of the three engines, listless and depressed as an unwanted salesman.

When the plane door opened, however, Hitler pumped up and was transformed: he stepped on stage as *der Führer*, posturing like Ludendorff, erect, squared, haughty. As the roar of welcome rose, he switched into his second mode, the wide-eyed Messiah. Delmer knew the light in his eyes from his school days: it was called the *leutseliges Leuchten*, the 'gracious shining' of the Hohenzollern emperors towards their devoted subjects. Now little Adolf Hitler in his belted mac was doing it for the age of the common man. What impressed Delmer was the range of German provincial dignitaries who had come out to greet the candidate: police, military, judicial, administrative, all calling him '*Mein Führer*'. Delmer watched Hitler switching his emotional magnetism on and off like an actor. Hitler's charisma did not win that April election; Field Marshal von Hindenburg got 53 per cent, but the Nazi vote went up to 36.8 per cent, and millions were now voting for Hitler. Three months later, in July 1932, the Nazis won 230 seats in the Reichstag to the Social Democrats' 133 and the Communists' 89.

On the first day of the flying tour, Delmer's overnight bag went missing at Königsberg. He was talking to Hitler and Goebbels on the railway platform when the press officer Putzi Hanfstängl drew him aside and indicated a smiling fellow in a pince-nez holding Delmer's bag. Its grateful owner was about to tip the little man from the lost property office a mark when Putzi hurriedly made the introductions. 'Mr Delmer, this is Herr Heinrich Himmler, the chief of Herr Hitler's security services.' Himmler had been searching the bag for bombs or assassin's kit.

As a journalist and information-monger, Sefton Delmer was drawn into Nazi intrigues and counter-intrigues, bluffs and double-bluffs, as they continued on their road to power. It all made good stories for the *Daily Express*. The Nazis wondered whether Sefton Delmer was a British spy with direct access to the British government; some Britons thought he was a Nazi spy, others later thought he was a Communist agent. Many people did not trust Delmer, but really good reporters are rarely trustworthy when they are on to a great story. Sefton Delmer was never a Nazi, just a 28-year-old newspaperman who had struck lucky.

Throughout 1932, the Nazis piled pressure on the Chancellor, Franz von Papen, who was replaced by Kurt von Schleicher in December. Street fighting between Communists and Nazis and the collapse of parliamentary government led the 85-year-old President von

Hindenburg to solve the problem by appointing Adolf Hitler as Reich Chancellor, *Reichskanzler*, the equivalent of German Prime Minister, on 30 January 1933. The diplomats told Delmer that Hitler was 'a Chancellor in handcuffs', because his power was balanced by that of von Papen, who was now the dictator of Prussia and enjoyed privileged access to von Hindenburg. But Hitler's henchman since 1922, Hermann Göring, was Prussian Minister of the Interior, and was Nazifying the entire police force. If Hitler was wearing handcuffs, his fat friend Göring held the keys.

On 27 February 1933, Sefton Delmer ran a mile and half from the *Daily Express* office to the parliament building. His tip-off had come from one of his petrol-station men: 'The Reichstag is on fire!' Delmer reached the Reichstag at 9.45 p.m., forty minutes after the alarm was given. Smoke and flame were funnelling up through the glass dome, and clanging fire engines arriving. An excited policeman told him they had got one of the men who did it, a man with nothing but his trousers on, who seemingly had used his jacket and shirt to start the fire, but they were still looking for any other accomplices. Delmer reached Reichstag entrance Portal Two just as Hitler went up the steps, two at a time, in his trenchcoat and black artist's hat, followed by Goebbels and the entourage. 'Mind if I come along too?' Delmer asked. 'Try your luck!' Hitler's bodyguard replied, grinning. Hitler just said '*Abend, Herr Delmer*', and that was his admission ticket.

Delmer stood listening avidly as Göring said to Hitler, 'Without doubt this is the work of the Communists, *Herr Reichskanzler.*' He went on to say that a number of Communist deputies had been in the building twenty minutes before the fire started. Hitler asked, 'Are the other public buildings safe?' Göring said he had mobilised the police to guard key spots. Delmer was sure that this was not just an act; Hitler and Göring really did fear a possible coup by the Communists.

They set off on a tour of the still burning parliament building, through pools of water and charred debris. Göring picked up some burned rag of curtain as evidence that 'they' had put cloths soaked in petrol over the furniture, but Delmer thought all of it could have been done by one man. They peered into the blazing furnace of the debating chamber, with flames roaring up into the cupola. Despite the firemen's hoses, the heat was like an oven. Hitler fell back to walk with Delmer. 'God grant that this be the work of the Communists. You are now witnessing the beginning of a great new epoch in German history, Herr Delmer. This fire is the beginning.' He stumbled over a hosepipe and recovered. 'If the Communists got hold of Europe and had control of it for but six months – what am I saying? – two months – the whole continent would be aflame like this building!'

On the first floor they met Franz von Papen, fresh from the *Herrenklub* where he had been dining with President Hindenburg. Delmer, thinking like an English public-school man, focused on the class difference between the two: von Papen very much the aristocrat, with a beautifully cut grey tweed overcoat over his dress suit, a black-and-white scarf round his neck and a Homburg hat in his gloved hand, Hitler the parvenu in his trench coat, soft black hat still on his head. Hitler was excitedly talking in his Austrian German about crushing the Communists. Von Papen withdrew his hand, which Hitler had shaken too hard, and said he was glad that at least the Gobelin tapestries and the Reichstag library had been saved. Hitler invited him to an immediate conference with Göring on what police measures to take, but von Papen declined politely, adding, as a final reminder of his higher authority, that he must first report to President Hindenburg.

Delmer expected congratulations from the *Daily Express* for his world scoop, but did not get any when he rang London. 'Is the story OK?' he asked, fishing for a compliment. 'OK I suppose,' said the sub-editor. 'But we don't want all this political stuff. We want more about the fire. United Press reports there are now 15 brigades on the spot and the dome has fallen in.'

Delmer did not believe that the Reichstag fire was set by the Communists, as the Nazis said, or by the Nazis themselves, as the Communists said. He thought the lone Dutch eccentric, Marinus van der Lubbe (later executed for the act), was probably responsible. But he was in no doubt that this was exactly the kind of excuse that Hitler needed to strike out against his enemies. Within hours the Communist Party HQ was raided for damning evidence. Communist parliamentarians and trade union leaders were rounded up by the police, along with left-wing doctors, lawyers, writers. On the morning of 28 February 1933, while newspapers blared 'Communist Plot' headlines, Hindenburg signed the emergency decree 'for the Defence of the People and the State' that Hitler and von Papen placed before him.

This is when the experiment of the Weimar Republic ends. The decree's abolition of free speech and the privacy of post and telephone was the death warrant of German democracy and marked the birth of a police state, for the police now had unrestricted right of search, arrest and confiscation. The Nazis had become unstoppable.

Sefton Delmer met Winston Churchill for the first and only time face to face in March 1933, in the aftermath of the Reichstag fire. As a Tory with Kiplingesque views, Delmer was an enormous admirer of Churchill. Nevertheless their encounter at Lord Beaverbrook's London home, Stornoway House, was not a success. The *Daily Express* leader writer, Frank Owen, introduced the two men after dinner. Churchill had a glass of brandy in one hand and a big cigar in the other, and he was red-faced and truculent as he talked with Owen about India. Churchill's views on India were reactionary and imperialist – he detested Gandhi as 'a seditious Middle Temple lawyer, now posing as a fakir', and once said that Indians were 'the beastliest people in the world, next to the Germans'. Delmer had no experience of the place, and so stayed silent. Then there came an opportunity. 'This aspect of the problem, sir,' he chipped in, 'is something on which I have frequently heard the views of Adolf Hitler . . . ' Churchill swung round, anger blazing in his light blue eyes, said, 'That imposhible fellow Hitler! I don't want to hear anything about him,' and stumped off. Delmer thought a statesman should be readier to listen to intelligence, particularly as he knew Churchill had tried unsuccessfully to meet Hitler when he visited Munich the year before. Churchill's son Randolph had phoned Putzi Hanfstängl to invite him to dinner, but Hitler declined several times. In the end the press officer came alone and was taxed by Churchill about Hitler's anti-Jewish views. 'Tell your boss from me,' said Churchill, 'that anti-Semitism may be a good starter, but it is a bad sticker.' After dinner, however, Churchill had asked Hanfstängl quietly what his chief might feel about an alliance between Germany, France and Britain against Russia.

Sefton Delmer left Germany because Beaverbrook, flattering him that he was the best foreign correspondent the paper had ever had, told him he would need the experience of working in Paris and New York to be truly international. Delmer knew he would miss his contacts in Berlin. Where else could his parrot Popitzschka leave its white droppings down the large dinner jacket of a heartily laughing Hermann Göring? On one of Delmer's final evenings in Berlin, Ernst

Röhm brought Heinrich Himmler to dinner and the subject of concentration camps came up. Joseph Stalin in Russia had already started massively expanding the system of work and prison camps he inherited from the Czars, and Hitler's regime (aping the Marxists whom in theory they loathed, as Victor Serge has pointed out) was building its own German gulag of concentration camps to deal with the enemy within. Heinrich Himmler was now chief of police in Bavaria and had opened a model institution at a place called Dachau. All those reports of brutalities were complete inventions, he insisted.

'If that is the case,' said Delmer, 'why don't you let me spend a few days there? Let me be treated as an ordinary internee and see what happens to me . . . What do you say?'

Himmler said it was a good idea and he would arrange it.

'It is a masterly idea,' laughed the jovial bully Röhm. 'We'll put you through the whole process from the initial beating up to the last bit where you get shot while escaping. *Prost!*' And roaring with laughter the guests downed another vodka, though Himmler only raised his glass primly.

Delmer could see the headline: 'Delmer in Dachau'. 'When can we do this, Herr Himmler? Could we make it this weekend?' Himmler demurred; after Easter would be best. When Delmer telephoned the Tuesday after Easter, an adjutant explained that the visit would have to be postponed because of an outbreak of cholera, and Delmer instantly envisaged a new headline: 'Cholera in Dachau'. Now the lads in black and brown were actually in power, his credit with them was running out.

He moved to Paris, but could not escape the pull of German affairs. In June 1934, he was in Venice for the historic first meeting between Hitler and Mussolini. There, he got a tip that Ernst Röhm had quarrelled with Hitler and was now plotting with General Kurt von Schleicher. He flew to London to brief Lord Beaverbrook, who confided that the former German Chancellor, Dr Brüning, had said on a secret visit to London that an attempt would soon be made to oust Hitler and substitute a Conservative government based on the army.

Delmer went back to Germany after a year away. His successor at the *Daily Express* Berlin bureau, Pembroke Stephens, had been expelled for finding out too much about Göring's secret rearmament schemes. The press were regularly shown a pasteboard *Reichswehr*, disarmed by Versailles, training with dummy tanks and wooden artillery, while the real guns and tanks were hidden. Delmer found that Beaverbrook's information was essentially correct: von Papen and the conservatives wanted to use Hindenburg and the army to overthrow Hitler and restore the Hohenzollern monarchy. Opposing them was the Göring–Himmler alliance, with Himmler now in control of the SS and the Gestapo. In the middle was Röhm's troublesome *Sturmabteilung* (SA, or Storm Troopers). Ernst Röhm wanted his three million brownshirts incorporated into the army under his command, but the army detested this idea as much as it did the homosexual Röhm and his undisciplined storm troopers. The stage was set for a showdown.

'The Night of the Long Knives', 30 June 1934, is like the scene in classic gangster movies when many scores are settled in one bloodbath. By personally turning on his old friend Röhm, massacring him and the SA leadership, Hitler got rid of a potential rival and at the same time placated the regular armed forces. But General von Schleicher and his wife were shot; so were von Papen's secretary and speech writer, as a warning to the military of just how ruthless Hitler could be. No one knows exactly how many murders and executions there were, but Sefton Delmer printed a list of 108 names in the *Daily Express*, and was asked to leave the country by the Gestapo.

When Hindenburg died on 2 August 1934, Adolf Hitler proclaimed himself *Führer und Reichskanzler*, effectively making Germany a one-party state under a dictator. From 4–10 September 1934, the NSDAP Reich Party Rally was staged in Nuremberg. It was a big story for the American reporter William L. Shirer, just returned to Germany and shocked by how many free-thinking Germans he had known in Weimar days had become fanatical Nazis. Shirer saw screaming crowds fainting in an almost religious ecstasy at the sight of Hitler. He heard proclamations that the Third Reich would last for a thousand years, listened to the execration of Jews and Bolsheviks, and watched 50,000 members of the new labour corps, the *Arbeitdienst*, doing military drill with shining spades instead of the guns denied them by the Versailles Treaty. Shirer hated – but was awed by – the way hysteria and joy could be induced in half a million people.

Dudley Clarke was also on the streets of Nuremberg that September, visiting Germany in the long vacation of his second year at the Army

Staff College in Camberley. He too was impressed by how the Nazis marshalled people – nearly 500 trains bringing in Party officials, the SS and the SA, the Labour Corps and the Hitler Youth from all over the country. He watched Hitler pass by 'standing up in a big Mercedes car going very fast and flanked on each side by car-loads of body-guards, alternatively driving past or falling back so that it would be almost impossible for a marksman on either side to get in a shot'. Leni Riefenstahl captured the drive-by sequence in *Triumph das Willens* or *Triumph of the Will*, her second attempt to film the Nazi party Nuremberg rally. She used a huge crew of 170 this time and multiple camera angles, cutting the choreographed conformity to create the ultimate Nazi propaganda film, where no one was out of step. Her earlier film had the embarrassment of the fat, sweaty, scarred figure of Ernst Röhm, now a non-person. The penultimate chunk of the new film was given over to a huge parade of SA men listening to Hitler's speech absolving them of blame for the 'dark shadow' of Röhm. At the end of the rally Rudolf Hess shouted 'The Party is Hitler, and Hitler is Germany, just as Germany is Hitler!'

*Triumph of the Will*, staged like a pseudo-religious Busby Berkeley movie within a huge set built by Albert Speer on the Nuremberg Zeppelin Field, had its glittering premiere at Berlin's Ufa-Palast am Zoo on 28 March 1935. It is Leni Riefenstahl's love song to Adolf Hitler, and a masterpiece of propaganda. Earlier that month, the Nazis defiantly tore up the Versailles Treaty's military restrictions. Germany revealed an air force equal to Britain's, and announced compulsory military service.

--

Even when he was covering the Spanish Civil War in 1936–7, Sefton Delmer could never quite escape the Germans, who were secretly assisting the rebel Nationalists, Generals Franco and Mola. Delmer had set out for Spain from Paris in his Ford V8 convertible (together with his typewriter, his wife Isabel, her paints and easel and their Siamese cat, Pilbul) the day after Franco announced his right-wing rebellion against the left-wing Spanish Republic in July 1936. At first, Delmer covered the campaign of Mola's Nationalists. He saw burnt-out vehicles, charred corpses and blasted body parts littering the Somosierra pass, white-faced trembling prisoners being led off to be shot, and the looting of San Sebastian and Irún. But he was soon expelled, because the Germans were coming. The Nationalists thought he was a British spy; they did not want an expert on Germany like him around when the German Condor Legion secretly landed, in defiance of Non-Intervention.

When Delmer arrived in Madrid by bus to report from the other, Republican, side at the beginning of the winter of 1936, the elected Republican government had fled to Valencia, and Madrid was waging its own exhilarating fight for survival against Franco's Nationalists, who were besieging the city. Once again Delmer was seen as suspect, for the opposite reasons: he knew personally the Führer whose air force planes had just begun dropping bombs on the Spanish capital. 'He's a bloody Nazi,' said one zealot in the International Brigades. Another Briton, who knew just how generous Delmer was with food, drink, cigarettes and money to the Brigadistas, replied: 'If you're a good Communist, I'd sooner be a bloody Nazi like him.'

Delmer moved early in 1937 into the Hotel Florida overlooking the Plaza Callao in the Gran Via, 'the friendliest, funniest, and most adventure-laden Hotel in which I have ever stayed'. In the bathroom Delmer installed a bar which he stocked with looted wine and spirits from King Alfonso XIII's cellar, purchased cheap from an Anarchist pub off the Puerta de Sol. Delmer liked to drink, though only after he had done his work. The American reporter Virginia Cowles remembered the Hotel Florida as a multinational haunt of 'idealists and mercenaries; scoundrels and martyrs; adventurers and *embusqués*; fanatics, traitors and plain down-and-outs', many of them in Tom's room with its electric burners and chafing dishes, a ham hanging from a coat-hanger, a litter of sardine tins and packets of crackers. The press who gathered there after eleven at night when they had finished the long wait to file their stories by phone from the Telefonica, the tallest building in Madrid, included Ernest Hemingway of the North American Newspaper Alliance and Martha Gellhorn of *Collier's*. When it was hot, Delmer would switch the light out and open the windows and play Beethoven's Fifth on his wind-up gramophone. The parties would go on till two or three in the morning, but they all ended when the room, luckily empty, was pulverised by a shell.

Delmer was putting on weight again after losing two stone at a German spa, and described himself as a kind of grinning fat boy of the Lower Fifth, in the dirtiest of shrunken and frayed grey flannels, a soup stained brown leather jacket over a khaki shirt and, if the sun warranted it, a wide brimmed straw hat on my head, of the kind worn by the Provençal peasants around Arles where I bought it.

This garb annoyed Constancia de la Mora in the Foreign Press Office who thought his scruffiness showed disrespect for the Spanish Republic. In her memoirs she said no one liked or trusted the *Daily Express* man, and she clearly did not share his sense of humour: 'Delmer always talked and behaved as though the Spanish were some tribe of strange and ignorant savages caught up in some absurd and primitive battle with bows and arrows.'

Sefton Delmer was honest enough to admit that he never saw the Nazi–Soviet pact coming. When he went to Russia with a British trade delegation in late March 1939, he misread the Soviet Union, whose chaotic muddle and inefficiency led him to believe it could neither fight nor supply its armed forces. The cynical Non-Aggression Treaty between Germany and the USSR on 23 August 1939 took Britain and France by surprise and doomed Poland. German troops invaded from one side on 1 September and a fortnight later Russians from the other. The Gestapo joined hands with OGPU, the forerunner of the KGB; as Evelyn Waugh wrote, 'east and west the prisoners rolled away to slavery'.

For Delmer the best thing about his Russian trip was that on the train journey from Warsaw to Moscow he made friends with the reporter representing *The Times*, a tall debonair type with a broken nose, whose name was Ian Fleming. The two journalists shared a suite at the National, the antique Intourist hotel opposite the Kremlin, drank vodka martinis and picked up a couple of girls from Odessa. It was through Fleming that Sefton Delmer would find his central role in political warfare and black propaganda when war broke out.

# 13

# Curtain Up

Clare Hollingworth always wanted to get away. The woman who became the doyenne of the Hong Kong Foreign Correspondents' Club determined early to escape from conventional Leicestershire to see the reality of the foreign lands whose maps she had clipped out from WW1 newspapers as a small child. She left the world of point-to-points and hunt balls first for Slavonic Studies at London University under Professor R. W. Seton-Watson (who had been a key figure in propaganda for Crewe House in WW1), and then to work for the League of Nations Union. She was 27 years old when she was hired in 1939 by the editor of the *Daily Telegraph* in Fleet Street to go as a correspondent to Poland, where she had previously worked in Warsaw and Katowice, helping refugees from Nazi Germany to get visas to escape. The Continent was on the cusp of another war, and Hollingworth's first professional newspaper assignment was about to get her one of the scoops of the century. With brand-new luggage from Harrods, she flew out the next morning from Hendon Airport to Warsaw, via Berlin.

It was Saturday, 26 August 1939, twenty-five years to the day since the British Expeditionary Force, retreating from Mons at the start of WW1, had briefly held up the invading Germans at the Battle of Le Cateau. It was also the anniversary of Crécy, one of the many battlefields to which Clare Hollingworth's father had taken her as a girl, and John Buchan's 64th, and last, birthday.

*Berlin* was the destination marked on the shabbier luggage of another passenger leaving London that day, on the boat-train from Victoria. 'Berlin?' remarked his chatty railway porter. 'Rum place to be going right now.' The owner of the cases, who bore a razor scar the colour of raw pork looping from his right ear to his mouth, was the fascist William Joyce, on his way to becoming the Nazi broadcaster nicknamed 'Lord Haw-Haw'. The Irishman was one jump ahead of the Metropolitan Police Special Branch, taking the ferry to Ostend.

Clare's flight to Warsaw took six hours. Noel Coward had flown the same route two months before. The entertainer had somehow imagined Warsaw as grey-stoned, medieval, twisting, but found it flat, wide and yellowish. He stayed at the Europejski Hotel, across the square from the impressive and well-guarded Polish Foreign Ministry building. In the same hotel, tiny Clare now met the *Daily Telegraph*'s man in the Polish capital, Hugh Carleton Greene, the younger brother of the novelist Graham Greene and a future director general of the BBC. Tall, thin, shambling Greene was a fluent German speaker who had been expelled as a correspondent from Berlin after ten years. He had visited Dachau concentration camp in 1933 and found the guards more criminal and brutal than their Communist prisoners. 'One of us has got to go to the border,' he said. Clare volunteered to take the train south to the German frontier. At railway stations along the line officials were posting notices of the mobilisation of the Polish Army.

Back in Katowice again she borrowed the official car of her friend the British Consul General, and the Union Jack fluttering from its bonnet helped her cross the closed frontier at Beuthen. Inside German Silesia she bought newspapers, aspirin, film, electric torches and bottles of wine, which were hard to obtain in Poland. She was driving back along the fortified frontier road towards Gleiwitz, the last town in Germany, when sixty-five military motorcycle dispatch riders, bunched together, overtook her. As she drove up a hillside towards the frontier, she found tarpaulins and screens of hessian erected along the road, concealing the valley on her left from view. But the wind blew on the afternoon of 29 August 1939 and lifted the curtain. Through the hole in the hide, the reporter Clare Hollingworth saw with her own eyes scores of German tanks lined up, ready to invade Poland.

---

Adolf Hitler told the commanders of his armed forces at Obersalzburg on 22 August: 'I shall supply a propaganda justification to bring about hostilities. It is of little consequence whether the reasons are believed. No one asks the victor whether he has told the truth.'

On 31 August 1939, two days after Clare Hollingworth's reconnaissance trip, a pair of black Ford V8s drove down the same road from Gleiwitz towards the Polish frontier. In early evening daylight they turned off into a clearing in the Ratibor Wood. The car boots were unlocked and all seven Germans changed into bits of Polish Army clothing and picked up black Luger pistols. A man called Karl put on headphones and squatted down by the other car boot to get a signal from the radio inside. At 19.27 he heard the code words '*Großmutter tot*' (Grandmother dead). They got into the cars and drove back towards the tall wooden radio transmitting tower that was visible for miles around.

Their leader, SS Sturmbannführer (Major) Alfred Helmut Naujocks, now looking entirely Polish, was the first man up the stairs and through the glass doors of Gleiwitzsender, the German radio station. He pistol-whipped a man in an office who fell heavily, knocking over a chair and a hatstand which crashed on to a metal filing cabinet. Then Naujocks was through into the studio where another of his team, the announcer Heinrich, had already taken a seat at the green baize table with the microphone on it, holding the fake script they had prepared with its lines praising independent Poland and denouncing Hitler and the Nazis.

Through the soundproof glass Naujocks could see Karl the radio engineer getting frantic in the cubicle as he failed to find the landline switch to Breslau to make any kind of broadcast. But eventually Karl banged 'go' on the glass. Heinrich read the phoney script rapidly and loudly. Then Naujocks faked interruption by shouting and firing four banging pistol shots inside the small studio. The pretend Polish rebels cut off the signal and ran out of the building.

Heinrich Müller of the Gestapo had also done his part of the job, operation KONSERVE, delivering what they called the 'tinned goods'. A fresh corpse now lay sprawled on the radio-station steps, dressed in civilian clothes. The dead man was tall and fair, about 30, with a strong, handsome face. This 'tin' was originally going to be selected from among the inmates of Sachsenhausen, the Nazi concentration camp north of Berlin opened in 1933, but indenting for an already processed prisoner would have left a paper trail through the ever-meticulous bureaucracy. So a local Polish-Silesian man with strongly patriotic views called Franciszek Honiok had been secretly arrested, drugged, dressed and killed with a bullet to the head, so he could be left on the steps, one of the apparent Polish perpetrators of the raid on the wireless station.

At 7 a.m. the next morning, Friday, 1 September 1939, Naujocks was sitting unshaven in the office of his boss, the 'Blond Beast', SS Obergruppenführer Reinhard Heydrich, chief of the *Reichssicherheitshauptamt* (RSHA), the Reich Central Security Office, being congratulated on a secret agent's job well done. Three weeks before, the ruthless and calculating Heydrich had told him, 'Actual proof of these attacks by the Poles is needed for the foreign press as well as for German propaganda purposes.' It had worked: the fake Poles had provided 'proof of these attacks' and the Chancellor himself, Adolf Hitler, had telephoned Heydrich at 5 a.m. to praise the provocation. No one would be allowed to find out the truth about the raid from abroad. On 1 September 1939 all Germans were banned from listening to foreign radio broadcasts, on pain of imprisonment or penal servitude. The penalty for spreading news from foreign broadcasts was death.

That same day, the front-page story of the *Völkischer Beobachter*, the Nazi party official paper, was all about the outrageous 'Polish' attacks of the 31 August, on a gamekeeper's house at Pitschen, on the customs post at Hochlinden – and particularly on the Gleiwitz radio station. 'Armed insurgents' had apparently managed to read out a propaganda statement in Polish and German before alarmed listeners to the broadcast could alert the local police. During the ensuing shoot-out, the newspaper said, one of the Polish 'bandits' had been killed. (This of course was Franciszek Honiok, the 'tinned goods'.)

And that was why, early in the morning of 1 September 1939, young Clare Hollingworth could be seen holding a telephone receiver out of the window of her hotel room in Katowice so that the British embassy in Warsaw could hear for itself the sound of the guns, tanks and planes from Generaloberst Gerd von Rundstedt's *Armeegruppe Süd*, invading Poland with the dawn chorus.

The use of propaganda and deception seen at Gleiwitz had long been a leitmotiv of Nazi behaviour. From the Reichstag fire, through the Berlin Olympics and the activities of Joseph Goebbels's Ministry for National Enlightenment and Propaganda, one can trace a deliberate policy of deceiving the outside world. Gleiwitz or Gliwice in southern Poland later became a satellite of the industrial extermination camp at Auschwitz-Birkenau, the site of a factory plant where forced labour packed the chemicals for Wehrmacht smokescreens. The

Nazis always believed in *Verschleierungsfähigkeit,* 'the obscuring power of smoke'.

The Gleiwitz raid was also a kind of tribute by imitation to similar British operations. Heydrich, the architect of the Nazi racist state despite persistent if ill-founded rumours of his own Jewishness, was just one among many Nazis who were, as the historian and wartime SIS officer Hugh Trevor-Roper pointed out, 'indefatigable readers of novelettes, especially about the British Secret Service – that Machiavellian institution which, they believed, had built up the British world-empire'. Heydrich wanted to be called 'C' too, like the head of the British Secret Intelligence Service. He went on to become the *Reichsprotektor* of Bohemia and Moravia, and was assassinated in Prague. The two roadsweepers who attacked his car with grenades in June 1942 were really Czech soldiers – but they were trained at Aston House in Hertfordshire by a British secret service, the SOE. The man who taught them to throw hand grenades like cricket balls was Alfgar Hesketh Prichard, a keen sportsman and crack shot just like his late father.

On the evening of Friday, 1 September 1939, the Prime Minister of Great Britain, Neville Chamberlain, lamented 'this terrible catastrophe' in the House of Commons. Of course, nobody could yet foresee the horrors that those German divisions were dragging behind them into Poland: destruction, looting, rape, torture and the murder of millions. Nor could anyone imagine the worldwide ripples of damage and death in the six years to follow. But because both Britain and France were pledged by treaty to defend Poland from aggression, and because the German Government did not withdraw its military forces by 11 a.m. British Summer Time on Sunday 3 September 1939, two decades of a dubious 'peace' came to a close.

Just after 11 o'clock on that Sunday morning, following a recorded talk on *Making the Most of Tinned Foods,* 70-year-old Neville Chamberlain came on the BBC wireless to announce from the Cabinet Room in Downing Street, in a tired and sad voice that sounded to one listener like 'stale digestive biscuits', that the ultimatum to the German government had expired, 'and that, consequently, this country is at war with Germany'.

Britain began the Second World War as it ended the First: scattering twenty million pages of propaganda over Germany. 'Truth raids', the

Air Minister, Sir Kingsley Wood, proudly called them. From ponderous Whitley bombers, the RAF dropped thousands of blocks of leaflets in German-style Gothic type that fluttered apart into separate sheets as they fell. What A. P. Herbert in *Punch* called 'bomphlets' or 'bomphs', Air Vice-Marshal Arthur Harris grumpily saw as tonnage of 'free toilet paper'. These first propaganda leaflets, printed by HM Stationery Office and following a line of thought emanating from Lord Halifax's Foreign Office, were aimed at encouraging the many 'good Germans' who, it was hoped and believed, opposed Adolf Hitler and his Nazi regime. Their tone was more like a highbrow newspaper editorial than a punchy popular advertisement, because they had been written by the author of a literary travel book, *The Road to Oxiana*, Robert Byron.

The very first sheet, dropped on 3 September 1939, was a note to the German people warning them that they had been deceived by their leaders: 'For years their iron censorship has kept from you truths that even uncivilized peoples know. It has imprisoned your minds in, as it were, a concentration camp.' Not only to German civilians were these sheets officially *verboten*. When an American war correspondent asked a censor at the British Ministry of Information for the text of the leaflet, he was refused on the grounds that 'We are not allowed to disclose information which might be of value to the enemy.' The journalist pointed out that the enemy now had two million copies of the sheet, so could he please have one as well. The hapless official blinked and agreed that there must be something wrong there. (This sort of embarrassment was only resolved in early November when the head of the secret department producing the leaflets began taking all the newspaper proprietors into his confidence.)

The Franco-British plan on the outbreak of war in September 1939 was much as it had been in 1914. Another two-corps British Expeditionary Force was soon over in Flanders, on the left of the French line, ready to block the Germans should they advance once again through Belgium. It seemed like a rerun with different names. Instead of Calais and Boulogne, the disembarkation ports were Cherbourg and Brest. Although Winston Churchill was soon again in charge of the navy as he had been from 1911 to 1915, the army corps were commanded this time by John Dill and Alan Brooke, and the commander-in-chief was the gallant Grenadier Guardsman Lord Gort, who had won the VC, three DSOs and nine mentions in dispatches in WWI, but had never had a major command.

In the murky linoleumed corridors and shabby offices of the War Office in London, the mufti of city suits and stiff collars vanished 'for the duration' on 1 September, and people tried not to show too much curiosity about each other's shoulder badges, buttons and ribbons, which told the stories of their previous wars. Dudley Clarke, now lieutenant colonel, carried on at his deputy assistant military secretary's desk. He noticed Lord Gort's daughter, one of the new generation of young women then appearing in Whitehall, looking trim in her new ATS uniform and black shoes. Another was the remarkable administrator Joan Bright, who became fond of the old War Office as a mausoleum:

its ancient ways, its unwashed walls, the uneven water-marks revealing the length of the office-cleaner's arm, the ceilings thick with dust and the dim evenings in blacked-out rooms which held the stale smell of scores of smokes and dozens of thick-cupped, thick-made teas . . . I liked the officers who were polite to women and the sturdy, loyal, flat-footed messengers who untiringly provided us with tea, cigarettes, drawing-pins and booty in the shape of pieces of carpet scrounged to cover the bare boards of our rooms.

In his BBC broadcast on 3 September, Neville Chamberlain admitted that the failure of his long struggle to bring peace was 'a bitter blow'. He concluded:

Now may God bless you all. May He defend the right. It is the evil things we shall be fighting against – brute force, bad faith, injustice, oppression and persecution – and against them I am certain that the right will prevail.

*God Save the King* was played, and all around the country many people stood up. Some of them were crying. Others sat in a doggedly British mood of grim apathy. This looked like the Kaiser's War all over again, though it was Poland that had been invaded this time, rather than Belgium. But how much worse would warfare be this time, after twenty-five years of scientific progress?

At that moment the air-raid sirens began their banshee wailing across southern England. Winston and Clementine Churchill watched the balloons go up: dozens of cylindrical barrage balloons or 'blimps' slowly rising to be tethered above the city skyline like fat silvery fish.

Police constables appeared on bicycles, blowing their whistles and sporting sandwich boards that read 'Take cover'.

Inside the War Office, overlooking Horse Guards' Arch, Dudley Clarke followed the procedure on his clipboard, storing papers, opening windows, gathering steel helmet, 'respirator, anti-gas' and emergency rations before descending to join the brass-hats in the newly strengthened basement. The Churchills too (accompanied by Inspector Walter Thompson, recently returned as personal armed bodyguard) made their way to the public shelter just down the street, with refreshments of brandy. 'Everyone was cheerful and jocular, as is the English manner when about to encounter the unknown,' Churchill wrote in *The Gathering Storm*, the first volume of his history of the WW2. But as he stood in that London street on a bright September morning, Churchill's imagination

drew pictures of ruin and carnage and vast explosions shaking the ground; of buildings clattering down in dust and rubble, of fire-brigades and ambulances scurrying through the smoke, beneath the drone of hostile aeroplanes. For had we not all been taught how terrible air raids would be?

Bombing haunted the 1930s. 'The bomber will always get through,' said Stanley Baldwin gloomily in 1932, and British planning for air-raids included mass evacuation by train, public and private shelter-building, as well as the manufacture of millions of gas masks and sand bags. HMSO published *Air Raid Precautions for Animals*, price 3d; art treasures were quietly removed from London or stored deep underground.

Dudley Clarke's memoir of the first year of the war, *Seven Assignments*, opens like a John Buchan adventure novel. It tells how Clarke, a 40-year-old soldier with a touch of Buchan's hero Richard Hannay, had been working late at the War Office on 31 August 1939 and came home after midnight to his cream and green bachelor flat. There he found, lying on the doormat, the visiting card of an aristocratic German staff officer, Lieutenant Colonel Gerhardt Count von Schwerin.

Clarke had first met von Schwerin at Easter 1939, when staying with his friend Major Kenneth Strong, the assistant military attaché at Berlin, and later that year in England, in July, when von Schwerin had been sent openly by the German General Staff to find out whether

Britain would really honour its obligations to Poland if the country were attacked. Clarke and von Schwerin were not exactly friends but they were of the same age and when they met in London they broached the subject of imminent war between their two countries. 'Send me a word of warning first,' Dudley had said, jocularly, in a taxi. Now, in a moment of pure Buchan, he found an embossed card, with *Auf wiedersehen* written on the back:

For a while I sat turning it over in my fingers, and then I went to the telephone by the window-seat. Outside the lights were still blazing in Piccadilly; but it was for the last time. At that very moment the forces of Nazi Germany were advancing into Poland.

Starting on Friday, 1 September 1939, three million reluctant and name-tagged people, including the very young, the pregnant, the disabled and the blind, were evacuated from cities like London and Glasgow, Birmingham and Liverpool to 'Safety Zone' towns and villages that did not really want them. The private British railways were placed under state control. Identity cards, based on the census register, were issued. The BBC's nascent television service to 25,000 viewers was cut off. All ham radio transmitting equipment was confiscated. Cinemas, dance halls and theatres were closed. All ARP wardens were mobilised, along with the Army, Navy and Air Force, and the first 'blackout' was instituted at sunset, 7.47 p.m. This meant that no street lighting was switched on and all householders were forbidden to let a single chink of light escape from their doors or windows. Drivers of motor vehicles could not use their headlights, lest bombers above should spot them. Even electric torches were to be smothered with crepe paper and always held downwards. From 'Black Friday' onwards, thousands of people would injure themselves stumbling about in dim homes and on dark wintry streets, and there were many more night-time car crashes and road deaths. But the thousands of hospital beds which had been cleared were not waiting for accidents like these, but for casualties imminently expected from enemy aerial bombing. Corporation swimming-baths had been drained and cardboard coffins stacked to await a multitude of corpses.

However, after half an hour of dread on the morning of Sunday the 3rd, with everyone braced for the Nazi bombers' 'knock-out blow', the sirens sounded the 'All Clear' and everyone left their air-raid shelters to find the sunny sandbagged streets as tranquil and undisturbed as before. The false alarm from the Thames estuary, instead of the ruthless thunder and lightning attack that everyone feared the Germans would deliver on the first day, was a fitting opening to that queer period of fighting and not fighting that Churchill called 'the pretended war' and then 'the Twilight War', but which impatient American journalists finally dubbed 'the Phoney War'. This lasted eight months, although the food rationing that was introduced then endured in some form for fourteen years. Evelyn Waugh's *Put Out More Flags* is the most acute satire on 'that odd, dead period before the Churchillian renaissance which people called at the time the Great Bore War'.

# 14

# Winston Is Back

That same Sunday, Churchill made his way to the House of Commons where, after the debate, Neville Chamberlain offered him a place in the War Cabinet, with the post of First Lord of the Admiralty. (The Board signalled to all Royal Navy ships 'Winston is back'.) That evening, on the wall behind his familiar old chair in the First Lord's office, Churchill found the wooden mapcase he had had fixed in the panelling twenty-eight years before. When he flung open the door, he saw the chart still marked with the disposition of the German Fleet on 23 May 1915, the day he had left.

Churchill came back to naval problems familiar from 1914–18. On the first day of war, as in WW1, the order was given to sever the two German undersea communication lifelines that connected Emden in Germany on the one hand with the Azores and the Americas, and on the other with Lisbon and Africa. This chopping of the telegraph and telephone cables left Nazi Germany relying on wireless systems or radiotelephony, whose codes and ciphers could be intercepted by the British 'Y' or listening service, and passed on to the codebreakers at Bletchley Park.

The old enemy was using new mines and better submarines to try and choke off the British Isles. Once again, U-boats were a headache and shipping became a prime German target. In WW1, Britain had lost half its merchant fleet, but in WW2, the Axis would sink 60 per cent or 11.3 million tons of British shipping, and kill over 50,000 British Empire merchant seamen, making theirs a more dangerous occupation than any of the armed forces. German Naval Intelligence cryptanalysts in B Dienst had been breaking British ciphers since the Abyssinian crisis of 1935 and thus knew exactly where many Royal Navy and Merchant Service ships were. War at sea was far from phoney.

In a reminder of Churchill's great fear of a quarter of a century before, on 14 October the German *U-47* crept past nets, booms, blockships, lookouts and patrols right into Scapa Flow, the Orkney-enclosed anchorage of the Royal Navy's Home Fleet. The U-boat torpedoed the huge 1914 battleship HMS *Royal Oak*, which sank in thirteen minutes with the loss of 810 officers and men. While struggling survivors in the water sang *Roll Out the Barrel* to keep their spirits up, the U-boat escaped. Four days later, Kapitänleutnant Günther Prien and his submarine crew surfaced, grinning, at a Nazi press conference, before being paraded through cheering crowds in Berlin. Churchill told the House of Commons that the Royal Navy had sunk 13 German U-boats and damaged some others, which was a better strike rate than in WW1. This was stretching the truth considerably, but Churchill felt he had to cheer the public up.

Prien had visited Scapa Flow before the war, posing as a tourist, and had also used aerial reconnaissance *Aufklärung* photos taken from directly above to find a narrow way in for his boat to sink the *Royal Oak*. Both high-flying Abwehr aircraft and some Lufthansa civilian flights had been secretly photographing tracts of Britain long before war broke out, and now Hermann Göring's Luftwaffe was filling in the gaps.

As First Lord of the Admiralty, Churchill went up to Scapa Flow, wept over the wreckage of the *Royal Oak*, and ordered more submarine defences. The death of one of his Dreadnoughts was a blow, making the nation seem vulnerable. Once again, when force had failed, he turned to fraud for protection. Churchill ordered the construction of dummy ships whose camouflage was good enough to deceive spotters in aeroplanes. He came back later to inspect the results. Mock-up warships were dotted about the huge Orkney anchorage. When Churchill pointed to one and said the Germans would never drop a bomb on it, he was told that it had convinced our own aerial reconnaissance. 'Then they need spectacles,' snapped the First Lord,

No gulls about her! You always find gulls above a living ship. But not around a dummy, unless you drop refuse from it too. Keep refuse in the water day and night, bow and stern of all these dummies! Feed the gulls and fool the Germans!

The British did manage to trick the German navy during the Battle of the River Plate in Montevideo, Uruguay. The German pocket-battleship *Admiral Graf Spee* was a fast armoured cruiser that could make twenty-six knots and carried six 11-inch guns with a range of seventeen miles. Accompanied by the auxiliary and prison ship *Altmark*, the *Graf Spee* continued the raiding practice of WW1, camouflaging herself as an Allied ship and sending false radio messages while sinking merchantmen in the South Atlantic and Indian Ocean. On 13 December 1939, lightly damaged after an engagement with the British cruisers *Exeter*, *Ajax* and *Achilles*, the marauder made for neutral Montevideo for repairs. British diplomats, naval attachés and secret agents worked together to delay the process; meanwhile they used the BBC news, diplomatic and dockside gossip and talk on telephone lines they knew were tapped to suggest that a large British fleet, including the aircraft carrier *Ark Royal*, was just over the horizon, when in fact it was five days away. The *Graf Spee*'s skipper, Captain Langsdorf, a decent man, was deceived into thinking he had no chance of escaping to the high seas. He released his crew, scuttled the *Graf Spee*, went ashore at Buenos Aires and, wrapped in his navy's ensign, shot himself.

—

'It's necessary to understand what real intelligence work is,' John le Carré once told George Plimpton in the *Paris Review*. 'At its best, it is simply the left arm of healthy government curiosity . . . It's the collection of information, a journalistic job, if you will, but done in secret.' The Naval Intelligence Division (NID) was not a secret service: it had no agents who did covert espionage. Though secrecy was its character, it was not its essence, for the organisation dealt in legitimate information from many sources. British Naval Intelligence started getting its act together again during the Spanish Civil War of 1936–9, when the German and Italian navies were actively assisting General Franco's rebellion against the Spanish Republic, and they had to be watched in the interests of the Royal Navy's blockade and rescue missions. Lessons had been learned from WW1: what was needed was an active, well-wired brain in an acute nervous system. The Admiralty was unlike the War Office or the Air Ministry in that its primary function was not administrative but operational: the Royal Navy's traditions were action and attack. What had gone wrong with the

Admiralty's cryptographic centre, Room 40, in WW1 was that its operational effectiveness was crippled by secrecy. Rigid compartments meant the brain's synapses did not always fire efficiently. Then, when cryptanalysts had information about the German fleet, there was no system to communicate that data swiftly to the British fleet. The brain could not connect with the hand. Hence the failure at the Battle of Jutland, when poor communication of knowledge meant the two fleets effectively missed each other.

When Rear Admiral John Godfrey became director of Naval Intelligence on 3 February 1939, he inherited the basis of an Operational Intelligence Centre (OIC) and was able to consult the most celebrated of his predecessors, Admiral Sir Reginald Hall, and to follow his advice in setting up systems to feed that centre. *Everything* was of potential interest to the director of Naval Intelligence. The net had to be spread wider than the technical world of signals and direction-finding, and the right staff found to process lots of information. Effective intelligence-gathering required curious, sceptical, energetic people. Accordingly, the NID (which expanded in WW2 from employing dozens to thousands) was leavened with civilian scholars, barristers, solicitors, writers and journalists, including the novelists Charles Morgan and William Plomer, Simon Nowell-Smith of the *Times Literary Supplement* and Hilary St George Saunders, the House of Commons librarian. Admiral Godfrey instituted a vitally important grading system for intelligence: the letter stood for the source, the number for the information, so A1 was first rate and D5 most dubious.

Admiral Godfrey made strong links with government, allied diplomats, the armed services and the secret ones, MI6, MI5 and Special Branch. He met Sir Roderick Jones of Reuters, the great press barons and their newspaper editors. Admiral Hall said the director of Naval Intelligence was entitled to enlist the help of anyone in the country from the Archbishop of Canterbury down, and he introduced Godfrey to Sir Montague Norman, governor of the Bank of England, and to various other powerful figures in the City of London. Hall himself had employed a stockbroker called Claude Serocold as his personal assistant in WW1 and advised Godfrey to do the same. The governor of the Bank of England and the chairman of Barings, Sir Edward Peacock, found Godfrey his right-hand man in May 1939. It was the suave chap who had drunk vodka with Sefton Delmer in the hotel opposite the Kremlin, the stockbroker Ian Lancaster Fleming.

Ian Fleming was an inspired choice to lead the coordinating section of Naval Intelligence, NID 17, where he had to supply ingenious ideas, create structural order, and liaise externally with the wider world. After Eton and Sandhurst, Ian Fleming had learned German and French, and worked for Reuters as a reporter before he became a stockbroker in the City. In Moscow in March 1939 he had told Delmer he was there as a favour to the editor of *The Times*. Years later, with the benefit of hindsight, Delmer claimed that

As soon as I saw [Ian Fleming], I knew he was on some intelligence job or other . . . he made such a determined show of typing away whenever the Russians were looking that it was clear he was no ordinary journalist.

On the train back to Warsaw, heading for the border of Stalinist Russia, Delmer had memorised his notes, torn them up and thrown them away. 'Why don't you swallow them?' mocked Ian Fleming, 'That's what all the best spies do.' But at Negeloroje, the customs officials went through Fleming's luggage with a toothcomb, stripped him and searched him. A carton of Russian contraceptives made of artificial latex which he was taking back to London to have the formula analysed was opened and each condom held up to the light. Fleming was already blushing scarlet when Delmer whispered, 'You should have swallowed them.'

Five months after this train journey, in the autumn of 1939, Delmer invited Fleming to lunch at his flat in Lincoln's Inn. Fleming turned up looking elegant in the dark blue uniform of the Royal Navy with the waved stripe-and-curl insignia of a lieutenant in the Volunteer Reserve.

'I thought you'd been to Sandhurst. What are you doing in the navy?'

'Oh, I've been given a special desk job at the Admiralty.'

Over coffee and brandy, Fleming announced that his boss would like to hear what Delmer had observed of the war in Poland. The reporter said he had seen no ships, only some bombing and air fighting, a lot of retreating and some excited people shooting fifth columnists. Nothing to interest the navy.

'Never you worry your head about that. Just do as I tell you.'

The next day, Delmer turned up at a door in the Mall behind the statue of Captain Cook and was escorted along sixty yards of bleak corridor to a transept in front of room 39. When the door opened he saw a crowded office with three tall windows looking out across the parade ground used for Trooping the Colour. A dozen men were working at desks and talking on the telephone, reminding Delmer of the back room of a Beirut bank, except the men were not clerks but section chiefs of British Naval Intelligence. Ian Fleming opened the door to room 38 and ushered Delmer in to meet Admiral Godfrey. There were other naval captains and commanders in the room as well as army and air force officers. They asked him a lot of questions and he told them, as he thought, 'little of interest'. But to the *Express* man the meeting was supremely interesting. He had discovered how important his friend Ian Fleming had become, 'nothing less than "17F", the personal assistant to the intelligence chief of the Senior Service.' Delmer also found himself curiously at home in a place where he would later do some of his greatest work.

Ian Fleming had the social confidence and forcefulness of character required to be an effective factotum to his energetic and horribly hard-driving boss. Fleming was not the wisest, but certainly the most vivid personality in NID 17: 'a skilled fixer and a vigorous showman', said Donald McLachlan, the historian of room 39. Fleming's Reuters training in good clear English made him first choice to draft his boss's replies to Churchill's imperious requests. Winston was back, and as demanding as ever.

--

The outbreak of war brought a frenzy of reorganising. The Ministry of Information (MoI) was one of two new British ministries that came into being. The other was the Ministry of Economic Warfare (MEW), tasked with disorganising the enemy's economic life so that Germany could not fight. Some thought the Ministry of Information was doing the same job on the communications of the Home Front. The Senate House of London University was taken over by 999 civil servants and new appointees. An officious law lord, Baron Macmillan, now had a brief spell as the first Minister of Information. As Hugh Macmillan, he had been John Buchan's assistant director of Intelligence in the old WWI Ministry of Information.

Sixty-four-year-old Buchan now bore the title Lord Tweedsmuir and was living in Ottawa. In his role as governor general, it had been his melancholy duty to sign Canada's declaration of war against Germany. 'This is the third war I have been in,' he wrote to an old friend, 'and no-one could hate the horrible thing more than I do.' He gave Macmillan advice from far away: no direct propaganda to America, no jibes at isolationism, 'no attempt to varnish', 'never deny a disaster'. He also suggested, 'Our news should follow the Reuter plan and be as objective as possible,' although this would mean battling the War Office and the Admiralty's 'passion for babyish secrecy'.

The MoI was meant to distribute news on behalf of all Government departments and the three fighting services, but this information policy (open doors and windows) conflicted with its second responsibility, which was censorship (close all doors and windows). It was possibly deformed by the secrecy of its gestation for future wartime needs, in late 1935, under the auspices of a standing subcommittee of the Committee for Imperial Defence. The MoI was meant to be in the public relations business, and its first director was intended to be Sir Stephen Tallents, the imaginative civil servant who wrote *The Projection of England*, a pamphlet that led to the founding of the British Council. But the appointment did not happen, as Tallents was replaced by Macmillan, and the ill-favoured MoI that shambled into the light looked askance at news or publicity. Fierce 'D' Notices were slapped on newspapers; the BBC was restricted. After four months, Macmillan was replaced by Sir John Reith, formerly of the BBC and Imperial Airways. In the MoI he found a lawyer running censorship, a man who saw his job as interpreting the regulations rabbinically rather than trying to woo a little more news from the Armed Services than they wanted to give. 'Also he thought an event wasn't news till the press got it!' John Reith wrote in his diary. 'Weird and infuriating. Save us from lawyers.' The Home Publicity Division of the Ministry attempted to boost morale, but failed. Its first poster, reading '*Your* courage, *your* cheerfulness, *your* resolution will bring *us* Victory' soon had cynical and disaffected members of the public asking who exactly 'you' and 'we' were in this equation.

The MoI was not responsible for the propaganda leaflets or 'bomphs' that showered over Germany at the start of the war. These were produced by a shadowy organisation in London called Department EH, whose letters stood for Electra House, a large building on Victoria Embankment where the chairman of the Imperial Communications Advisory Board had his office. He was the very same Sir Campbell Stuart who had run Crewe House in WWI and who had been invited to do the same job of Propaganda to Enemy Countries during 1938. Department EH's first job, at the height of the Munich crisis, was to help get German translations of speeches by Chamberlain, Daladier and Roosevelt rapidly broadcast on the powerful commercial station, Radio Luxembourg, which was close to Germany and audible by many there. This was a delicate matter. The BBC itself was actually trying to get Luxembourg closed down as a 'pirate' station because so many British listeners preferred its light music to BBC fare, so the corporation was not best pleased when its staff and facilities at Broadcasting House were hurriedly used by the British government to assist a rival broadcaster.

--

The head of the Secret Intelligence Service (SIS), also known as MI6, is traditionally referred to as 'C' after the initial letter of the surname of its first head in 1911, Mansfield Cumming. Before the war, SIS's reputation was probably a lot higher in spy fiction than it was in reality. Admiral Sir Hugh Sinclair liked secrecy, and being the only one 'in the know', for security reasons. Most SIS officers then operated abroad as Passport Control Officers in British embassies and consulates, but in late 1936, aware that this was well known to UK's enemies, Sinclair had secretly established an alternative organisation, the 'Z' network, to gather intelligence from Nazi Germany and to a lesser extent from Fascist Italy. 'Z' was run by Claude Dansey, and the cover story for the end of his career in the Secret Service was the plausible rumour that he had been sacked for dipping his hand in the till in Rome. Nobody else in SIS apart from Admiral Sinclair knew about 'Z', which operated from a suite on the eighth floor of the north-west wing of Bush House and used adjoining companies – Geoffrey Duveen & Co, Joel Brothers Diamond Company – as a front.

In March 1938 Sinclair had also asked Major Lawrence Grand, RE to look at the possibilities of creating a British organisation for covert offensive activities, looking at 'every possibility of attacking potential enemies by means other than the operations of military forces'. Grand was promoted to colonel and set up the Devices or Destruction Section of MI6 – Section D – with the innocuous camouflage name of the Statistical Research Department of the War Office, and he began establishing a network of agents in cities abroad, especially in the Balkans. He started studying undercover sabotage, the training of saboteurs and methods of countering sabotage, as well as producing and experimenting with ammunition and explosives. Secret services are supposed to keep quiet about what they do, but Section D was bound to produce a lot of noise, which conflicted with that. Section D, moreover, saw destructive sabotage against potential enemies as not just physical, but also moral, mental, and verbal. But these were also in the spheres of propaganda and radio, which brought in EH, the Overseas Department of the Ministry of Information, and MI7 in the War Office, leaving plenty of room for confusion and muddle.

From March 1939, Sinclair's Secret Service money was also paying for a small War Office Military Intelligence Research unit called MI(R), headed by a fiery, chain-smoking Royal Engineer called Colonel J. C. F. Holland. Jo Holland was already immersed in the study of guerrilla warfare, and wrote a joint paper with Grand in March 1939 on possible guerrilla operations against Germany. MI(R)'s brief was to keep studying unconventional warfare for the uniformed services, to draw up a Field Service Regulations Handbook for guerrillas and organised irregular bands, and to investigate any destructive devices that could be produced to help them. Holland appointed two grade II staff officers to assist him. Major Colin Gubbins of the Royal Artillery researched and wrote two pamphlets, *The Partisan Leader's Handbook* ('Surprise is the most important thing in everything you undertake') and, with Holland, *The Art of Guerrilla Warfare*, which looked particularly at T. E. Lawrence in WW1, the IRA, the Arab rebellion in Palestine and the North-West Frontier of India. The other grade II officer Colonel Holland appointed, a bang-happy sapper called Major Millis Jefferis, who later invented the 'sticky bomb', wrote *How To Use High Explosives*. ('If distributed today,' Patrick Howarth remarked in his book, *Undercover*, in 1980, 'they would probably be described as terrorists' handbooks.'). MI (R) also headhunted especially adventurous types from among the linguists, explorers, writers and executives already earmarked by the director of military intelligence. After being interviewed by Major Gerald Templer of military intelligence in War

Office room 365, they would be sent in plain clothes to Cambridge for a course (later informally known as 'The Gauleiters') whose lectures covered guerrilla warfare, resistance, sabotage, subversion and clandestine wireless communication.

Sinclair had been 'C' since 1923, but he did not just control the spies who gathered human intelligence (HUMINT). He was also the director of the Government Code and Cipher School (GC&CS), which dealt in the signals intelligence (SIGINT) that became so vital to victory in WW2. Sinclair himself had formed the School in 1919, when he was still the director of Naval Intelligence, by combining the remnants of Room 40 at the Admiralty with the War Office's equivalent. The school – its name was a camouflage; it taught nothing – was under the nominal control of the Foreign Office, and stingily paid for out of the Foreign Office vote. GC&CS continued wartime signals intelligence in peacetime by intercepting and interpreting the cable and wireless communications of both hostile and friendly powers. Much material was received under the 1920 Official Secrets Act that Sinclair had helped to frame, whereby all cable companies operating in Britain were legally obliged to hand over copies of all telegrams sent or received within ten days. There was also a chain of radio intercept or 'Y' stations in key points at home and abroad. During the 1920s, when Bolshevik subversion was seen as the great threat, GC&CS, under its deputy director Alastair Denniston, had decrypted all the codes and ciphers of Soviet Russia. Unfortunately, British politicians in Stanley Baldwin's government in 1927 could not resist publicly boasting about this covertly acquired information. The Soviets then began sending their diplomatic and commercial wireless traffic using 'one time pads', which were impenetrable. This disastrous loss of intelligence taught a hard lesson about secrecy which was not forgotten in WW2: never let your enemy know what you know, or how you know it.

Denniston was a little Scot who had husbanded the section through lean times. There were only sixty-six staff in 1919; in 1935, there were still no more than 104 on the payroll. Two years later, Admiral Sinclair told Denniston to start earmarking and recruiting more 'men of the professor type' from universities who would be ready to start cryptanalysis when war broke out. In that eventuality, the GC&CS would move fifty miles from Broadway Buildings in London to a large ugly mansion that SIS had purchased in 1938 in Buckinghamshire, a house called Bletchley Park. Among the 12,000 men and women who eventually worked at what was known as 'Station X' were eccentric geniuses, including an untidy young Fellow of King's College, Cambridge, called Alan M. Turing, often described as the founder of modern computer science. His work led to the building of the world's first programmable digital computer, Colossus, which helped break German top-secret codes and produced the top grade of special intelligence known as ultra.

When the German Navy appeared in the Mediterranean in 1936, giving military support to General Franco's insurgent forces during the Spanish Civil War, they were signalling in cryptograms that the British could not read, communicating through Enigma electro-mechanical enciphering machines that looked like a typewriter in a wooden box, complicated by variable plugs and rotor wheels. In 1937, GC&CS were analysing the traffic, without understanding the messages, of the German air, army and police forces who were also using Enigma ciphers. As war approached in 1939, the Polish cipher bureau contacted its British and French counterparts and asked them to visit. On 24 July 1939, a month before Clare Hollingworth made the same journey, Alastair Denniston, with two GC&CS colleagues, Alfred Knox and Humphrey Sandwith, flew from Hendon to Warsaw. The next day they were driven to Pyry, a village in woods south of Warsaw, and shown into a room with lumpy objects on a table covered by a sheet. Their Polish hosts then uncovered their *pièces de résistance*: three Enigma machines that they had built themselves, with the rotors correctly wired in the German way. The French and the British were given one each for further technical research, to be shipped via Paris.

The British Enigma machine reached Victoria Station on the 'Golden Arrow' boat-train late on 16 August 1939, in a large diplomatic bag, camouflaged by the mountainous luggage of the singer Sacha Guitry and his wife. It was met on the platform by Sinclair's deputy, Colonel Stewart Menzies, wearing black tie for an evening engagement, with the *Légion d'honneur* rosette in his buttonhole. Three months later, a fortnight after Sinclair's death, Menzies was anointed 'C', given the secret ivory emblem of his office by King George VI, and became the keeper of the golden eggs of ULTRA.

SIGINT saved SIS from the disasters of their HUMINT. On 9 November 1939, two SIS agents were captured at Venlo on the German–Dutch border. Major Richard Stevens, a passport control officer, and Captain Sigismund Payne Best, a monocled cove working for the 'Z' network, had believed they were going to recruit an anti-Nazi dissident, a 'good German' high up in the Luftwaffe who was leading the resistance to Hitler. In fact the whole thing was a deception operation by German counter-espionage, *Reichssicherheitshauptamt IVE*, an elaborate sting to roll up all the British intelligence networks in Holland. Among the thugs who captured them was the same Alfred Naujocks who had led the fake raid on Gleiwitz radio station by the Polish border at the very start of the war. For this new trick, Hitler personally gave Naujocks the Iron Cross. Under interrogation, Stevens and Best told everything. Much of their information about the British secret services went into a Gestapo publication called *Informations-heft GroßBrittanien*, prepared by SS Major General Walter Schellenberg as the invasion handbook for the Nazi forces who would assault, occupy and purge the United Kingdom of Great Britain after Göring's air force dominated the skies.

In the summer of 1939, Arthur Watts, the president of the Radio Society of Great Britain, was approached by the War Office to find amateur radio enthusiasts prepared to listen for clandestine enemy morse code signals on short wavebands. Thousands of 'radio hams' became part-time V. I.s or Voluntary Interceptors, organised by groups and sectors into the Radio Security Service (RSS), one of the nine wartime secret services. They found no spies transmitting in England, but they did pick up weak Morse signals from Europe that came in coded groups. They turned out to be wireless messages directed to German secret service agents abroad. In its new headquarters at Arkley View, North Barnet, the RSS – which included Oxford dons like Hugh Trevor-Roper, Stuart Hampshire and Gilbert Ryle – collated and classified the intercept logs, before passing the messages on to Bletchley Park for deciphering. This ability to read German intelligence messages was crucial to later deception operations and to the 'Double Cross' system.

The connection between double agents and radio began before the war, with a character code-named snow, who in real life was a Welsh electrical engineer called Arthur Owens. Owens first contacted the

Abwehr, the Wehrmacht intelligence service, in Germany in 1936, and they later gave him a wireless transmitter set. Owens thus became one of only six known German spies in Britain. But he also reported what he had done to the British authorities. At the outbreak of war Arthur Owens ended up in Wandsworth Prison in the custody of the Security Service, the British secret service which does counter-espionage and is also known as MI5. Owens pleaded with Major T. A. 'Tar' Robertson of 'B' Division, MI5, to be allowed to radio supervised messages back to Hamburg. This began in September 1939, but snow at first sent mainly weather reports.

Later, snow was allowed to travel to Belgium with his new radical Welsh Nationalist colleague 'G. W.' (a retired police inspector called Gwilym Williams) to meet his Abwehr controller, Major Nikolaus Ritter. From this expedition, MI5 learned the whereabouts of three more spies, what military and naval matters the Germans were interested in, their sabotage techniques and plans, and their secret communications via postage stamps and microdots. Most important of all, the RSS's monitoring of snow's German wireless control station led in April 1940 to the first breaking of the Abwehr codes. The decrypts of Abwehr signals that circulated among the secret services became known as Illicit (or Intelligence) Series Oliver Strachey (ISOS), after the section head at Bletchley Park, brother of Lytton Strachey, who had been breaking codes, including Japanese ones, since WW1.

Reading German intelligence signals enabled MI5 to arrest twenty of the twenty-one German spies who arrived in Britain by parachute, small boat, seaplane and U-boat between September and November 1940. (The twenty-first ran out of food and money and shot himself in despair in an air-raid shelter in Cambridge.) Twelve of the spies were interned for the duration of the war, five were executed and three became 'double agents' for Tar Robertson and his section, B1A. This meant they would continue to work for, and be paid by, the Germans, but all their reports and movements were carefully controlled by the British. Some volunteered to do this, others 'were volunteered' by the prospect of a grim alternative. Thanks to the breakthroughs from snow, the British Security Service controlled all German spies in Britain during WW2, knowing exactly who they all were and when and where new ones arrived. Moreover, by supplying all the information the spies communicated to Germany, Britain gained many opportunities for deceiving the enemy.*

After a row between MI6 and MI5 about who had jurisdiction and control over these channels, a coordinating committee evolved to make sure that credibility was maintained and wires did not get crossed. The initial W or Wireless Board, chaired by Admiral John Godfrey, brought together the heads of all the intelligence services. Later this task was devolved to a lower level organisation. Named officially the Twenty Committee and known colloquially as the Twenty Club, it was really the 'Double Cross Committee', since the Roman numeral for 20 is XX. The Double Cross Committee first met at Wormwood Scrubs prison on 2 January 1941 and then every Wednesday for the next four years and four months that the war in Europe lasted.

Naval Intelligence's usual representative on the XX Committee was a Jewish barrister, Lieutenant Commander Ewen Montagu, RNVR, the son-in-law of Solomon J. Solomon. The day before the first XX meeting, Montagu had been seen, in uniform, drinking and chatting about yachts with a smooth-looking man in the American bar of London's Savoy hotel. This was a Yugoslav called Dusko Popov who, after twelve days in England being questioned and checked out by MI5 and MI6, was on the eve of flying back on the first of many regular trips to Lisbon to meet his Abwehr controller, who went under the name Ludovico von Karsthoff. In the Savoy bar, Montagu gave Popov a friendly letter that he could show as proof of their acquaintance. Every detail of what they talked about would be passed on, because Dusko Popov was starting his career as one of the most successful double agents of the war, code-named TRICYCLE by the British, and IVAN by the Germans. He fed false information to the Germans and brought back true intelligence on German rocketry and strategy.

---

* This holds equally true for the Germans too. From the spring of 1942 to the autumn of 1943 Oberleutnant H. J. Giskes of the local German Abwehrstelle controlled all the SOE agents dropped into occupied Holland, as he let London know, quite openly, on April Fool's Day 1944. The whole excruciating story of how the British were hoodwinked is told in *Between Silk and Cyanide: A Codemaker's Story 1941–1945* by Leo Marks, and in *SOE in the Low Countries* by M. R. D. Foot.

Both German and British scientists had been working with invisible electromagnetic radio waves. British boffins, led by Robert Watson Watt, had been developing 'radiolocation' since 1935, bouncing radio pulses back and using the blips to locate the range, height and bearing of approaching enemy aircraft. When the Home Chain of Radio Direction Finding aerials was linked to the RAF Ground Control Interception system, it helped win the Battle of Britain in 1940. Radar (RAdio Direction And Ranging) equipment fitted in RAF aircraft helped defeat the night blitz in 1941, and when installed in Royal Navy ships detected faraway enemy vessels and aided the laying of accurate gunfire at distance. Short-wave radar in ships and planes proved devastating against U-boats in the Battle of the Atlantic in 1943.

German research centred on transmitting beams that aided bomber navigation, and using intersecting beams to indicate a bombing target. From scraps of intelligence – captured German pilots talking, a photographed aerial array, equipment on a crashed German aeroplane – the young scientist R. V. Jones worked out what the Germans were doing, told Churchill's Cabinet about it, and sent up a plane over Derby (where the vital Rolls-Royce aero-engine works were a key Lufwaffe target) to discover that the narrow directional beam was where two thicker radio beams, one of dots, the other of dashes, overlapped. Radio countermeasures could then be set in force, the *Knickebein* of beam bombing bent, the *X-Gerät* of precision bombing annulled. The British radio guidance system, known as 'Gee', was developed in 1942 to guide fleets of British bombers to the Ruhr. After commandos were sent to capture parts of a German Würzburg radar at Bruneval in France in 1942, R. V. Jones and Joan Curran devised a way of 'spoofing' German radar with a smokescreen of deceptive reflective material dropped from aeroplanes. These strips of paper and aluminium foil, now called 'chaff', were code-named 'WINDOW' in WW2.

# 15

# Hiding the Silver

The population of Great Britain had been alarmed, among other things, by the sci-fi film written by H. G. Wells, *Things to Come*, in which fleets of bombers devastated an English Everytown. For them, rearmament meant building more fighter aircraft and anti-aircraft guns, and civil defence meant air-raid precautions, concealment and camouflage. Winston Churchill, however, thought that public anxiety about aerial bombing was roused both by pacifists eager to promote their anti-war cause, and by Air Ministry officials exaggerating out of self-importance.

On 30 March 1939, the same day the government pledged 8 per cent of its £25 million Civil Defence budget to 'obscuration of glare, and camouflage', *The Times* had run a long and authoritative article on the subject. 'Camouflage: Nature's Hints to Man' stated: 'In view of the revolutionary methods of modern aerial warfare, concealment has now assumed a new and most vital function.' The author of the piece, anonymously credited as 'a Scientific Correspondent', was the zoologist Dr Hugh Cott of Cambridge University, then completing his masterly study *Adaptive Coloration in Animals*, which Methuen published in 1940. Cott gave a biologist's overview of counter-shading and 'dazzle', disruptive patterns contradicting structural features, and, pointing out some errors in modern military camouflage, made a plea for more science to be applied to the art, which was 'still in its infancy – a child suffering from arrested development'.

Cott's piece prompted a slew of letters to *The Times* throughout April 1939. Camouflage was no longer secret, and the sort of people who write to newspapers had lots of points to make about its usefulness or its earlier history. A. J. Insall wrote from the Imperial War Museum, mentioning 'armoured snipers' posts made to represent natural tree trunks' and other 'excellent examples' of camouflage from the sniping schools of WW1. But the debate re-aroused old rivalries. Norman Wilkinson wrote to defend the culture of dazzle painting, saying that its purpose was not 'diminishing visibility' as Cott alleged, and pointing out that 5,000 wartime vessels had been successfully painted. Professor Sir John Graham Kerr then wrote from the Athenaeum to point out that camouflage was biological, not cultural, and that he had been the first to lay out the distractive functions of 'dazzle' in a memorandum of September 1914. What the Government needed now, he said, was 'a special Department, presided over by someone possessing high scientific qualifications' (not a million miles from himself, perhaps), to guide the camouflage activities of civilians and military on land, at sea and in the air. This spurred Wilkinson on to state that ship disguise was *not* biological camouflage, and tartly to remind 'Mr Graham Kerr' that when the Royal Commission of Awards to Inventors had thrashed out the history of 'dazzle', Wilkinson was the only one to receive an award. (This was a twenty-year-old battle: Wilkinson and Kerr's first print spat had been in *Nature* in 1919.) In May 1939, Graham Kerr also popped up in Parliament to ask the Secretary of State for Air whether he was using the biological principles of camouflage to diminish the conspicuousness of aircraft hangars, and once again in *The Times* where he reminded readers that a recent picture of a panda's 'patchwork of violent contrasts' constituted the kind of 'dazzle' that was potentially useful in war camouflage.

The British military was already thinking about this. In December 1937, a Camouflage Research Establishment (CRE) had been set up at the Royal Aircraft Establishment at Farnborough. Its director was Lieutenant Colonel Francis Wyatt, formerly of the Royal Engineers, the man who had taken command of Wimereux 'Special Works Park' in 1916. The man who appointed Wyatt, Reginald Stradling, had also been with the RE in WW1, when he won the MC and two mentions in dispatches. Stradling was a civil engineer with a doctorate in building materials, and an expert on steel structures. He became head of ARP research inside Sir John Anderson's Department of Civilian Defence, looking particularly at how to protect from bomb damage. From April 1939, when the Ministry of Supply that Churchill had been pleading for was finally established, the CRE helped with technical advice on camouflaging 'Military Establishments, Fixed Permanent Defences,

Royal Ordnance Factories (excluding agency factories run by civil firms) [and] Ministry of Supply Establishments'. The Royal Engineer Board's 'E' Committee was also concerned with camouflage, and developed steel wool (now used for scouring) as a useful material to attach to rolls of wire netting. Steel wool could be painted, and did not rot or burn.

The Air Ministry in London had also been concerned about the possible bombing of its own aircraft factories and aerodromes as well as other prominent civil structures. At the end of 1938, it opened a camouflage design section at the ministry offices, Adastral House, on the north-east corner of the Aldwych where it meets Kingsway, opposite the large American-style office block, Bush House. Captain Lancelot Glasson, MC, who had lost a leg as one of Wyatt's *camoufleurs* in WW1, was in charge there. Himself a painter, Glasson started gathering visual artists who knew something about flying. Among them was Captain Gilbert Solomon, a former pilot in the RFC who happened to be the nephew of the first British *camoufleur* Solomon J. Solomon (who had died in 1928). Another was Richard Carline who, with his brother Sydney, had also flown in the RFC and made sketches from the air over the Middle East which they later turned into superb oil paintings for the Imperial War Museum. (Their sister Hilda married the artist Stanley Spencer.) In May 1939, Tom Monnington, a future President of the Royal Academy, joined, and Leon Underwood, first inducted by Solomon J. Solomon at Wimereux, came forward with a sculptor's three-dimensional ideas. Lieutenant Colonel C. H. R. Chesney, DSO, also formerly of Wimereux, started writing *The Art of Camouflage* (Robert Hale, 1941) partly to continue his old disagreements with Solomon about the effectiveness of paint in camouflage, and to stress the importance of deception. Remnants of the old team were re-assembled to face the new threat.

Adastral House had a large studio and a viewing room, rather like Wilkinson's former studios at the Royal Academy, where models of hundreds of 'key points' in Britain's industrial infrastructure – factories, power stations, gasworks, oil tanks, water reservoirs, docks, railways, etc. – could be painted and looked at from various angles. Special matte paints in fourteen prescribed tones were manufactured. In July 1939, ARP Handbook No. 11 recommended two painting techniques that could be combined for large buildings: distortion of form, and imitation of surroundings. Thus, aided by a 50 per cent government grant, a factory roof could have its symmetries disrupted and imitations of neighbouring streets, houses and back gardens painted upon it. Clause 36 of the 1939 Civil Defence Bill gave the Government the power to require factories and public utilities to be camouflaged, so a new industry sprang up to satisfy the need. The Ministry of Supply also started encouraging 'groups of fishermen and their families' to braid fishing nets for camouflage purposes: 'Nets are needed for obscuring guns, ammunition wagons, tanks, buildings, stores, and many other things which it is desirable to conceal from enemy aircraft.' A Camouflage Advisory Panel including the distinguished artist Paul Nash and the biologist Hugh Cott was set up on 2 August 1939 to identify key areas for emergency camouflage. (Cott's piece in *The Times* had been noticed.)

The paint manufacturers were, for their own commercial reasons, obviously keen to encourage everyone to paint everything everywhere, and non-governmental organisations and individuals also stepped forward with ideas for 'misleading the bomber'. Among the consultancies was the Industrial Camouflage Research Unit, who worked from the architect Ernö Goldfinger's offices in Bedford Square. Despite the grand name, this was actually just some young painters scuffling about trying to do something after the outbreak of war. Among them was Julian Trevelyan, a British Surrealist and Mass Observer who had taken hallucinogens three times under medical supervision to get a revelation of universal beauty: 'I had, under Mescalin, fallen in love with a sausage roll and with a piece of crumpled newspaper from out of the pig-bucket.' He had been at the opening of the Spanish Republic's pavilion in Paris in June 1937 when Picasso's *Guernica* was first shown, and thought it the shining peak of the Spaniard's genius. He had marched with the Surrealists on May Day 1938 in a top hat and a Neville Chamberlain mask, wearing the sign 'Chamberlain Must Go'. Trevelyan was on the last peacetime ferry out of St Malo to Southampton, together with the beautiful photographer Lee Miller and her future husband, the Surrealist, artist, writer and collector, Roland Penrose, who had just been staying with Max Ernst in Avignon and Pablo Picasso in Antibes. Their train arrived at Waterloo Station on Sunday morning, 3 September, to the wail of the first air-raid sirens and the sight of ARP wardens in white tin hats carrying wooden football rattles to sound in case of gas. Watching the silver barrage balloons rising into the blue sky, Trevelyan wondered if Surrealism could outdo the oddity of war. In fact, the arcane literary and cultural movement barely survived the outbreak of hostilities.

Trevelyan and Penrose joined the printmaker Bill Hayter and the engraver Buckland Wright in offering their services as *camoufleurs*. 'In those early days it was easy to sell any kind of camouflage,' wrote Trevelyan. Penrose confirmed: 'It was thought by many people that camouflage was simply a question of painting stripes over an object.' A rash of squiggly green patterns was breaking out across the country, along with a fanciful realism – palm trees on gasworks – of laughable conspicuousness. Trevelyan saw it as a kind of magical ritual, doing your bit to ward off harm. The apotropaic rites included criss-crossing paper strips across windows to stop glass blasting into splinters, and daubing dung-coloured wiggles and splotches on the house roof or the garden shed or over the car in order to aroint aircraft. Trevelyan admitted that few members of the Industrial Camouflage Research Unit had flown much and they knew little more about the planes they were trying to deceive than most amateurs.

There was no shortage of artists applying to do civil and industrial camouflage. By 5 October 1939 *The Times* was announcing 'No More Camouflage Workers Needed'. Sufficient candidates 'have been examined by a selection committee, with a view to the compilation of a section of the central register which is now being prepared for the national service department of the Ministry of Labour'. In the House of Commons on 25 October, A. P. Herbert asked the Prime Minister to consider setting up a Department of the Arts, to maintain artistic effort and education, and to use artists' powers fully and effectively 'for the purposes of war'. Neville Chamberlain did not think that was necessary, but he was pleased to note that a Central Institute of Art and Design had recently been formed to achieve such ends.

The Central Institute of Art and Design was set up by a panel that included Kenneth Clark, who had been director of the National Gallery and surveyor of the King's pictures since 1934, and Jack Beddington, the Jewish publicity manager of Shell-Mex and BP Ltd, an industrial patron who used artists of the calibre of Edward Bawden, Paul Nash, Ben Nicholson, John Piper, Graham Sutherland and Rex

Whistler in his poster and press campaigns to market oil and petrol, and gave the director Paul Rotha his first leg-up as a documentary film-maker. Jack Beddington had positioned Shell cleverly in the 1930s, choosing not to deface the countryside with fixed advertisements but to decorate Shell's mobile lorries and tankers with colourful slogans instead, and had been canny enough to take up an idea from John Betjeman and start the popular Shell County Guides in 1934. Shell-Mex and BP thus in effect camouflaged themselves with 'green' credentials, good practice for Beddington's later PR work in the national interest.

The declaration of war meant a cabinet reshuffle, and Sir John Anderson became Minister for Home Security, responsible for civil camouflage. In March 1940, Home Security reorganised it all under a Directorate of Camouflage, headquartered in Leamington Spa, under Wing Commander T. R. Cave-Brown-Cave. In May 1940 the War Office set up a Camouflage Development and Training Centre that eventually ended up at Farnham Castle, near Aldershot, not far from Wyatt's group of *camoufleurs* at Farnborough. The man put in charge of the War Office Camouflage Centre was Colonel Frederick Beddington, Jack's brother, who had been a sniper in the Great War before training at the Slade. From Beaumetz in France, he had been running the British Expeditionary Force's camouflage, which featured in the first dispatch by the first BBC radio war correspondent, Richard Dimbleby, on 11 October 1939. Everything worked through personal contacts in those days. Frederick Beddington's chief instructor was Colonel Richard McLean Buckley who had also worked with Solomon J. Solomon in the Great War. Another prospective *camoufleur*, Geoffrey Barkas, had been in the British film industry for years making features (*Palaver, Q-Ships, Tell England*) and documentaries (*Tall Timber Tales, Wings over Everest*) in Africa, India and Canada, before getting a job with Jack Beddington touring a Shell promotional show called 'How Your Motor Car Works'. When Barkas telephoned Beddington to try and get into filming the war, Jack recommended him to his brother Freddie. Thus Geoffrey Barkas and Julian Trevelyan were picked to become Camouflage Officers in the Royal Engineers.

It was all something of a boys' club and much of the early work was amateurish. The artist who best catches some of the absurdity of British camouflage at this time was the graphic genius William Heath

Robinson, whose son Oliver Heath Robinson taught camouflage at Farnham. The outbreak of WW2 had brought 67-year-old Heath Robinson back to *The Sketch*, for whom he had done drawings in WW1, and his subject matter now became the English defence of the Home Front through ridiculously serious camouflage and seriously ridiculous deception.

# 16

# A Great Blow Between the Eyes

'The beginnings of any war by the British', wrote General Archibald Wavell, 'are always marked by improvidence, improvisations, and too often, alas, impossibilities being asked of the troops.' The Phoney War turned real and bloody in Norway. While no one could doubt British courage and phlegm in what Churchill called 'the first main clinch of the war', the Germans managed to deceive their intelligence, wrong-foot their diplomacy and outmanoeuvre and outgun their armed services. The Norwegian campaign would end in retreat and evacuation.

All Scandinavia was neutral at the start of WW2. In January 1940, Britain began secret planning to violate that neutrality by stopping German ships, mining the waters and seizing the port of Narvik in Norway to prevent the winter export of Swedish iron ore for Germany's military industries. Britain also planned to send an Anglo-French expeditionary force to help the Finns in their fight against the Russians. But on 9 April 1940, Germany jumped the gun by suddenly invading Denmark and Norway. The Germans seized Narvik by a Trojan-horse deception: German iron-ore ships apparently waiting peacefully in the harbour suddenly disgorged hidden troops. (The British later did the same at Namsos in Norway, disembarking troops at night, hiding all traces from reconnaissance aircraft by day.) In the silky words of Joseph Goebbels, both Denmark and Norway were 'taken under the protection of the Reich to forestall Allied occupation'. The German Foreign Minister von Ribbentrop added, 'Germany has preserved Scandinavia from destruction and will be responsible for true neutrality in the North until the end of the war.'

In mid-April 1940, after a successful, violent naval action off Narvik, Churchill was urging the British general on the ground, Pat Mackesy, to attack the Germans in Narvik directly. Given his available forces and the local conditions, Mackesy thought he could not mount a direct assault from the sea but would have to take the surrounding fjords first. Under Churchill's goading he finally snapped and wrote an irate message that included the line 'Are snows of Narvik to run red as sands of Gallipoli?' This would have got him sacked had it been sent. But an alert young staff officer read it and held up its transmission by pretending that wireless communications were 'out' to Catterick, until tempers had cooled and the message could be moderated. Churchill did sack Mackesy the next month, but at least it was not for this cable. That junior staff officer was Captain J. T. Rankin of the Hallamshire battalion of the York and Lancaster regiment, my father.

In Norway, the Germans were deploying a new kind of mobile infantry – *Fallschirmtruppen* – dropped by parachute, and the Luftwaffe commanded the air. They used tactics developed in Spain, close air support of attacking infantry, and incendiary bombs, to devastating effect. Yet the two-month Norway campaign turned out to be a long way from the complete fiasco it seemed. The Germans had lost over 5,000 men and 242 aircraft, and the *Kriegsmarine* proportionately suffered more than the Royal Navy. Meanwhile, King Haakon VII and the Norwegian government, no longer neutral, moved to London, to be joined the Dutch royal family and government. Three million tons of Norwegian merchant shipping (the fourth largest fleet in the world) came over to the Allied side, as did remnants of the Norwegian navy and air forces. Because Major Vidkun Quisling of Norway was believed to have helped the Germans invade, his surname entered the English dictionary as a new and loathsome synonym for 'traitor'. In 1943, the British-trained Norwegian resistance succeeded in preventing Nazi scientists from getting deuterium oxide or 'heavy water' for atomic bomb research from the Norsk Hydroelectric Plant at Vemork in the Telemark region. And the mountainous, fjord-riven kingdom of Norway managed in time to tie the hands and feet of some twelve divisions of German soldiers, nearly half a million men.

Most important of all, the two-day debate in the House of Commons on the conduct of the war in Norway on 7 and 8 May 1940 sealed the fate of Neville Chamberlain as British Prime Minister. Leo Amery and Lloyd George openly urged Chamberlain to go, and in the division the PM's Conservative majority shrank from over 200 to 81.

Seventy-one-year-old Chamberlain realised that he himself was the main stumbling block to the establishment of a vigorous, all-party coalition, and he resigned two days later. On 10 May 1940, Winston Churchill, the 65-year-old who had started and impelled the whole Norway campaign but miraculously escaped its avalanche of consequences, now acquired 'the chief power in the State'. He would be Prime Minister for the next five years and three months of world war. Churchill the historian later smoothed his luck into Fate:

I cannot conceal from the reader of this truthful account that as I went to bed at about 3 a.m. I was conscious of a profound sense of relief. At last I had the authority to give directions over the whole scene. I felt as if I were walking with destiny, and that all my past life had been but a preparation for this hour and this trial. Ten years in the political wilderness had freed me from ordinary party antagonisms. My warnings over the last six years had been so numerous, so detailed, and were now so terribly vindicated, that no one could gainsay me. I could not be reproached either for making the war or with want of preparation for it. I thought I knew a good deal about it all, and I was sure I should not fail.

Finale of *The Gathering Storm*, vol. 1 of *The Second World War*.

It was a dramatic day in European history. At 3 a.m., Germany launched *Fall Gelb* (Operation YELLOW), its offensive in the West, a simultaneous land and air attack on the Netherlands, Belgium and Luxembourg. Neutrality and obsequious diplomacy were no defence against Nazi contempt for protocol; water and wire offered no protection against blitzkrieg. Neutral Switzerland mobilised fully, and Eire began to panic: 'The fact that we want to keep out of war', said Eamon de Valera to a Fianna Fail convention, 'will not, or may not, be sufficient to save us.'

The onslaught was underway; standing up to it a tremendous task. When Churchill told the House of Commons on 13 May, 'I have nothing to offer but blood, toil, tears and sweat,' he was almost quoting the opening chapter to volume 5 of *The World Crisis*, thinking back to what he had described as 'incomparably the greatest war in history', the Eastern Front of the First World War, which destroyed three empires and involved 'the toils, perils, sufferings and passions of millions of men. Their sweat, their tears, their blood bedewed the endless plain . . .'

The alarming idea of soldiers dropping from the skies jolted civilian Britain into action. The day after Anthony Eden, the Secretary of State for War, appealed for Local Defence Volunteers – what we think of now as 'Dad's Army' – to guard against enemy parachute landings, a quarter of a million men joined. Eden was still speaking on the BBC after the nine o'clock news on 14 May 1940 ('You will not be paid, but you will receive uniforms and will be armed . . .') when listeners started phoning their local police stations in order to sign up in 'the parashooters'.

As the United Kingdom pulled itself together in 1940, the working classes were quicker than some of the upper classes – many of whom loathed and distrusted Churchill – at putting their shoulders to the wheel. On 13 May, the Labour Party conference had backed the new national government, which had Labour ministers in the War Cabinet. The cartoon by the New Zealander David Low in the *Evening Standard* the next day, 'All behind you Winston', showed an army of politicians and people striding in step with Churchill, all rolling up their sleeves. Suddenly, the country was organising for Socialism; the mood was swinging leftwards.

The Minister of Labour and National Service was the formidable Transport and General Workers Union leader and ex-docker Ernest Bevin, who soon had the Trades Union Congress and the British Employer's Confederation working together. When Herbert Morrison, the Minister of Supply, asked all contractors to 'work at war speed', in shifts covering twenty-four hours a day, seven days a week, providing 'more shells, more tanks, more guns', the TUC sent a message: 'Men of the fighting forces, we salute your courage and determination. We are unitedly resolved that all our resources shall be used to the full to provide the arms and munitions you need.' 'Go to it!' exhorted the Ministry posters.

In his 'Grand Coalition' cabinet, Churchill made himself Minister of Defence. The Air Ministry was separated from a newly created Ministry of Aircraft Production, which was put under the newspaper magnate Lord Beaverbrook. The appointment was contentious. King George VI sent a worried message to Churchill, because on his 1939 Canadian tour he had heard from John Buchan, the governor general, how many Canadians distrusted Beaverbrook. But malevolent or not, Max was energetic. Within a week the Nuffield Aircraft Factory and

Vickers Supermarine were under the control of one management and working round the clock. And as a newspaperman, Beaverbrook understood the power of symbolic gestures. When he appealed to the housewives of Britain in July to donate their aluminium cookware to help build aircraft frames, the total amount collected might have only equalled one day's supply, but every woman who gave a pot or a pan could imagine a bit of her kitchen in that Hurricane or Spitfire zipping through the burning blue.

The firebombing of Rotterdam from the air frightened other Dutch cities and Holland capitulated. Curiously, a Luftwaffe officer charged with improving the camouflage of his airfield had accidentally exposed the earlier German secret plans to invade Holland and Belgium. When Erich Hoenmanns, lost in fog on his way to Köln, crash-landed on 10 January 1940 near Mechelen-sur-Meuse in Belgium, he was carrying a passenger, Major Helmuth Reinberger, with all the invasion plans in a dispatch case that he failed to destroy. The Allies thought it was only a 'haversack ruse'; Guy Liddell of MI5 reckoned it 'part of the scheme for the war of nerves', and John Colville, a secretary at No. 10 Downing Street, recorded in his diary on 15 January that 'the landing itself, the ostentatiously ineffective efforts of the pilot to burn the papers, and subsequently to commit suicide, are suspiciously like a "put-up job".' In fact, they were all wrong, for the case held the real plans at that time.

When German Army Group B (two armies of five infantry corps) invaded Holland and Belgium on 10 May, it seemed like the Schlieffen Plan of 1914 again, and the British and French armies moved forward seventy-five miles to block the right hand. This was exactly what the Germans wanted them to do – charge the matador's cloak – because their main weapon was in the other hand. The French army had believed the fortified Maginot line stretching from Switzerland to Longuyon, 'a battleship built on land', was impregnable; the forested terrain of the Ardennes was surely impassable to tanks. Yet it was through here that von Rundstedt's German Army Group A – four armies, including four armoured corps in twelve divisions – delivered a body blow to France, the *Sichelschnitt* or sweep of the scythe round behind them.

France was unready. Their ageing, corpulent generals had become complacently defence-minded; the dash of Maréchal Ferdinand Foch had gone. General Gamelin's nickname was 'Gagamelin', and morale was low among the sullen, bored troops. Goebbels's people, meanwhile, dropped leaflets on the French lines suggesting that Britain was ready to fight to the last Frenchman, and that *les Tommies*, all looters and lechers, were already busy in their homes with their wives and sweethearts. From the radio, the traitor Ferdonnet oozed contempt for the French ruling class. France's confident front was an elaborate *pâtisserie*, concealing a stew of *je m'en foutisme*, defeatism, confused communism, and demoralisation. Churchill felt the French were 'rotted from within before they were smitten from without'.

The *Schwerpunkt* or concentration point of the German attack was at Sedan, 120 miles north-east of Paris. An RAF reconnaissance Spitfire saw the columns of enemy vehicles stretching back for miles. The German armoured divisions punched through, crossed the Meuse and the Oise, and drove a 'Panzer corridor' towards the French coast, along the Somme, through the British rear. Marc Bloch, the medieval historian who later died a hero of the French resistance, tortured and shot by the Gestapo in June 1944, believed that the 'strange defeat' of France in 1940 was in part intellectual. The French high command simply could not conceive of a new kind of war, waged by Germans with 'methodical opportunism' and an utterly different rhythm and faster tempo. 'It was much more terrifying to find ourselves suddenly at grips with a section of tanks in open country. The Germans took no account of roads. They were everywhere.' Bloch noted how the Germans 'relied on action and on improvisation. We, on the other hand, believed in doing nothing and in behaving as we had always behaved.'

Apart from aircraft, the Germans did not have superiority in numbers of men or vehicles, but violence and speed in their use won the day. Because the shocked and dazed French did not counter-attack, the BEF became cut off from its reserves, and had to bend its line backwards as it was squeezed and pushed towards the coast of northern France. Fighting a rearguard action towards the Channel ports, making use of successive rivers and canals, now became the BEF's only option. The stubborn retreat towards the 'defensive perimeter' of Dunkirk began.

Meanwhile, Dudley Clarke was having a busy war. He had already accomplished a four-month-long trip to Africa, reconnoitring the entire 3,000-mile overland route from Kenya to Egypt as the supply line to reinforce the Middle East should Italy enter the war and choke off sea lanes. He was just back from two missions to wintry Norway in three weeks, when he was suddenly given a new job at the War Office in May 1940. Three days before, German Panzers had reached Abbeville at the mouth of the Somme, and the tip of the *Sichelschnitt* scythe was now moving up the coast towards Boulogne and Calais, preparing to cut the BEF to pieces. Boulogne was hastily defended by the 20th Guards Brigade and then mostly evacuated by the Royal Navy.

Calais was a different story. On 22 and 23 May a small British force disembarked there, ill-equipped and with conflicting orders from different sources. It soon became clear what the 30th Infantry Brigade had to do: sacrifice themselves to save others. Dudley Clarke passed on the dismal signal to Brigadier Claude Nicholson and his beleaguered garrison of 3,000 English riflemen and 800 French soldiers. He dictated the message on the red phone to the Duty Captain at the Admiralty; it was flashed from a mast on Horse Guards Parade to a destroyer in the English Channel; and from there to Nicholson's signallers at Calais. It told them that they were to hold the citadel of Calais 'for the sake of Allied solidarity' and to help the rest of the BEF get safely to Dunkirk, where they were going to be evacuated.

Nicholson's men, already short of food, water and ammunition, denied the chance of evacuation themselves yet refusing to surrender, courageously fought off aeroplanes, tanks, artillery and infantry for three more days. Only 440 men from the garrison managed to escape death or capture, the last four dozen getting out on the British minesweeper *Gulzar*. Second Lieutenant Airey Neave was already wounded by the end:

The last stand was made among the wagons-lits and in the sand-dunes. A man shot himself with his own rifle in an archway which housed the regimental aid post; beside me a young soldier was crying quietly. A field-grey figure appeared shouting and waving a revolver. Then a huge man in German uniform and a Red Cross armband put me gently on a stretcher. I was a prisoner of war.

Next, Dudley Clarke undertook a hush-hush mission to Eire, which was determinedly neutral but anxious not to be annexed and 'protected' by the Nazis. (This mission, from 24–27 May 1940, is so elliptically described in Clarke's censored book that his biographer David Mure thought the country might have been Spain.) In his first message to US President Roosevelt on 15 May, Churchill had written that 'we have many reports of possible German parachute or air-borne descents in Ireland'. Dudley Clarke flew out from Hendon aerodrome in plain clothes, accompanying two Irishmen, a civil servant, Joseph Walshe, and a military intelligence officer, Colonel Liam Archer, in the twin-engined Flamingo (which had an armchair with a huge ashtray) often used by Churchill. They landed at Belfast in Northern Ireland, and travelled, separately, by train, the next day to Dublin. At the Shelbourne Hotel a mysterious man went through all of Clarke's kit removing anything that might incriminate him if anyone furtively searched his luggage. Then Clarke was driven to an evening meeting in an underground conference room with men who did not shake hands or introduce themselves, but who included the Irish politician, Frank Aiken, and the Irish chief of staff, General McKenna. Clarke informed them that General Huddleston, General Officer Commanding, Northern Ireland, had a mobile column standing by, ready to travel south into the Republic to help the Irish if the Germans invaded. There were more conferences the following day, with visits to the docks, Phoenix Park and Baldonnel aerodrome, where ways of stopping paratroopers and gliders were discussed, as well as the British anxieties that the IRA (which had been conducting a bombing campaign in England since 1939) could become a German Fifth Column.

The Flamingo quietly returned Clarke from Belfast to London on Monday, 27 May. Behind him, the Irish army was being reinforced and placed on a war footing, the Army Reserve and the Volunteers called up, and a National Defence Council established. Within days, in an imaginative, bold gesture, Churchill offered de Valera a United Ireland as the prize of joining the war on the Allied side. But Eamon de Valera wanted neutrality more than unity; Northern Ireland could see nothing beyond itself, and the historic opportunity passed.

On the steps of the War Office, back in uniform, Clarke bumped into General Haining, who told him that another of Clarke's former bosses, Sir John Dill, had just become Chief of the Imperial General

Staff (CIGS), and Clarke was now in the 'Head Office' team. Clarke had liked Dill at the Staff College and in Palestine, and knew him to be energetic and intellectual. Dill had achieved his lifetime's ambition in becoming Britain's top soldier, but he was doing so at a dark hour in the nation's military history, with the Royal Navy about to evacuate the British Army from France. The winning chalice felt poisoned. There are two views on Dill. Some say he was an exhausted has-been and that Churchill's nickname for him, 'Dilly-Dally', was appropriate. However, the historian Alex Danchev has argued persuasively that John Dill was a good man who paved the way for the next CIGS, Alan Brooke, to stand up to Churchill in his bullying mode, and that he firmly cemented Anglo-American relations when he went to Washington DC as head of the Joint Staff Mission and senior British member of the Combined Chiefs of Staff, where he died in post.

Clarke was appointed chief staff officer of Dill's personal staff, warding off unnecessary business from his harassed chief. That night a telephone line was established from the War Office to the BEF's new GHQ, which was situated on the beach of La Panne, just inside Belgium, east of Bray Dunes and Dunkirk, within the perimeter where the BEF was either going to resist, surrender or evacuate. This was where Lord Gort, the commander-in-chief of the BEF, had received the news on 27 May that King Leopold of Belgium had capitulated to the Germans without telling his government or his allies. The German air force was bombing the oil tanks and docks at Dunkirk and a pall of black smoke was visible for miles. That day over 7,500 men were evacuated from the harbour, and on the 28th over 25,000 from the beaches as well. Clarke spent much of 29 May on the telephone, because one of his duties was to listen to all conversations between Dill and Gort in order to record decisions and notify them to the operations staff and other departments. Clarke lived the war by proxy, imagining the perimeter shrinking in on the retreating army whose next job would likely be to defend England itself.

The epic of Dunkirk and 'Operation Dynamo', organised by Rear Admiral Bertram Ramsey, the miraculous evacuation of a British army by the Royal Navy between 27 May and 4 June 1940, while the RAF fought off enemy planes above, still makes Britons blink with pride.

'Knowing some of the difficulties,' wrote John Masefield in *The Nine Days Wonder*, 'I should say that the Operation was the greatest thing this nation has ever done.' But like Corunna and Gallipoli, it was also a debacle, defeat on an enormous scale. 'Wars are not won by evacuations,' Churchill warned the House of Commons on 4 June. 'What has happened in France and Belgium is a colossal military disaster.'

King George VI sent Gort a message: 'The hearts of every one of us at home are with you and your magnificent troops in this hour of peril.' Gort asked for more fighters to keep the German bombers and dive-bombers off the ports and beaches. Then German artillery began to shell Dunkirk. Soon there were reports of unruly French units retreating from Belgium, and the roads to the coast piling up with transport and marching men.

Among them, not ten miles away, was my grandfather, Major Geoffrey Page, DSO, who had been in the retreat from Mons and Le Cateau in 1914. He was now a brigade major in a car with two motorcycles, moving slowly along the south side of the Canal de Bergues, on a narrow road clogged with French horse-drawn carts and vehicles. Where needed, the officers ordered men to push cars and lorries into the canal. At the guarded bridge, the only wheeled vehicles being allowed over and inside the British perimeter were staff cars like theirs, and artillery guns. Dunkirk, six miles off, was marked by an enormous mushroom of smoke. 46th Divisional HQ ordered Page's brigade to hold the line of the Canal de Moeres from Teteghem to Dunkirk:

Our first job was to collect our brigade from the mass of British troops of all arms who, without transport or organization, were flocking down towards Dunkirk. The B.E.F. had started the war with a mania for incognito and in the crowds of men in battledress it was impossible to recognise our troops. I stood at the road fork near Teteghem shouting '138 Brigade, to the right'. I remember a crowd of about 200 men led by a Chaplain, who was shouting 'Which is the way to Dunkirk?'

On 30 May, Churchill wrote out an order to Gort, telling him to continue to defend the perimeter, the 'Corunna line'. He knew that Gort was likely to stay until the last man had left, so Churchill instructed him to hand over to a corps commander and return home 'when your effective force does not exceed the equivalent of three divisions . . . It would be a needless triumph to the enemy to capture you.'

When Gort rang back to protest and found that Dill had gone to a

1 Winston Churchill (with cigar, as Minister of Munitions) in the British Army School of Camouflage's training trenches in Kensington Gardens, 1917. He is inspecting a new armoured dome for a machine-gun post, which is concealed by painted cloth.

2 Dummy dead German from the Camouflage School. Some fakes were used as sniper hides; others, in captured uniforms, were placed in no-man's-land and their limbs moved by wire. Anyone deceived into rescuing such 'wounded' men would be ambushed.

3 Australian troops carrying a wood and canvas dummy British Mark 1 Tank in France in 1917. Decoy tanks in disruptive pattern camouflage were used to draw enemy shell-fire or to add a false impression of strength to a feint attack.

4  The legendary 'Lawrence of Arabia', Captain T. E. Lawrence, photographed above Feisal's camp at Wejh early in 1917, wearing Arab clothing and a western wrist-watch. The archaeologist and intelligence officer was a key liaison between British forces and the Arab revolt against the Ottoman Turks, pioneering new forms of guerrilla warfare.

5  Propaganda in war. The German execution by firing-squad of Nurse Edith Cavell in Belgium was used in this 1916 Canadian poster as an emotional spur to recruitment. She was tried for the crime of helping British, Belgian and French soldiers escape into neutral Holland. But was the Norfolk spinster quite what she seemed, and was her death actually an atrocity?

& 7 American 77th Division soldiers learning field camouflage at a British school for
couts, observers and snipers in France, May 1918. The instructor with the walking stick has
rought them right up to a concealed sharpshooter who reveals himself in the second picture.
Stacks of silhouette figures used in the fake or 'Chinese' attacks of trench warfare.
When erected by pulley in half-light they diverted enemy attention and helped to waste his
mmunition. The papier-mâché heads behind were painted and dressed realistically, then
aised above a trench parapet in order to lure enemy snipers into giving away their
ositions by shooting at them.

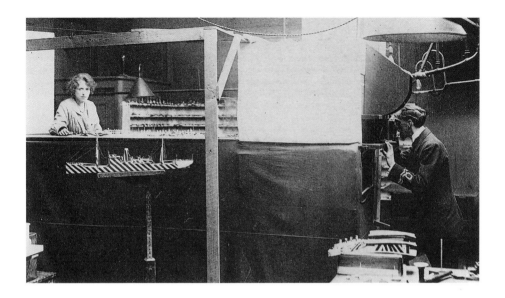

9 A special studio set up by marine artist Norman Wilkinson at the Royal Academy of Arts, Burlington House, London, during WW1. Models of 'dazzle-painted' ships were placed on a turntable and then viewed through a submarine periscope to test whether their visual effect was deceptive. Successful patterns were then transferred by 'artist-officers' to real ships in dockyards.

10 'The object of camouflage is to give the impression that your head is where your stern is.' SS *Olympic*, painted in 'razzle-dazzle', is seen from the port stern quarter. German U-boats were taking such a toll on shipping in 1917 that anything which might divert the aim of a submarine commander firing a torpedo had to be tried.

11 1940: Prime Minister Winston Churchill inspecting coastal defences against Nazi invasion near Hartlepool. The sand-bagged gun emplacement has been camouflaged as an innocent seaside roundabout.
12 WW2 British Local Defence Volunteer emerging from cover, about to throw a bomb. Amateur defenders might have been over-whelmed had the German army invaded in 1940/41.

13 1940: Adolf Hitler jigs for joy after signing the armistice at Compiègne. The total surrender of France was sweet revenge for the humiliating defeat of Germany in 1918.
14 Cartoon from *The Sketch*, 10 September 1941. The national treasure W. Heath Robinson (1872–1944) imagined wonderfully absurd camouflage and deception schemes in both world wars.

15 The 'Prop-Shop' at the Special Operations Executive's Station XV in the Thatched Barn roadhouse near Elstree film studios. SOE's camouflage section disguised secret agents as well as their special devices for sabotage and subversion in Axis-occupied territories.

BUILT TO DECEIVE:
16 Dummy British aircraft at El-Adem air-base near Tobruk, part of an 'A Force' deception scheme in the Mediterranean.
17 Dummy landing craft moored in North Africa, purportedly for the invasion of Greece in 1943.
18 Dummy tank to fool German and Italian observers in North Africa in 1943, made by Royal Armoured Corps engineers from painted canvas stretched over a metal framework rigged on a lorry chassis.

19 Lt Col David Stirling, founder of the SAS, with patrol commander Lt Edward McDonald (with Fairbairn Sykes Commando dagger) and Cpl Bill Kennedy. The famous Special Air Service regiment was at first a wholly imaginary unit, originating in a Dudley Clarke deception that there were British paratroops in North Africa in 1941, when there were actually none.

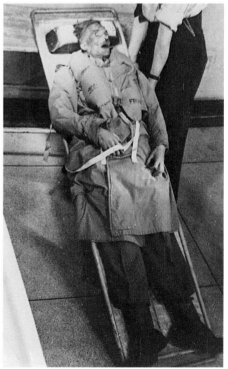

20  Lt Col Dudley Clarke, head of 'A Force' in Cairo and Britain's top deceiver, was arrested in drag in Madrid in October 1941, while disseminating false information to reach German agents in Spain.

21  Fluent German-speaking *Daily Express* journalist Sefton Delmer became the maestro of British 'black' propaganda broadcast to Nazi Germany in WW2.

22  The genuine corpse of 'The Man Who Never Was', before shipping by submarine to Spain in April 1943, with a false identity and a briefcase full of forged papers intended to deceive the Germans about the forthcoming invasion of Sicily.

23  A still from the film *I Was Monty's Double*, in which actor M. E. Clifton James impersonates Field Marshal Bernard Montgomery in Gibraltar, just as he did in late May 1944. The impersonation allowed the real 'Monty' to prepare for the 6 June invasion of Normandy.

24 Barcelona-born Juan Pujol García, the most successful double agent of WW2. Codenamed GARBO by the British (because he was such a good actor), he was decorated by Germany with the Iron Cross for leading the extensive ALARIC spy ring in Britain, which was in fact wholly imaginary and coordinated with Allied strategic deception.

25 Juan Pujol's crucial message as received by teleprinter at German HQ on 9 June 1944. As top German V-Man or Agent ARABAL he suggested that the D-Day landings were a diversion from the real attack, which convinced Hitler to hold back nineteen divisions at Calais, thus saving the Normandy invasion.

26 Prime Minister Winston Churchill sets foot in liberated France on 12 June 1944, six days after D-Day. The Normandy beachhead was not yet wholly secured and his wizards had to keep their spell working for many more weeks to achieve the triumphant apotheosis of British military deception in the twentieth century.

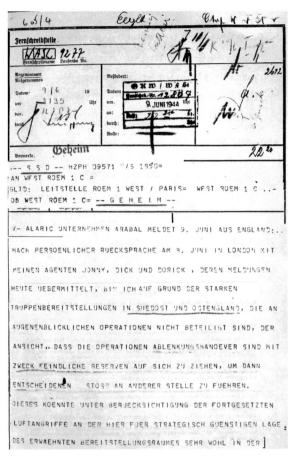

meeting with Churchill, he asked for his call to be transferred. Clarke walked over to the Admiralty to make the connection, and was shown into the War Room where the chiefs of staff were gathered with several cabinet ministers. The Prime Minister was striding up and down, wreathed in cigar smoke, and the atmosphere was tense. Clarke explained his mission and was waved to a red telephone on a desk in the window. Gort was put through to Dill. Every so often Dill turned to repeat the news over his shoulder: the figures of men clear from the beaches, now over 120,000, the number of those waiting, around 80,000, the weather, the perimeter. Then Dill and Gort discussed the rearguard: it was agreed that Major General Harold Alexander and 1st Corps would stay to the end. When the First Sea Lord, Admiral Dudley Pound, took over the telephone to talk naval matters to Gort, Dill stepped back, suddenly pale.

Mr Churchill noticed it instantly, and interrupted the pacing to step to his side. 'Something wrong?' he asked; and the answer came very quietly: 'My son is with the 1st Corps.' Nothing more was said. The Prime Minister took the CIGS by the arm in an impulsive gesture of sympathy that was better than any words, and in a moment he was himself again.

Eventually General Dill picked up the receiver again. Gort was protesting at his recall, but the Prime Minister was insistent. 'Tell him from me that he is to return according to instructions. That is an Order, and it is not to be questioned – an Order of His Majesty's Government!' Later, Churchill took the seat by the red telephone, and gave such a sympathetic message of encouragement to General Gort that Clarke wished it had been recorded for future inspiration.

The PM began speaking guardedly to Gort about future co-operation with the French. Next day he was in fact planning to fly to Paris to meet the French Prime Minister Paul Reynaud. Clarke knew the skies were not safe from the Luftwaffe, yet here was Winston Churchill saying, on a not wholly secure cross-Channel telephone line, 'I am going to fly with the CIGS to . . .' Dudley Clarke therefore put out his hand and switched off the call. This took considerable courage. The Prime Minister looked up with that bulldog glare that Karsh of Ottawa later caught in his famous photo-portrait by daring to snatch Churchill's cigar away. 'Please, Sir, it is not safe to speak of these plans on the telephone,' said Clarke. Churchill stared at him, then Dill intervened and Gort was soon back on the line. 'There's somebody here who tells me I mustn't say what I was going to say,' the Prime Minister grumbled, with a twinkle, and the conversation continued more discreetly.

And so they came, sailing across the English Channel to Dunkirk in France, the 'Mosquito Armada' of nearly 900 big and little ships – barges and beach-boats, coasters and corvettes, dinghies and destroyers; luggers, minesweepers, oyster-dredgers, skiffs, *schuits*, sloops, smacks; trawlers and tugs, whalers and wherries, yachts and yawls – to pluck the British soldiers and the French *poilus* off the broken mole and the bombed beaches and get them back safely to England.

At 6 p.m. on Friday, 31 May, General Gort was taken off the beach a mile west of La Panne and out to the minesweeper *Hebe*. During the heaviest Luftwaffe air raid of the day, the General sat calmly in a corner of the bridge in a borrowed tin hat looking through binoculars at the wrecked and burning beach. Men were shouting, bombs exploding in the water, the 4-inch high-altitude gun on the fo'c'sle just below the bridge was banging away noisily. 'Won't you go below and take cover, Sir?' 'No, thank you, I'm quite happy where I am.'

Major Geoffrey Page reached the sands of Malo-les-Bains on the Saturday night. Burning vehicles lit up a long column of troops, six or seven abreast, stretching out into the darkness towards the mole. In front was the sea, and out to sea faint lights. The 46th Division men decided to try their luck there and walked into the water, first to their waists, and then up to their chests. There was a swell and some of the men were beginning to get mouthfuls of brine when out of the dark appeared the first rowing boats. Aboard a paddle steamer, Major Page was plied with whisky and sandwiches, had his clothes dried in the engine room and was delivered to Harwich at noon on Sunday, 2 June, the twenty-first birthday of his only child, my mother. Five days later, my father, Captain Rankin, was evacuated from Harstad in Norway.

Dunkirk would be put to brilliant propaganda use. 'What began as a miserable blunder,' J. B. Priestley said on the wireless the following Wednesday, 'a catalogue of misfortunes and miscalculations, ended as an epic of gallantry. We have a queer habit – and you can see it running through our history – of conjuring up such transformations.'

Priestley's talk focused on the little pleasure boats and paddle steamers that were called away from the 'innocent foolish world' of British seaside holidays 'to sail into the inferno . . . to rescue our soldiers'.

Great Britain had sent nearly half a million men and all their equipment into France, and it ended in a shambles. The collapse of France demolished the strategy, and the material losses were horrendous. The RAF lost over 1,500 aircrew and more than 900 aeroplanes, half of them fighters. The Royal Navy lost over 200 ships, including six destroyers, eight personnel carriers and seventeen trawlers. The British Army had over 68,000 casualties: nearly 12,000 men killed, the rest wounded, missing or captured. In addition, the BEF also left behind on the smouldering beaches and wrecked fields of France all their tanks, armoured cars and carriers, tractors, lorries and cars: 120,000 vehicles. They lost virtually all their artillery: 2,472 guns out of 2,794. The waste was prodigious, as German newsreel footage shows. The British had to abandon, throw away or destroy half a million tons of stores, supplies and ammunition, 165,000 tons of petrol, 20,000 motorcycles, 90,000 rifles, 8,000 Bren guns, hundreds of anti-tank rifles, hillocks of grenades; a tragic litter of papers, possessions, clothes, kit, in a great scattering of men's stuff.

'The Dunkirk spirit' transforms such a defeat into a victory. But this particular blow, devastating in its losses and psychic impact, stunning in its revelation of vulnerability, also galvanised Britain into new ways of protecting itself. Necessity is always the mother of invention, and the glaring weakness of 1940 required a bigger show of strength; hence more camouflage, more deception, more propaganda.

# 17

# Commando Dagger

On Tuesday, 4 June 1940, Harold Nicolson MP wrote to his wife Vita Sackville-West, wishing her courage and hope: 'This afternoon Winston made the finest speech I have ever heard. The House was deeply moved.' The famous Dunkirk speech to the House of Commons begins as an historical narrative of the events leading to Boulogne, Calais and Dunkirk, praises the armed services and everyone involved in the miraculous deliverance of 335,000 men 'out of the jaws of death and shame' and then turns to the imminent threat of military invasion, a reality unknown for a thousand years. Its peroration is justly renowned, '. . . we shall fight on the beaches, we shall fight on the landing-grounds, we shall fight in the fields and in the streets, we shall fight in the hills; we shall never surrender . . . ' and it ends by invoking the New World. But there are less well known elements in it. Readying British citizens for Nazi invasion, Churchill said:

When we see the originality of malice, the ingenuity of aggression, which our enemy displays, we may certainly prepare ourselves for every kind of novel stratagem and every kind of brutal and treacherous manoeuvre. I think no idea is so outlandish that it should not be considered and viewed with a searching, but at the same time, I hope, with a steady eye.

Immediately after making his speech, Churchill suggested to General Ismay, the head of the military wing of his Cabinet War Secretariat, that 'we should immediately set to work to organise raiding forces on these coasts . . . How wonderful it would be if the Germans could be made to wonder where they were going to be struck next, instead of forcing us to try to wall in the Island and roof it over!'

With Britain on the back foot, everyone was discussing what the country was going to do, including people like Dudley Clarke in the CIGS Secretariat at the War Office. In *Seven Assignments*, Clarke says that he began thinking of historical parallels of countries overrun, like Spain under Napoleon from 1808 to 1814, when the bandit-like resistance movement gave the world the name 'guerrilla'. Ninety years later the Boers had found the same solution, harrying the British Army with loosely organised groups of horsemen, the Boer 'Commandos'. This idea of the irregular armed band, led by a man like Smuts, took Clarke on to his own experience in 1936 of unorthodox warfare in Palestine where 'a handful of ill-armed fanatics' had run rings round an army corps of regular troops. Perhaps the Spaniard, the Boer and the Arab had something to teach the stodgy British Army. Archibald Wavell had talked about this, observing that attack was the easiest form of warfare, and that it paid to be more aggressive when you were weaker. Mobility was crucial to all guerrillas, and the British had the advantage of their traditional element, the sea, with an enemy stretched out along thousands of miles of European coastline from Bodø to Biarritz. Clarke first sketched out the idea of an amphibious 'commando' in three hand-written sheets dated 30 May 1940.

On Wednesday, 5 June, General John Dill inspected some soldiers who had returned in good order from Dunkirk. Their morale was surprisingly high, and he was keen to harness their 'offensive spirit'. Clarke says that it was then that he suggested the idea of 'commandos', and the CIGS said he would put it to the Prime Minister the next day. He asked Clarke to rough it out on paper, bearing in mind two limitations: no existing unit could be diverted from Home Defence, and thrifty use was to be made of the meagre stock of arms remaining after the losses in France.

Dill took Clarke's idea to the chief of staffs meeting and the cabinet, and before lunch that Thursday summoned Clarke to tell him 'Your Commando scheme is approved, and I want you to get it going at once. Try to get a raid across the Channel mounted at the earliest possible moment.' The same day, 'flushed with a sense of deliverance', Winston Churchill wrote a vigorous minute on organising the Striking Companies:

Enterprises must be prepared, with specially trained troops of the hunter class, who can develop a reign of terror down these coasts, first of all on the 'butcher and bolt' policy . . . I look to the Joint Chiefs of Staff to propose me measures for a vigorous, enterprising and ceaseless offensive against the whole German-occupied coastline . . .

Clarke was to drop his staff duties and take charge of a brand new section in the Directorate of Military Operations responsible for every sort of raiding. That afternoon, under the direction of Brigadier Otto Lund, Clarke began setting up MO9. For recruits, he began looking to the 'Independent Companies' which had been formed in April 1940 from Territorial Army volunteers, lightly armed and equipped to operate behind enemy lines or on his flanks for several days. Five of the independent companies (each with twenty officers and 270 other ranks) were then fighting in Norway under Colin Gubbins, but another five were facing disbandment in Scotland, and from these he could have a free choice.

The next problem was transport. On 7 June, Dudley Clarke talked to the assistant chief of the Naval Staff at the Admiralty, who, he said, 'fairly bounced from his chair' with enthusiasm at the idea of this piratical enterprise and introduced Clarke to the Second Sea Lord. The Second Sea Lord said he had the very man, G. A. Garnon-Williams, at the time busy sinking blockade ships in Zeebrugge harbour. Clarke took the sleeper to Scotland, found two officers at the independent companies' camp outside Glasgow, and asked them each to pick a hundred volunteers for 'independent mobile operations' in a unit to be known for the moment as 'No. 11 Commando'. They were to bring them south to go on a raid in the next two weeks.

Clarke later explained the principle behind this in a November 1948 BBC Home Service talk, 'The Birth of the Commandos':

Normally, in any regular army, a soldier is placed, quite at random, under the command of a particular officer, whom he is compelled to obey by law . . . The guerrilla, on the other hand, more often than not chooses his own officer. He joins up with a band because he reckons it has a good leader, who is going to give him the best run for his money. In the same way, the leader only takes the men he wants, and soon gets rid of those who fail to live up to his standard. It seemed to us very important that this 'leadership' principle, in some form or another, should be introduced into the Commandos from the start.

On Monday, 10 June, Clarke met Garnon-Williams, who suggested the Hamble river near Southampton was a good place to collect men and motor boats. On the 12th the soldiers arrived from Scotland for the first of several exercises, invading Hampshire against a friendly infantry battalion. The boats were not good and Garnon-Williams came up with a solution: six RAF 'crash-boats', the kind that T. E. Lawrence had worked on, designed for the speedy rescue of airmen who came down in the water. The date for the first raid was set for 24 June 1940. By now the raiding concept was gaining ground elsewhere too, and Major General Alan Bourne of the Royal Marines was being put in charge of a new Combined Operations HQ. Rumblings of organisational discontent and flashes of service rivalry marked the opening skirmishes of a Whitehall turf war.

Dudley Clarke got permission to go on the first raid, operation COLLAR, but was forbidden to land. Four boats set out from Dover, Newhaven and Folkestone with 120 men, faces blackened with make-up from Willie Clarkson's theatre shop in Wardour Street, heading for points on the enemy coast south of Boulogne. They were not equipped for exact navigation, the compass was unreliable and they were delayed by buzzing RAF fighters who had not been told about them. At two o'clock in the morning of 24 June their boat hissed into the beach and the men waded ashore and disappeared into the sand dunes.

Clarke says it was alarming just waiting in the lapping craft. A German plane flew directly overhead and a German patrol boat hovered ominously nearby. Then firing broke out a mile down the coast: machine guns and rifles, the thump of grenades. A German cyclist patrol came along the beach. Ronnie Tod, the officer, dropped the drum of his unfamiliar .45 Thompson sub-machine gun and the noise caused the German patrol to open fire wildly into the darkness. Clarke felt a stunning blow to his head and fell to the deck, sore and shaken. He had a pain in his hip where he had crushed his silver tobacco box. The implication in his account is that he was hit by a stray bullet, but Ernest Chappell (who was also there) says there was no firing. Perhaps Clarke's injuries were caused by falling over in the boat in the darkness.

The German patrol had made off, but it was possible they would return with reinforcements. Then sinister dark figures appeared on the beach and started wading quietly towards the boat, bayonets out-thrust. In a voice strangled with fear, Clarke asked for a password and got an unintelligible answer. Was it the enemy? But the ordinary English name of an NCO dissipated the intense excitement. They backed carefully out into deep water and edged away to the north. It was growing light by now and when the German plane came back they cut the engine to show no wake. Then the mysterious German boat moved off, and they headed for home. Daylight showed the left side of Dudley Clarke's head, neck and coat to be caked with dried blood, and his left ear almost severed. Patched up, he was treated as a hero when the commandos returned to Dover. Only one of the four boat crews had actually done anything, landing at Le Touquet, clumsily killing two sentries and throwing some grenades.

This first attempt of the commandos was small beer, but the Ministry of Information put out a slim communiqué the next day that hinted at great games afoot.

### British Raid on Enemy Coastline

Naval and military raiders, in co-operation with the RAF, carried out successful reconnaissances of the enemy coastline; landings were effected at a number of points and contacts made with German troops. Casualties were inflicted on the enemy, but no British casualties occurred, and much useful information was obtained.

The second commando raid, on occupied Guernsey on 14 July, was not very successful either, once again owing to faulty compasses and inadequate boats. One party landed on the wrong island and two weak swimmers were left behind to be captured. But Churchill was keen on the concept and three days later appointed the fire-eating Admiral of the Fleet, Sir Roger Keyes, who had led the famous raid on the U-boat bases of Zeebrugge and Ostend in April 1918, to be the new head of Combined Operations Command. Keyes cancelled the 'little-and-often' policy and prepared for a big commando raid. On 25 August 1940, Churchill wrote to Anthony Eden, his Secretary of State for War, urging the development of 'storm troops' on German lines.

The defeat of France was accomplished by an incredibly small number of highly-equipped *élite*, while the dull mass of the German Army came on behind, made good the conquest and occupied it. If we are to have any campaign in 1941 it will be amphibious in its character . . . which will depend on surprise landings of lightly equipped, nimble forces accustomed to work like packs of hounds . . . For every reason therefore we must develop the storm troop or Commando idea. I have asked for five thousand parachutists, and we must have at least ten thousand of these small 'bands of brothers' who will be capable of lightning action.

Dudley Clarke stayed with MO9 from June to November 1940. One of his key ideas was to get away from the comfortable staples of army 'herd' life, instead making individuals responsible for their own food, transport and shelter. In addition to his pay, each commando was given an allowance of six shillings and eightpence a day (an officer received double, thirteen shillings and fourpence), and how he spent it was up to him. If the order was 'Assemble tomorrow at Dover docks at 10 a.m.', no transport was laid on, but every man had to make his own way there. Thus the individual commando was encouraged to be self-reliant and self-confident, free to put forward a good idea or a new technique, but also to pair up: one assault course was called 'Me and My Pal' because it was far harder to do alone. These men were physically fit, could swim, climb or march, and had been trained with every kind of portable weapon, including the enemy's. Majors Fairbairn and Sykes, formerly of the Shanghai Police, taught close-quarter combat and developed the commando dagger. Wavell had said in the 1930s that he saw the ideal infantryman as a mixture of cat-burglar, poacher and gunman, and in 1940 Clarke wanted the amphibious commando to be a cross between an Elizabethan pirate, a Chicago gangster and a Frontier tribesman. These cinematic references did not always go down well with traditional authority, but they were a powerful lure to recruits.

Once again Clarke's love for showbiz and the theatrical came up trumps. He managed to enlist the actor David Niven, who had been an officer in the Highland Light Infantry but had to resign in disgrace after raising his hand to ask, at the end of a long and boring lecture by a visiting major general, 'Can you tell me the time, please? I have to catch a train.' He had then escaped close arrest by charming his captor, going on to become famous in Hollywood. It was a brilliant idea of Clarke's to make the debonair star of *The Dawn Patrol* the link between the MO9 office and the hard nuts of the Commandos, because it made 'special service of a hazardous nature' seem even more appealing and amusing to the men. According to his colourful 1971 reminiscences *The Moon's a Balloon*, Niven once came down to London from hush-hush commando training at Lochailort Castle and was arrested by Special Branch and questioned by MI5 about the telegram he had sent to a female foreign national: ARRIVING WEDNESDAY MORNING WILL COME STRAIGHT TO FLAT

WITH SECRET WEAPON. He explained that his urgent mission was fornication, not espionage.

Niven suggested to Clarke that his new uncle by marriage, Bob Laycock, would do well in the commandos. Indeed, after they got him in, Major General Sir Robert Laycock rose to become chief of Combined Operations from 1943–7. On his way to the top, and largely selecting men out of White's Club in London's St James's Street, Laycock raised No. 8 Commando, which included David Stirling as well as the writer Evelyn Waugh, whose experiences in Crete with 'Layforce' would become central to his great three-part novel of WW2, *Sword of Honour*.

In November 1940, Waugh wrote to his wife that she need have no misgivings about his status. 'Everyone in the army is competing feverishly to get into a commando . . . The officers are divided more or less equally into dandies and highly efficient professional soldiers.' In Waugh's comic novel *Put Out More Flags*, the army commandos became the inevitable destination not only of the adventurous cad Basil Seal but of his dim and decent friend Sir Alastair Digby-Vane-Trumpington:

> Then Alastair said, 'Sonia, would you think it bloody of me if I volunteered for special service?'
>
> 'Dangerous?'
>
> 'I don't suppose so really. But very exciting. They're getting up special parties for raiding. They go across to France and creep up behind Germans and cut their throats in the dark.'
>
> 'It doesn't seem much of a time to leave a girl,' said Sonia, 'but I can see you want to.'
>
> 'They have special knives and tommy-guns and knuckle-dusters; they wear rope-soled shoes . . . They carry rope-ladders round their waists and files sewn in the seams of their coats to escape with. D'you mind very much if I accept?'
>
> 'No, darling. I couldn't keep you from the rope ladder. Not from the rope ladder I couldn't. I see that.'

From amateurish beginnings, the commandos grew into a professional force, raiding north to Spitzbergen and south to the shores of Libya. Following raids on Vaagso, Bruneval, St Nazaire and Dieppe, they became such an irritant to the Third Reich that Adolf Hitler issued an edict in October 1942 ordering all commandos to be shot out of hand, without parley or pardon.

By then, Dudley Clarke was otherwise engaged. On 13 November 1940, General Haining summoned Clarke to say that the CIGS had had a personal signal from General Sir Archibald Wavell, commander-in-chief, Middle East, saying that he wanted to form 'a special section of intelligence for deception of the enemy' in Cairo, under a General Staff officer, grade 1, and asking specifically for Lieutenant Colonel D. W. Clarke.

# 18

# British Resistance

The British people mobilised not just to work in mine, field and factory or Dig for Victory, but also to fight. 'The one desire of all the males and many women was to have a weapon,' wrote Churchill. By the end of May 1940 there were 400,000 men in the Local Defence Volunteers, and 1.4 million registered by the end of June, nearly half of whom had served in WW1. Yachtsmen and motor boaters joined in too: the Upper Thames Patrol guarded 125 miles of river. Young men aged 18 to 19? who were too young for national service could join the Home Defence Companies, and there was also a Non-Combatant Corps (nicknamed the 'Norwegian Camel Corps') where conscientious objectors could do useful work with pick and shovel, brush and trowel. Some went into bomb disposal, so 'conchies' were certainly not cowards. On 5 June all these Home Forces were bolted together with the Field Army and the Anti-Aircraft Units into what were briefly called the 'Ironsides', who used, among other things, Bren-gun-carrying armoured motor cars bodged out of boilerplate. The idea was supposed to be linked to Oliver Cromwell, the Protector of England, as 'Ironsides' had been his Roundhead nickname, and the new commander-in-chief, Home Forces, was also called Sir Edmund Ironside. On 5 June 1940, CROMWELL became the code word for all troops to take up their battle stations.

In the days when the threat came from Napoleon, only the 'fencibles' or the volunteer militia near the coast would have rallied. But the advent of the parachutist in Hitler's war meant that every village in England might be endangered, and every citizen had to be *en garde*. In June, at General Ironside's instruction, a massive campaign of deception at a local level began. All direction and road signs across the country were taken down so as not to aid *Fallschirm-Infanterie* dropping from the sky. Tradesmen's vans and shop signs were painted over. The names on railway station platforms were reduced to three inches high, which made arrivals and departures on crowded, jerky, smoky train journeys even worse. Lost military drivers learned to nip into red Gilbert Scott telephone boxes, whose address might still be written down inside by the 999 instructions, or to navigate by the corporation names on manhole covers.

By June, church bells could only be rung to warn of enemy landings, and all bank holidays were cancelled. Travel to within twenty miles of the coast from the Wash to Portland was prohibited, while a new evacuation moved thousands of children back away from the shores. For the few who had cars then, petrol was already rationed to essential use. 'Is Your Journey Really Necessary?' posters appeared. Even telephoning was discouraged. The country was concealing itself, crouching down.

Britain was locking up, too. From 1 September 1939, German and Austrian males between the ages of 16 and 60 had been required to register with the police as enemy aliens, and most had been put in Category C: sympathetic to Britain, to be left at liberty. But now, between 12 May 1940 and the end of July, some 27,000 'friendly aliens' (including 4,000 Italians after 10 June) were rounded up and interned on the Isle of Man or deported by sea. After the British ship *Arandora Star* was torpedoed by Günther Prien's *U-47* on 2 July en route to Canada, killing over 700 German and Italian internees, 450 traumatised survivors were brought back and then immediately shipped off again to Australia on the *Dunera*. The prison camps became mini-universities, centres of art, music and learning, because many internees were anti-Nazi Jewish refugees, including scholars, scientists and intellectuals who had been fleeing Hitler over the last seven years.

All this came from a moral panic about the possibility of enemy deception on a grand scale. Was there a 'Fifth Column' of hidden traitors and spies lurking ready to dash out and help enemy parachutists? The term Fifth Column came from early in the Spanish Civil War, when Franco's Nationalists had four military columns besieging Madrid but also claimed to have a fifth one, operating clandestinely inside the Republican defences. In mid-May, the *Daily Express* and *Daily Telegraph* were claiming that such deceptive agents or 'Quislings' had opened the gates of Norway and Holland to the invading Nazis. Were there similar people here in the UK – foreign women, refugees, dissidents, blackmailed Jews – perhaps waiting to assist Germans parachuting down dressed as policemen, vicars or air-raid wardens? Heath Robinson was having none of it. His cartoon in the *Sketch* of 3 July 1940, *The Sixth Column at Work: Here they come. Disguised parachutists receive a warm welcome*, shows hordes of patient Englishmen in flat caps and hats cycling about with large tubs and baths of hot water ready to catch parachuting beldames and bishops as they float to the ground.

This moment of national alarm and emergency felt rather like 1914 again. Ordinarily sane people became obsessed by carrier pigeons, strange chalk marks on telegraph poles (actually left by 'bob-a-job' scouts and guides indicating where they had visited), mysterious lights, buzzing wirelesses, suspicious conversations, funny-looking strangers. General Ironside's diary of 31 May records 'Fifth column reports coming in from everywhere. A man with an arm-band on and a swastika pulled up near an important aerodrome in the Southern Command . . .' Following these reports, he posted pickets everywhere. 'Perhaps we shall catch some swine.' General Ironside was a source of alarmism, but he was also the victim of an alarmist whispering campaign that linked him to the British fascists. The British authorities' problem became how to separate innocent from guilty amid rising hysteria. 'Collar the lot' was the view of Major General Sir Vernon Kell, Director General of the Security Service, MI5, which, while trying to keep tabs on real German spies, was drowning in erroneous information about possible subversion by Communists, Fascists, Indian Nationalists, the IRA and Pacifists.

At the end of the day, the only deception was self-deception. There was no evidence of any plans for espionage, sabotage or 'Fifth Column' activities among the foreign internees. If there was an enemy within, it was native, not foreign; noble, not *arriviste*. Too many high-born Britons were attracted to the ideology of the vigorous if vulgar Nazis; even the abdicated King Edward VIII, now the Duke of Windsor, was not immune to the flattering blandishments of Hitlerism – the Windsors had visited Nazi Germany in September 1937 and were photographed, smiling, with the Führer.

Parliament passed a new Treachery Act in May 1940 under which grave cases of espionage and sabotage carried the death penalty, and after the discovery inside the US embassy of a right-wing spy, Tyler Kent, passing secret documents to Italy and Germany, the Home Secretary, Sir John Anderson, overcame his sceptical liberal scruples to take action. On 23 May, under the hurried amendment 1A (adding mere associates and sympathisers to those criminalised under 18b of the Defence Regulations for 'execution of acts prejudicial to the security of the State'), the Metropolitan Police raided the London headquarters of the British Union of Fascists and arrested thirty-four members, including the Blackshirts' leader, Sir Oswald Mosley, who had been insisting this was 'a Jews' war'. Two more noted anti-Semites and pro-Nazis, John Beckett of the racialist British People's Party and Captain Maule Ramsay, president of the Right Club and Conservative MP for Peebles, were also taken into custody in Brixton prison. Around 750 fascists were detained. Aristocratic appeasers like Lord Tavistock, the Duke of Buccleuch and the Duke of Westminster who had once been involved in the Nordic League and the Anglo-German friendship association, The Link, now ducked their ducal heads and went to ground on large estates. On 28 May, Churchill appointed Lord Swinton as head of the Home Defence (Security) Executive 'to co-ordinate action against the Fifth column'. General Kell, who had founded and led MI5 for thirty years, was retired. In the House of Commons on 4 June Churchill said:

We have found it necessary to take measures of increasing stringency, not only against enemy aliens and suspicious characters of other nationalities, but also against British subjects who may become a danger or a nuisance should the war be transported to the United Kingdom. I know there are a great many people affected by the orders which we have made who are the passionate enemies of Nazi Germany. I am very sorry for them, but we cannot, at the present time and under the present stress, draw all the distinctions which we should like to do . . . There is, however, another class, for which I do not feel the slightest sympathy. Parliament has given us the powers to put down Fifth Column activities with a strong hand, and we shall use those powers, subject to the supervision and correction of the House, without the slightest hesitation until we are satisfied . . . that this malignancy in our midst has been effectively stamped out.

On that day, three *Evening Standard* journalists, Michael Foot, Frank Owen and Peter Howard, finished a swiftly written 40,000 word polemical attack on the appeasers, disarmers and Chamberlainites of the 1930s whose folly and sloth had led to the disaster of Dunkirk. It was published a month later by Victor Gollancz under the title of *Guilty Men* by 'Cato'. Even though it was not available in W. H. Smith's chain of shops it became notorious and sold 200,000 copies.

Yet by the middle of August 1940, Churchill was declaring to the House of Commons that he 'always thought' the Fifth Column danger 'was exaggerated in these islands'. If that was true, he had nevertheless put the 'Fifth Column' idea to most effective propaganda use.

---

On 10 June, scenting wounded prey, Fascist Italy boldly declared war on the Republic of France, and promptly captured 200 yards of the Riviera. Paris was declared an 'open city', which was a formal statement that it would not be defended. Now it became a ghost town as shops, offices and hotels closed, traffic vanished and the government slipped away from the advancing Germans. Sefton Delmer was one of the last reporters left in Paris, with Edward Ward of the BBC, Robert Cooper of *The Times* and Walter Farr of the *Daily Mail*. By lunchtime on 14 June, the swastika was fluttering from the radio mast on top of the Eiffel Tower and *Kavallerie* were trotting down the Champs Elysées. Reynaud then resigned, and the new Prime Minister, white-moustached Maréchal Pétain, spoke on 17 June: 'It is with a heavy heart that I say we must cease the fight,' adding, in another broadcast three days later, 'Not so strong as twenty-two years ago, we had also fewer friends, too few children, too few arms, too few allies. This is the cause of our defeat.' France had fought and lost, with nearly 100,000 dead, 120,000 wounded, and over a million and half soldiers captured.

On 21 June, in the Forest of Compiègne, Hitler in his grey uniform relished the ritual of French abasement. Bill Shirer of CBS watched the Führer strutting among toadies, 'his face afire with scorn, anger, hate, revenge, triumph'. Generaloberst Keitel read out the Nazis' propagandist, self-righteous preamble, clearly written by Hitler:

Trusting to the assurance given to the German Reich by the American President Wilson and confirmed by the Allied Powers, the German Defence Forces in November 1918 laid down their arms. Thus ended a war which the German people and its Government did not want, and in which in spite of vastly superior forces the enemy did not succeed in defeating the German

Army, the German Navy or the German Air Force . . . Broken promises and perjury were used against a nation which after four years of heroic resistance had shown only one weakness – namely, that of believing the promises of democratic statesmen.

On September 3, 1939, twenty-five years after the outbreak of the World War, Great Britain and France declared war on Germany without any reason. Now the war has been decided by arms. France is defeated . . .

'What General Weygand called the Battle of France is over,' Churchill had said on 18 June. 'I expect that the Battle of Britain is about to begin . . . Hitler knows that he will have to break us in this island or lose the war.' The mood became even more exalted and grim. Upper-class Harold Nicolson and Vita Sackville-West prepared suicide potions, but ordinary people felt strangely exhilarated, even liberated. The playwright and broadcaster J. B. Priestley felt a new mood in the country, of courage and hope: 'Our people began to show the world what stuff they're made of, and the sight was glorious.' Churchill commanded some of that mood with superb morale-boosting language: 'Let us therefore brace ourselves to our duties, and so bear ourselves that, if the British Empire and its Commonwealth last for a thousand years, men will still say, "This was their finest hour." '

It was great rhetoric. But how was Churchill himself behaving under the strain? At times he was impatient and inconsiderate. There is only one letter extant between Winston and Clementine Churchill in all 1940, dated 27 June, a loving wifely letter telling him something she feels he ought to know, that 'there is a danger of your being generally disliked by your colleagues & subordinates because of your rough sarcastic & overbearing manner'. She confessed to noticing 'a deterioration' in his manner; 'you are not so kind as you used to be'. She could not bear

that those who serve the Country & yourself should not love you as well as admire and respect you – Besides you won't get the best results by irascibility & rudeness. They *will* breed either dislike or a slave mentality – (Rebellion in War time being out of the question!)

It was time for writers, artists and film-makers to man the propaganda barricades. 'Arm the people,' wrote George Orwell pugnaciously to *Time and Tide* on 22 June, urging the distribution of hand grenades, shotguns and all the weapons in gunsmith's shops. The author of *Homage to Catalonia* had just joined the Primrose Hill platoon of the 5th (London) Local Defence Volunteers battalion, but his letter to *Time and Tide* sounds as if he were still in the Partido Obrero de Unificación Marxista militia with whom he fought on the Republican side in the Spanish Civil War. Later, Orwell saw June 1940 as a revolutionary moment:

Had any real leadership existed on the Left, there is little doubt that the return of the troops from Dunkirk could have been the beginning of the end of British capitalism. It was a moment at which the willingness for sacrifice and drastic changes extended not only to the working class but to nearly the whole of the middle class, whose patriotism, when it comes to the pinch, is stronger than their self-interest. There was . . . a feeling of being on the edge of a new society in which much of the greed, apathy, injustice and corruption of the past would have disappeared.

Orwell had had a dream one night back in August 1939 that helped convince him of his true feelings. He dreamed that the war had already started, and this had two lessons for him. The first was that he would be relieved from dread when it did happen, and the second was the sure knowledge that he was patriotic at heart, that he would not sabotage his own side, and that he would support the war and fight in it if possible. The next day he had read in the newspapers about the Nazi–Soviet pact. By March 1940, Orwell had reached George Bernard Shaw's position:

When war has once started there is no such thing as neutrality. All activities are war activities. Whether you want to or not, you are obliged to help either your own side or the enemy. The Pacifists, Communists, Fascists etc are at this moment helping Hitler.

In April 1940, reviewing Malcolm Muggeridge's *The Thirties*, Orwell recognised something more sympathetic to him in its closing chapters than the self-righteousness of the leftwing intelligentsia:

It is the emotion of the middle-class man, brought up in the military tradition, who finds in the moment of crisis that he is a patriot after all. It is all very well to be 'advanced' and 'enlightened', to snigger at Colonel Blimp and proclaim your emancipation from all traditional loyalties, but a time comes when the sand of the desert is sodden red and what have I done for thee, England, my England?

'George Orwell' was the nom de plume of a writer who often features in discussions of what Englishness is. The name camouflages the old Etonian and former Burma Police officer Eric A. Blair under the regal Christian name of England's warrior saint, St George, joined to the name of the river Orwell that winds through Suffolk to the North Sea. Although a democratic Socialist and a man of the Left (with what a Special Branch surveillance report called 'advanced communist views'), Orwell wrote shrewdly on Kipling and understood the spiritual need for patriotism and the military virtues among the English people. 'The English are not intellectual,' he observed in 'The Lion and the Unicorn: Socialism and the English Genius', 'but they have a certain power of acting without taking thought . . . Also, in moments of supreme crisis the whole nation can suddenly draw together and act upon a species of instinct', like cattle facing a wolf. Orwell says that the gentle, hypocritical English can be martial, but hate militarism. The British Army don't goose-step, he points out, because the people in the street would laugh.

Another war veteran who wanted a better world and had just had his first spell with the Local Defence Volunteers was the novelist J. B. Priestley, keeping watch and ward at night on a high down near his home at Godshill in the Isle of Wight. As he described the experience in his third broadcast talk on Sunday, 16 June, Priestley found himself back in a world not of the Flanders trenches or of Spanish revolutionaries but of Thomas Hardy, out among Wessex country people, 'ploughman and parson, shepherd and clerk', talking about 'what happened to us in the last war, and about the hay and the barley, about beef and milk and cheese and tobacco'. He felt a powerful sense of community and continuity with those gone before, like the men who stood ready for Napoleon's *Grande Armée*:

But then the sounds of bombs and gunfire and planes all died away. The 'All Clear' went, and then there was nothing but the misty cool night, drowned in silence, and this handful of us on the hilltop. I remember wishing then that we could send all our children out of this island, every boy and girl of them across the sea to the wide Dominions, and turn Britain into the greatest fortress the world has known; so that then, with an easy mind, we could fight and fight these Nazis until we broke their black hearts.

Balding, bespectacled Tom Wintringham, *Daily Worker* journalist and founder of the International Brigades in Spain, was, like Orwell, a veteran wounded in the Spanish Civil War who thought that 1940 was the moment to make 'an army of the people'. Born bourgeois in Grimsby, Wintringham was an ardent public-school Communist who went on to be a motor-cycle dispatch rider with the kite balloon section of the RFC in WW1. He had commanded the British battalion of the International Brigades at the Battle of the Jarama in February 1937 when 150 of them got killed. Wintringham, once a loyal Stalinist hack, was expelled from the Communist Party in October 1938 over his affair with the 'undesirable' American reporter Kitty Bowler, but his card remained marked by MI5. Orwell likened him to G. A. Henty with Marxist training.

In May 1940 Wintringham was writing vigorously not just for *Tribune* and the *New Statesman* but also for the *Daily Mirror* and *Picture Post*, which were read by millions. His slogan for an article in the *Daily Mirror* addressed to the newly appointed Edmund Ironside, was 'An Aroused People, An Angry People, An Armed People': he was preaching popular guerrilla warfare. 'How to Deal with Parachute Troops' filled a page of *The War Weekly* on 7 June and was soon followed by two more long pieces for *Picture Post*: 'Against Invasion' and 'Arm the Citizens'. The War Office printed 100,000 copies of the second article and distributed them to Local Defence Volunteer units. His biographer Hugh Purcell says in *The Last English Revolutionary* that his pieces

gave practical instructions for a people's war based on his experience in Spain: how to destroy tanks and bridges, capture or kill German parachutists, fortify a village, make and throw hand grenades, engage in street fighting.

All Wintringham's articles were worked together into a longer text, *New Ways of War*, published as a Penguin Special in July 1940, which sold 75,000 copies in a few months. In it he points out that German 'Blitzkrieg' evolved to escape trench deadlock, giving a certain amount of autonomy to front-line soldiers, and that the German tactic of 'infiltration' also worked through individual initiative. He criticises snobbish and rigid military thinking in Britain and says what is needed is 'an army of free men' who can think for themselves. Scanning history to find 'democratic' forces who prevailed against more powerful autocratic societies, Wintringham looks to, among others, ancient Greeks against Persians, the Roman republic versus Carthage, and the bowmen of Crécy and Agincourt against French knights. For examples of a 'People's War', he cites the Dutch Republic, Garibaldi's Italy, Japanese-occupied China and the old native Anglo-Saxon tradition of the *fyrd*, the freemen's shire militia. *New Ways of War* is a rousing piece of propaganda that mixes bomb-making tips, guerrilla tactics and battle lore with his radical argument for a 'People's War':

There are those who say that the idea of arming the people is a revolutionary idea. It certainly is. And after what we have seen of the efficiency and patriotism of those who ruled us until recently, most of us can find plenty of room in this country for some sort of revolution, for a change that will sweep away the muck of the past.

*Picture Post*, a pioneering illustrated magazine from the moment it first appeared on 1 October 1938, strongly supported Wintringham's ideas. The most popular news magazine ever published in Britain, it was later imitated by *Signal* in Germany, *LIFE* in the USA, *Paris-Match* in France, *Drum* in South Africa. The proprietor was the Conservative Edward Hulton, the son and grandson of newspaper proprietors, and in June 1940 the left-leaning Tom Hopkinson became its editor. Hopkinson was dining with Wintringham at Hulton's one night in late June, talking about the frustrations of the LDVs having to do drill with broomsticks when they were itching to learn to fight. On 20 June, in the Secret Session of the House of Commons, hadn't Winston Churchill said that 'the essence of the defence of Britain is to attack the landed enemy at once, leap at his throat and keep the grip until the life is out of him'? Then the idea clicked. Why couldn't *Picture Post* help provide the training?

It was all set up by midnight. Hulton rang a friend, the Earl of Jersey, who came round straight away. He owned a large mansion to the west of London set in the spacious grounds of Osterley Park. Jersey was happy to allow the grounds to be used for a training course, but said he'd rather the house wasn't blown up, as it had been in the family for some time. 'Can we dig weapon pits? Loose off mines? Throw hand grenades? Set fire to old lorries in the grounds?' asked Wintringham. 'Certainly. Anything you think useful.'

Wintringham became Director of Training at Osterley Park School which started its first course on 10 July. By now Churchill was overruling Eden and insisting on changing the name 'Local Defence Volunteers' to 'Home Guard'. Churchill found 'Local', as in 'Local Government' uninspiring, and he absolutely hated the name 'Communal Feeding Centres', telling the Ministry of Food: 'It is an odious expression, suggestive of Communism and the workhouse. I suggest you call them "British Restaurants". Everybody associates the word "restaurant" with a good meal, and they may as well have the name if they cannot get anything else.'

Members of the Home Guard came to Osterley from all over the country for two-day courses in irregular warfare at Hulton's expense. Every week there were three courses with sixty men on each. By the end of July there were a hundred men on every course and more clamouring to get in. In August 1940 over 2,000 attended. *Picture Post* appealed to the USA for private citizens to donate weapons: they sent a serviceable shipload to Liverpool, including six-shooters, buffalo guns, hunting rifles and .45 calibre gangsters' 'Tommy guns' which were duly distributed to the Home Guard. *Picture Post* also showed how to manufacture heavier weapons in the garage. 'Make Your Own Mortar for 38/6d' was one feature, with instructions for milling home-made black powder. This sort of thing perturbed the conventional authorities.

The Communist Party of Great Britain had not been proscribed nor the *Daily Worker* banned at this time. Their line of 'revolutionary defeatism', which had emerged from the peculiar contortions of having to explain the Nazi–Soviet pact in Marxist–Leninist terms, had now evolved into shouting loudly for the conscription of all wealth, the arming of the workers and a new government. The hard-line Security Executive was worried by CPGB propaganda and suggested in July that the Home Office frame a new defence regulation making it an offence 'to attempt to subvert duly constituted authority'. The Permanent Under Secretary at the Home Office, Sir Alexander Maxwell, resisted this in an admirable minute, dated 6 September 1940:

There would be widespread opposition to such a regulation as inconsistent with English liberty. Our tradition is that while orders issued by the duly constituted authority must be obeyed, every civilian is at liberty to show, if he can, that such orders are silly or mischievous and the duly constituted authorities are composed of fools or rogues . . . Accordingly we do not regard activities which are designed to bring the duly constituted authorities into contempt as necessarily subversive; they are only subversive if they are calculated to incite persons to disobey the law, or to change the Government by unconstitutional means. This doctrine gives, of course, great and indeed dangerous liberty to persons who desire revolution, or desire to impede the war effort . . . but the readiness to take this risk is the cardinal distinction between democracy and totalitarianism.

Wintringham's 'The Home Guard Can Fight', spread across several pages of *Picture Post* on 21 September 1940, was probably the apogee of the revolutionary militia idea, before the War Office and GHQ Home Forces began taking Osterley Park's staff and facilities under their control. The article shows the Home Guard shooting at model aircraft, stalking, sniping, attacking behind smokescreens and blowing up vehicles with mines. Pictures of the staff include a moustached surrealist poet, painter and art collector who was teaching camouflage, Captain Roland Penrose.

When Penrose's *Home Guard Manual of Camouflage* was published by Routledge in October 1941, he was described as 'Lecturer to the War Office School for Instructors to the Home Guard, formerly lecturer at the Osterley Park School for Training of the Home Guard'. Penrose, former leading light in the British Surrealist movement, was now working full-time at the South Eastern Command Fieldcraft School at Burwash in Sussex. The course, near Kipling's home, Bateman's, at Burwash taught men to use the countryside for cover and camouflage when dealing with German parachutists:

Your fieldcraft training can be summed up as being a way of learning by practice many of the things which animals do by instinct . . . We have forgotten to do these things largely because we live in civilized communities, where custom, law, and policemen have put a padded cushion between us and the raw struggle for existence. When the Nazis come the cushion will be removed. The nearer we are to animals, the better we shall be prepared: but we can only regain the wisdom of animals through our brains, through thinking, learning and practising.

The study of hedges and ditches, woods and roads, fields and streams, how cattle behave when someone is in their vicinity, what cock pheasants, wood pigeons, magpies, jays, lapwings and blackbirds do when disturbed, how to move silently or under fire, how to freeze, how to become inconspicuous, how to bivouac, the dangers of shape, shine, shadow and silhouette, the power of Nazi field glasses, the best way to use your ears, eyes and nose at night, reconnaissance and message carrying were all part of the Burwash syllabus. Like Osterley, Burwash pioneered some of what later army battle-schools would do, teaching platoon skills that would be useful when the British were fighting in Normandy hedgerows in 1944.

At Burwash, Roland Penrose's lantern slides certainly caught the eye of his Home Guard students. Some were from Cott's *Adaptive Coloration in Animals*, but one that flashed up was a colour photo of his beautiful wife Lee Miller, daubed in light green camouflage paint, lying naked on a lawn under a soft-fruit net and some tufts of raffia. Solomon J. Solomon would have appreciated Penrose's *Manual of Camouflage* which assimilates every hard-earned point from WW1, and begins with the need for surprise in defending Britain.

To an old soldier, the idea of hiding from your enemy and the use of deception may possibly be repulsive. He may feel that it is not brave and not cricket. But that matters very little to our enemies, who are ruthlessly exploiting every means of deception at the present time to gain their spectacular victories. They can only be stopped by new methods, however revolutionary these may appear . . .

Penrose's first point is the need to escape observation from all angles. A parachutist can land behind you, the enemy pilot can spot you from the air. 'From directly above there is no dead ground, and trenches, wire and tracks show up as though drawn on the map.' The Home Guard would have to learn to conceal their fixed positions, quarters and transport by fitting in with the patterns of the landscape, using nets and screens or tarred sandbags covered with earth and rubbish. It is all very much in the tradition of Hesketh Prichard's disguised loopholes and sniper suits cut out and stitched from panels of hessian sacking in WW1, but what was then an obscure art was now becoming part of accepted military thinking. Penrose made the Home Guard conscious of their black shadows and white faces when taking cover, and taught the usefulness of cow dung. The aim was not just concealment:

Deception is the active counterpart and is of great importance in counter-attack and guerrilla warfare . . . By using decoys and dummies we shall be able to draw the enemy's attention away from our vital points . . . [while] by the use of camouflage and smoke screens our real positions and movements can be hidden from him.

The British were rallying after the German victories across Europe, knowing they were underdogs. Penrose saw that the whole point of camouflage was to make the weaker party stronger. 'It is useless in warfare to be merely brave, if bravery means presenting oneself as a useless target to the enemy. It is far better to employ intelligence and concealment, so as to induce *him* to present a target.'

*Camoufleurs* were getting into their stride all over the country, with plenty to conceal. To stop the invader getting a foothold, arrays of defences were being engineered around the south-east English coast, many of them guarded by seaward-facing gun batteries salvaged from ageing ships and obsolescent tanks. Metal scaffolding was erected in the sea, hundreds of mines were buried in beaches, there were miles and miles of coiling barbed wire, thousands of concrete pillboxes for machine guns and hefty anti-tank blocks. More defences were erected inland along 'stop lines' designed to slow down enemy armoured forces pushing towards London. In Norfolk, for example, the stop-lines ran along the courses of the rivers Ant, Bure, Wensum, Yare and Ouse, whose bridges were rigged with explosives, and there were 30,000 men wearing the brassard of the Local Defence Volunteers ('Look, Duck and Vanish'), mostly with bicycles, whose role was to observe, patrol and protect, as well as help obstruct. Anti-tank obstacles included ditches, deepened canals, slots in the road into which iron girders were socketed, and dragon's teeth of concrete blocks, at least five feet high. Open fields had carts put in them to prevent planes landing, or poles put up to rip the wings off gliders.

Every village had its 'strong post' to defend and eight different kinds of concrete pillbox could be erected at key spots. Some of these were hideously prominent, poorly designed deathtraps. Other pillboxes conformed better to the cliffs, fields or hedges and were camouflaged afterwards with earth and undergrowth. The best were built into existing structures – barns, haystacks, lighthouses, medieval ruins, sheds, windmills, yacht clubs – so as not to be seen straight away. As the war went on, subsequent *camoufleurs* competed with ingenious disguises for strongholds: beach huts, bookstalls, bungalows, bus stops, cafés, chalets, garages, ice-cream parlours, Regency pavilions,

267

*[handwritten marginalia: point. WWI was reactive WWII became proactive / parachuter – surroundings – defenses in battlefield]*

railway signal boxes, gentlemen's toilets, half-timbered Tudor tea-shoppes and twee thatched cottages. The *camoufleurs* did their best, but it did not alter the fact that there were no tanks available to fight the Germans if they invaded, and no armour-piercing anti-tank weapons. Ironside's plan was that householders would drop home-made Molotov cocktails from upper windows on to enemy vehicles.

They were certainly ready to have a go. The LDV managed to shoot and kill their first four motorists, none of them Germans, in separate locations on the night of 2/3 June, and there were more fatal accidents to come. Battle of Britain pilots who ejected over England faced a real danger of being riddled with bullets by their own countrymen as suspected German parachutists. In the summer of 1941, the Home Guard was organised into battalions affiliated to county regiments, with military ranks, but some people did not take them seriously. A. J. P. Taylor wrote in *English History 1914–1945* that

The Home Guard harassed innocent citizens for their identity cards; put up primitive road-blocks, the traces of which may delight future archaeologists; and sometimes made bombs out of petrol tins. In a serious invasion, its members would presumably have been massacred if they had managed to assemble at all. Their spirit was willing though their equipment was scanty.

This is the line also followed by the affectionate and popular BBC TV comedy series *Dad's Army*, which ran for ten years from 1968, written by David Croft and Jimmy Perry. It was in keeping with wartime skits by comedians like Robb Wilton, and other con-temporaneous jokes. 'And what steps would you take if the Germans invaded, my man?' asks the inspecting officer. 'Big long 'uns, sir!'

And yet perhaps the Home Guard and the *camoufleurs*, in their determined efforts against as yet unrealised enemies, were taking part in a huge effort of psychological warfare. John Colville's diary, *The Fringes of Power*, records a dinner at Chequers on Friday, 12 July, with Generals Paget, Auchinleck, and Ismay as well as Duncan Sandys at the table. The day before, Churchill had toured the defences in Kent, inspecting pillboxes and troops from Dover to Whitstable. The discussion turned to invasion. There was 'an argument about encouraging the populace to fight. If they meet the invader with scythes and brickbats they will be massacred.' Churchill pointed out that 'here we want every citizen to fight and they will do so the more if they know the alternative is massacre. The L.D.V must be armed and prepared . . .' According to Colville, Churchill thought the invasion scare was 'keeping every man and woman tuned to a high pitch of readiness. He does not wish the scare to abate therefore, and although personally he doubts whether invasion is a serious menace he intends to give that impression, and to talk about long and dangerous vigils, etc., when he broadcasts on Sunday.'

In that talk on 14 July 1940 Churchill said that the war would be long and hard, but he also delivered a magnificent morale-boosting piece of rhetoric that praised both the armed forces and the British Home Guard:

Behind these soldiers of the regular Army, as a means of destruction for parachutists, air-borne invaders, and any traitors that may be found in our midst . . . we have more than a million of the Local Defence Volunteers, or, as they are much better called, the 'Home Guard'. These officers and men, a large proportion of whom have been through the last War, have the strongest desire to attack and come to close quarters with the enemy wherever he may appear. Should the invader come to Britain, there will be no placid lying down of the people in submission before him as we have seen, alas, in other countries. We shall defend every village, every town, and every city . . . we would rather see London laid in ruins and ashes than . . . tamely and abjectly enslaved . . .

This is no war of chieftains or of princes, of dynasties or national ambitions; it is a War of peoples and of causes. There are vast numbers not only in this island but in every land, who will render faithful service in this War, but whose names will never be known, whose deeds will never be recorded. This is a War of the Unknown Warriors . . .

The Home Guard, like so many things in WW2, was not actually all that it seemed. Its great secret was camouflaging the Auxiliary Units, one of Britain's nine wartime secret services. The Auxiliary Units wore ordinary Home Guard uniforms but were actually guerrilla cells. They had been set up by Lawrence Grand's secret organisation Section D, who supplied weapons and explosives, acting in unison with MI(R), whose field was guerrilla warfare. The Auxiliary Units' job would only begin after the 'stop lines' and fixed positions had been overrun. They were the 'stay-behind parties' that Major General Thorne had requested of the War Office, when he was in command of XII Corps, whose job was defending Sussex and Kent against the first thrust of the imminently expected German invasion, code-named *Seelöwe* or SEALION. General Ironside had then tasked his former ADC, Colonel Colin Gubbins, the professional soldier who had just been in Norway with the Independent Companies, to organise this resistance. Gubbins chose other enterprising officers to set up cells of Auxiliary Units all around the British coast.

Ian Fleming's older brother, the writer and explorer Peter Fleming, from the Grenadier Guards like Thorne, was 33 years old when he went to assist him in setting up the 'XII Corps Observation Unit' in Kent. The glamorous and good-looking travel writer and journalist, author of *One's Company, Travels in Tartary* and *Brazilian Adventure*, had already written a paper for MI(R) on the possibilities of guerrilla warfare with irregular cavalry in China. He had also shown a talent for deception, proposing a fake document to alert the USA to Japanese ambitions in the Pacific and South-East Asia. But now, in May 1940, just back from reconnaissance activities in Norway, Peter Fleming helped to create a real partisan force.

Basing himself at a farm near Bilting, north of Ashford in Kent, and assisted by a detachment of Lovat Scouts 'of WW1 fame' and a formidable sapper called Michael Calvert (who later won renown with Wingate's Chindits in Burma and the SAS in north-west Europe), Fleming picked countrymen – foresters, gamekeepers and poachers who knew the lie of the land – and trained them to hide up by day and come out to sabotage at night. This was just the job for the adventurous Fleming, who would end his days as a country squire in Oxfordshire, happiest out shooting.

The Auxiliary Units dug underground lairs in the woods, one of which was an expanded badger sett in a derelict chalk-pit, and skilfully concealed all sign of them. These hideouts were like the den made by the hunted hero of Geoffrey Household's 1939 novel *Rogue Male*, or as Fleming himself wrote in his book *Invasion 1940* (reissued as *Operation Sea Lion*) 'the Lost Boys' subterranean home in the second act of *Peter Pan*'. Fleming was also inspired by the English mythic hero Robin Hood, acquiring half a dozen longbows and encouraging men to use them not only to kill quietly, as guns could not, but also to carry fire and noise to the enemy by shooting arrows fitted with incendiaries or detonators. Thinking about it later, Fleming reckoned that the greenwood game could work in the summer, but with the leaves off the trees, the English resistance outlaws would soon be tracked down and eliminated by *Einsatzkommandos*. German plans make it clear that armed insurgents in occupied Britain would have been shot out of hand, and there would have been ruthless reprisals against civilians, some of whom might have given vital information away to save themselves. Some Auxiliary Units seriously considered the assassination of collaborators in the event of invasion.

Colonel Gubbins chose another veteran of MI (R) and Norway, the young Arctic explorer Andrew Croft, to organise Suffolk and Essex. Croft, who had been at Stowe with David Niven, was the son of the vicar of Kelvedon, and he based himself at home, storing explosives and weapons in his father's coach house. Then he began enlisting the locals: farmers and fruit-growers, a game warden, a master of fox-hounds, a butcher and some poachers and smugglers, in order to create two dozen small patrols each working out of their well-hidden Operational Base (OB).

By the end of 1941 the Auxiliary Units had 534 concealed OBs in the UK, with 138 more on the way. Over 3,500 men served in the Auxiliary Units and most went for their resistance training to Coleshill House, the Inigo Jones-designed home of the Earl of Radnor, situated north-east of Swindon. The address was GHQ Auxiliary Units, c/o GPO Highworth, Wiltshire; the formidable postmistress Mabel Stranks acted as a gatekeeper. Coleshill was the professional version of Osterley Park. Every weekend, two dozen Auxiliary Unit members came for courses given by regulars: close-quarter combat, weapons, explosives and fieldcraft by day, and silent exercises at night. Auxiliary Units got first pick of scarce weapons like Thompson sub-machine guns, semi-automatic pistols and Springfield rifles from the USA, and Section D's technical establishment near Stevenage supplied the Auxiliary Units with spigot mortars, incendiary fougasses, and blocks of the new yellow plastic explosive, together with the delayed-action chemical fuses known as time-pencils.

The English *maquis* learned hands-on how not to over-egg the pudding. One group watched in amazement as the car they merely wanted to disable rose high into the air and crashed in the next field. Readying Kent for invasion, they buried explosives in milk churns under bridges, and booby-trapped the large mansions that the Germans might commandeer as headquarters. 'Mad Mike' Calvert also blew out the centre sections of the piers at Brighton, Worthing and Eastbourne.

Adrian Hoare's *Standing Up To Hitler* (2002) gives details of some of the forty-five Auxiliary Units in Norfolk from 1940 to 1944. Fewer than 300 men were recruited, cautiously and quietly, and signed the Official Secrets Act. They got no medal, just a small badge with 202, the number of their battalion. The Royal Engineers dug several of the Auxiliary Units' OBs, but they made others themselves at night and weekends. These had camouflaged entrances and exits, and some were booby-trapped. All the Norfolk men had Smith and Wesson .38 revolvers, a military pass and three morphine pills, one for severe pain, two for unconsciousness and three for suicide. Their copy of the *1938 Norfolk Calendar* concealed a saboteur's handbook with instructions how to blow up tanks, planes, trucks, armoured vehicles, railway lines, ammunition dumps etc. They had magnets to clamp explosives to metal, thin wires to trip-wire spigot mines and to disable motorcyclists, and a wide variety of destructive ordnance. They trained at Coleshill but also at Leicester Square Farm, Syderstone, North Creake, where they had to run and shoot wearing gasmasks, and crawl under live machine-gun fire a few inches above their heads while instructors threw thunder flashes at them. They also studied German military tactics, routines and sentry procedures, and were taught stalking by Lovat Scouts and unarmed combat by ferocious commandos. Most of their exercises were carried out at night.

The British resistance was best at secrecy and stealth. Few knew, fewer talked, and no one saw them on their black-face night manoeuvres as they slipped around the regular forces, probing and penetrating their defences. When General Bernard Montgomery took over the corps that was defending Kent and Sussex in 1941, he was taken by Peter Fleming's successor, Captain Norman Field, to a sloping pasture above the village of Charing with a magnificent view south to the English Channel. Field suggested that the general sit a while on a weathered sheep trough. Montgomery, the future victor of Alamein, stared aggressively towards occupied France, turned to speak to Field, and found himself alone with turf-munching sheep. Then he heard a voice beneath him and Field's head popped up from a rectangular aperture that slid open in the wooden trough. A ladder led down to a two-man observation post stocked with food and water. Its windows were genuine rabbit holes, now weatherproofed and glazed. The Auxiliary Units had made the hide by placing an anti-aircraft gun on top, and its 'crew' had filled the protective sandbags around it with the chalk spoil they secretly dug out.

As David Lampe was the first to point out in his revealing 1968 investigation, *The Last Ditch*, the greatest legacy of the Auxiliary Units was the practical experience it gave its overall supremo, Colonel Colin Gubbins, in organising, equipping, supplying, camouflaging and inspiring clandestine armies for resistance and liberation. On 22 July 1940, Hugh Dalton's Ministry of Economic Warfare was given a new striking force 'to co-ordinate all action, by way of subversion and sabotage, against the enemy overseas'. This was the SOE, whose unofficial mandate came from Churchill: 'Set Europe ablaze!' Gubbins, now brigadier, became SOE's director of operations and training from 18 November 1940 until the Labour government closed down its worldwide operations on the last day of 1945.

The novelist Graham Greene was working in the Ministry of Information, persuading other authors to come on board the war effort, when he wrote a short story about the British resistance which was published in *Collier's* magazine in June 1940. In 'The Lieutenant Died Last', a squad of German parachutists land near an English village called Potter, lock up the inhabitants in the only pub and set off to blow up the nearby London to Edinburgh railway line. They are foiled by a shabby old tramp and poacher who has a Mauser rifle from when he fought the Boers and who knows the land better than they do. In 1943, Mike Balcon's Ealing Films – the scriptwriters were Angus McPhail, John Dighton, and Diana Morgan – turned Greene's story into *Went The Day Well?*, one of the really interesting British films of the war, directed by Greene's friend Alberto Cavalcanti, the gay Brazilian maverick and avant-garde documentarist who had helped make the brilliant *Night Mail* for John Grierson's GPO Film Unit.

In the film *Went the Day Well?* a party of Royal Engineers arrives in an English village called Bramley Green. Slowly their deception is unmasked: they are actually English-speaking German parachutists in disguise, and the local squire is a Fifth Columnist helping them. The Germans in British disguise kill the vicar in his vestry when he starts to ring the church bell, and massacre the Home Guard on their bicycles.

An evacuee boy and a sailor on leave save the day, but it is the homely English village women who break their codes of domesticity and hospitality to start killing the enemy. The post mistress uses an axe, the vicar's daughter shoots the treacherous squire dead with a pistol, the land girl snipes with a Lee-Enfield .303, and the bossy lady from the manor house whom we thought was a prattling middle-class fool heroically scoops up a German grenade to save the working-class evacuee children, but not herself.

> Went the day well?
> We died and never knew
> But well or ill,
> Freedom, we died for you

As early as January 1937, the Air Ministry had recognised that the British film industry might be useful in war time. After an approach by Alexander Korda of London Films, RAF officers saw the potential of Denham Studios' workshops, lighting rigs and versatile craftsmen:

These men are specialists at 'make believe' and deception in defeating both the eye and the camera. They possess the workmen, material and shops to build jerry constructions for deception purposes.

Alexander Korda was born Sándor László Kellman in Hungary and had escaped from poverty and anti-Semitism in the flat plains by dreaming big and living extravagantly. Korda realised early that if you tipped waiters and doormen at the best restaurants and hotels generously in cash you could actually hold off paying the big bills until someone turned up with a business proposition. After making films in Budapest, Vienna, Berlin, Hollywood and Paris, he came to England in 1931. His first film was with another Hungarian (acting under the name Leslie Howard) whom he would later make a star in *The Scarlet Pimpernel*.

Korda became more British than the British, living north of Regent's Park in Avenue Road (with a camp butler called Benjamin), acquiring a chauffeured Rolls-Royce, and buying shoes for his tiny, dainty feet at Lobb, Homburg hats at Locke, double-breasted suits in Savile Row. He became a British subject in 1936, and was knighted in 1942. Korda founded London Film Productions in 1932 (its logo was Big Ben) and had a huge international success in 1933 with his first big film, *The Private Life of Henry VIII*, starring Charles Laughton (who won the Best Actor Academy Award) and Merle Oberon, Korda's discovery and future wife. This was the first British feature film to conquer world markets. Suddenly all sorts of people saw that money could be made from cinema and there was a mini gold-rush to finance more British films and film studios. Sir Connop Guthrie, chairman of the wealthy Prudential Insurance Company, encouraged by Sir Robert Vansittart of the British government, enabled Korda to build Denham Studios, which he wanted to be the best outside Hollywood.

In Winston Churchill, Korda recognised a fellow showman, and in September 1934 he offered Churchill £10,000 to write the script of a film on King George V for the Silver Jubilee in 1935. The film was never made, but Korda also wanted Churchill to 'advise' on various historical and imperial movies he planned, including one on Lawrence of Arabia that the subject himself personally dissuaded Korda from attempting. Ten years later, Korda offered £20,000 for the film rights to Churchill's *Life of Marlborough* and paid £50,000 for *A History of the English-Speaking Peoples*. According to Korda's nephew Michael, it was Churchill who stocked the pond outside Korda's office at Denham Studios with swans (having gained royal permission).

The British secret services were also among those drawn to the possibilities of the British film industry, in particular Claude Dansey, who saw that London Films could be useful 'cover' for persons from his 'Z' network travelling to foreign countries. Moreover, Korda had a double debt of honour to Britain, not only because he had been financed by Dansey's rich friends but because a shadowy figure from the secret services known as Brigadier Maurice had once saved his life. Just before WW2, Korda made the patriotic spy-film, *Q Planes,* and on the outbreak of war, he swiftly made the first propaganda film about the RAF, *The Lion Has Wings* (with Ralph Richardson), which was in cinemas by November 1939. Korda also helped his adopted country by making films which boosted Britain, whether set in the imperial past (*Sanders of the River, Elephant Boy, The Drum, The Four Feathers*) or the idealistic future (*Things to Come, The Man Who Could Work Miracles*). Churchill sent Korda to Hollywood during the war to continue making propaganda films that were also romantic entertainment. In 1941, *Lady Hamilton* (known in the USA as *That Hamilton Woman*) starred Laurence Olivier and Vivien Leigh, then in the full limelight of their adultery, as Admiral Lord Nelson and Emma Hamilton. Winston Churchill adored the film. And why not? It was Churchill himself who wrote Nelson's stirring speech on why Napoleon Bonaparte must be resisted.

Film-set crews from Shepperton started building fake aircraft factories to decoy bombers away from the real ones, Short's at Rochester, De Havilland's at Hatfield, Boulton & Paul's at Wolverhampton and the Bristol aircraft Company at Filton. Technicians from many British film studios – Gaumont-British, Sound City, Green Bros – were building dummy RAF aircraft – Battles, Blenheims, Hurricanes, Wellingtons, Whitleys – for dispersal on real and dummy airfields. The decoy airfields, which had lit-up landing paths at night, were known as 'Q sites' after the 'Q ships' of WWI. One RAF veteran recalled them looking 'pathetic' in daytime, just some poles, wires and bulbs, but at night the effect was amazing. 'It was quite impossible to tell a fake from the real thing.' In *Trojan Horses: Deception Operations in the Second World War* (1989) by Martin Young and Robbie Stamp, the electrician Geoff Selwood described how his unit experimented with visual effects for fake airfields. They made what they called a 'Running Rabbit', a long curve of wires which a light could run along. Then they built a railway out of barrage balloon cables and a series of small steel towers to carry a lighting rig on a trolley with a little rocket engine:

So what happened, we powered this trolley along the rails with the rocket, then we would switch the lights on and the running Rabbit would take over. The whole effect was just like a plane coming in to land very fast. It would slow down and then the Running Rabbit took over and for all the world it looked as if a plane had landed and then turned off the runway to the dispersal area.

In a further ingenious development, 'K sites' were built, more elaborate than the 'Q sites'. They simulated small aerodromes and attempted to draw the enemy away from the real ones. To look authentic by daylight, the 'K sites' needed more personnel than the 'Q sites', to move dummy aircraft about, to fake tracks, rearrange supplies and even man machine guns. 'K sites' were normally laid out on the line of enemy approach, that is, east of the real target, and were connected to it by telephone. Seymour Reit's *Masquerade: The*

*Amazing Camouflage Deceptions of World War II* (1978) records one K-Area call overheard during the Battle of Britain that seems almost too good to be true:

FLIGHT SGT. (agitated): Sir! We're being attacked!

PILOT OFFICER: Splendid, Sergeant. Good show.

FLIGHT SGT.: They're smashing the place to bits!

PILOT OFFICER: Yes, excellent. Carry on.

FLIGHT SGT.: But, sir – we need fighter cover! *They're wrecking my best decoys!*

# 19

# Fire over England

*If the Invader Comes*, 'issued by the Ministry of Information in co-operation with the War Office and the Ministry of Home Security', was a 1940 government leaflet whose authors were determined that no British citizen should be taken by surprise as the citizens of Poland, Holland and Belgium had been. In the event of invasion, there were seven rules. The first was to stay put, and not move. 'If you run away . . . you will be machine-gunned from the air . . . and you will also block the roads.' The seventh was to think before you act: 'But think always of your country before you think of yourself.'

There is another method which the Germans adopt in their invasion. They make use of the civilian population in order to create confusion and panic. They spread false rumours and issue false instructions. In order to prevent this, you should obey the second rule which is as follows: –

(2) DO NOT BELIEVE RUMOURS AND DO NOT SPREAD THEM. WHEN YOU RECEIVE AN ORDER, MAKE QUITE SURE THAT IT IS A TRUE ORDER AND NOT A FAKED ORDER. MOST OF YOU KNOW YOUR POLICEMAN AND A.R.P. WARDENS BY SIGHT, YOU CAN TRUST THEM. IF YOU KEEP YOUR HEADS, YOU CAN ALSO TELL WHETHER A MILITARY OFFICER IS REALLY BRITISH OR ONLY PRETENDING TO BE SO. IF IN DOUBT ASK THE POLICEMAN OR A.R.P. WARDEN. USE YOUR COMMON SENSE.

For Churchill, almost all means were valid in self-defence. The man who was prepared to spray mustard gas on German troops from massed aircraft if they got ashore obviously saw rumour as another useful weapon against them. John Baker White, an officer in the overlap between EH and Section D and then later in the Political Warfare Executive (PWE), wrote a book in 1955 about British psychological warfare, which he called *The Big Lie*. 'We used the Big Lie when we were weak and the Great Truth when we were strong,' he stated. 'By rumour and deception we built another wall about Fortress Britain, when we stood alone.' This was the dark corollary to Churchill's speeches, using words not to hearten and lift up your own people, but to discourage and depress the enemy.

Baker White's own most notable contribution to the 1940 war effort is recorded in his book's first chapter – 'The Sea Is On Fire!' Were the German Army (quite unused to amphibious operations) ever to invade England, it would have to cross the English Channel in some kind of landing craft. Baker White's notion was to make them fear they might be cooked alive in the process. He wanted them to imagine a frightful double death, first blazing hot and then drowning cold. This was not a wholly new idea: 'Greek fire', a kind of napalm, had flared in Byzantine times, and burning ships were used by the British against both the Spanish Armada and the Napoleonic fleet.

Baker White's immediate source may have been Dennis Wheatley. The best-selling author of *The Devil Rides Out* and *They Found Atlantis* had found no war job to date but had been busy penning the lurid (and ludicrous) adventures of his secret agent hero, 'lean, Satanic-looking' Gregory Sallust, 'the man the Nazis couldn't kill!' in thrillers like *The Scarlet Impostor*, *Faked Passports* and *The Black Baroness*. However, Wheatley was close to the Security Service, and his second wife was working as an MI5 driver in May 1940. When one of her counter-intelligence passengers told her he was a bit stuck for ideas to resist invaders, Joan Wheatley suggested trying her imaginative husband. In twenty-four hours from 27 to 28 May 1940 Wheatley produced a 7,000-word paper, *Resistance to Invasion*, fizzing with forty-five ingenious ideas of how to 'undermine enemy morale', which was circulated in Whitehall. One proposal was for an oil ship to be detonated from shore, 'so that flaming oil will spread over the water and ignite the enemy craft'.

As it happened, there was an excess of petroleum products in Britain owing to the disruption of Shell and BP's trade with Scandinavia and Europe. General Ironside had noted it on 27 May: 'There is far too much petrol in the coast areas, most of it unguarded.' But Maurice Hankey, Minister without Portfolio, thought that these fossil fuels could be used 'for defensive purposes', so various flame-thrower trials and experiments were carried out along the south coast at places like Dumpton Gap and Dungeness, which were in full view of occupied France and passing German aircraft. John Baker White saw an experiment at St Margaret's Bay in Kent, when oil was pumped from bowsers down the cliff along pipes buried in the beach to ignition flashpoints, and he found the flame barrage's red tongues in clouds of black smoke 'a frightening spectacle'. It was, in fact, soon found impossible to set the sea on fire. Notwithstanding the truth, Baker White was determined to use the idea of a burning ocean to terrify the enemy.

Rumours were an established part of psychological warfare. The British called them 'sibs' – from the Latin *sibilare*, meaning 'to hiss' or 'to whistle' – and their purpose was to disturb enemy soldiers. In September 1940 Baker White put his idea of 'setting the sea on fire' through the three committees that vetted all sibs before release, in case they were genuinely true or compromised actual operations.

There was a tangled bureaucracy to negotiate. In July 1940, the Special Operations Executive (SOE) had been set up under Hugh Dalton, who was also head of the Ministry of Economic Warfare, founded when war broke out. SOE effectively took over the co-ordination of Section D (the Sabotage Service, formerly under the Foreign Office), MI(R) (guerrilla warfare, formerly under the War Office) and EH (Electra House, Propaganda to Enemy Countries, formerly under joint control of the Foreign Office and the Ministry of Information). Initially the idea was to split the new SOE itself into words and deeds, SO1 (Subversion/Propaganda) and SO2 (Sabotage/Operations). The dichotomy was, in fact, disastrous. Ministry fought with ministry. Only after many months of tedious Whitehall inter-departmental turf wars did a new secret service emerge separately out of SO1. This was the Political Warfare Executive (PWE), chartered on 20 September 1941 and no longer under the Ministry of Economic Warfare. PWE's cover name was Political Intelligence Department (PID) of the Foreign Office, and its job was all forms of propaganda against the enemy, including overall control of the BBC. As John Baker White put it: 'A deception – a Big Lie – was a military operation. Political warfare was the machine used to project it to the enemy.'

A big pearl of a lie is best seeded with a grain of truth. Whereas the sib that the British had imported 200 man-eating sharks from Australia to release in the English Channel was strikingly short of supporting evidence, there really were some burned Germans to support Baker White's sib about setting the sea on fire. Churchill told the secret session of the House of Commons that the Germans had gathered 'upwards of 1,700 self-propelled barges and more than 200 sea-going ships' in occupied ports, ready for invasion. When the RAF attacked this German shipping in harbours from Emden to Le Havre with incendiaries and high explosive in September, the flames could be seen from Kent. Injured German soldiers were transferred to Paris hospitals and the story spread that they had been burned in a failed invasion. French wags began to stand behind the German soldiers occupying their country and pretend to warm their hands on them. Belgians swore they knew nurses who had tended hundreds of moaning Germans with burns.

The RAF dropped leaflets and a *Short Invasion Phrasebook* with handy phrases for 'The Water's on fire!' in German, French and Dutch: (*Hier brennt sogar das Wasser! Même l'eau brûle ici! Hier staat waarachtig het water in brand!*) On the BBC German Service, Sefton Delmer gave mock English lessons: '*Das Boot sinkt* . . . the boat is sinking' with useful verbs '*Ich brenne* . . . I burn, *Du brennst* . . . you burn, *Er brennt* . . . he burns . . . And if I may be allowed to suggest a phrase: *Der SS Sturmführer brennt auch ganz schön* . . . The SS Captain is al-so bur-ning quite nice-ly.' Other broadcasts in German gave out the names of captured German seamen saying they had been 'rescued' from the sea while the fate of their unfortunate companions was not known.

All along the Atlantic coast German soldiers put two and two together and made four hundred. Captured Luftwaffe pilots had all heard the story, Wehrmacht personnel wrote home with lurid versions of it. The burning-sea story also spread through Britain (whether by accident or design) almost as fast as the Russians-with-snow-on-their-boots story had in WW1. From Dorset to Dover and from Sandwich to Shingle Street in Suffolk there were stories of dozens, scores, hundreds, no, *thousands* of German soldiers hideously charred and incinerated in a seaborne invasion that failed horribly sometime around the weekend of 14–15 September 1940.

In fact, only thirty-six German dead bodies were washed ashore in Britain that year, mostly Luftwaffe pilots and air crew. If they were burned, it was because they had been shot down in flames. However, the first crack at 'The Big Lie' was astonishingly successful. Perhaps Hitler was right when he sneered sarcastically on 4 September 1940 that 'the British should not forget to raise their most important general to the rank of Field Marshal of the Empire. I mean General Bluff. That is their only reliable ally.' (Hitler also made use of bluff, keeping the invasion threat going long after he had decided to turn on Russia instead.)

The German invasion came not by water but by air, because Hermann Göring promised Adolf Hitler that his airmen would win command of the skies before the army and navy crossed the sea. In 1937, while entertaining Lord Trenchard, boasting of the superior powers of his secretly rebuilt Luftwaffe, Göring took his guest outside for a magnificent firework display in the chilly night. Loudspeakers blared out an amplified recording of an artillery barrage, mixed with the whine of dive-bombers swooping to drop their whistling loads of explosive bombs. This was barely two months after the destruction of Guernica. 'That's German might for you,' Göring shouted. 'I see you trembled. One day German might will make the whole world tremble.' 'You must be off your head,' the founder of the RAF angrily replied. 'I warn you, Göring, don't underestimate the RAF.'

From July to October 1940 the Luftwaffe and the RAF clashed above southern England in the series of air combats that became known as the 'Battle of Britain'. Some doubt if there was ever a coherent German plan; bombers would simply bash Britain until it gave up, which it surely had to. But the illogical British stubbornly refused to surrender, and what ensured was the mythic battle of which Churchill said, in August 1940, 'Never in the field of human conflict was so much owed by so many to so few.' The 'few' were British, Canadian, Czech, Polish and South African pilots.

London's Croydon Aerodrome was attacked on 18 August, when the Home Guard managed to shoot down a Dornier with 180 rounds of rifle fire. Central London and the City were first hit by the German air force on the night of the 24th. Then it became a war of tit for tat. RAF Bomber Command bombed 'military targets' in the German capital Berlin. Major Nazi reprisal bombing started at teatime on Saturday, 7 September 1940. A huge armada of enemy aircraft flew up the Thames estuary in broad daylight, over 300 bomber planes with more than 600 fighters protecting them. The journalist Virginia Cowles, weekending in the country, saw them in the distance like a swarm of insects. They were heading for the wharves and warehouses of the East End of London where the riches of the British Empire were unloaded and stored at the sprawling docks. The Port of London Authority Docks had forty-five miles of quays and moved a third of the UK's imports and a quarter of its exports. Three big railway stations – London Bridge, Waterloo and Charing Cross – were knocked out, water mains and sewage pipes were broken, gas and electricity cut. Two hours after the 'All Clear' sounded, another 250 bombers returned at night, their path clearly lit by flames, to add their tonnage of explosives and incendiaries to the inferno below. More than sixty major blazes and a thousand smaller ones made it the worst conflagration since the Great Fire of London. From five miles away, you could read the *Evening Standard* by it in the blackout.

Colonel John Fisher Turner, in charge of creating dummy airfields, was put on his mettle by the Blitz. His job was to improvise and co-ordinate the military and civil bombing decoy systems from his office at Shepperton Studios. From November 1940 he set in motion the decoy fires of the QF sites, and the Special Fires of the SF or 'Starfish' sites. As Colin Dobinson showed in *Fields of Deception: Britain's Bombing Decoys of World War II* (Methuen, 2000), these decoy fires were set in country outside urban areas where further bombs would do no harm. The idea was to simulate the different visual effects of burning buildings; the aim of these second-degree decoys was to catch the attention of the second and third waves of Luftwaffe bombers who would then unload their bombs uselessly on what they thought were the right targets. Cans of creosote and roofing-felt generated what the manual calls the 'large spluttering fire', while big drums of creosote with rolls of roofing-felt stuffed in end-on led to 'heavy initial smoke fire'. The 'dull basic fire' was a brazier of coal, ignited by flare cans. With the addition of a header tank of oil and a structure of pipes, this became the 'coal drip fire', which flared up dramatically when fuel was sprinkled on it. Eventually forty-two towns had 130 'Starfish' systems helping to protect them. As German use of incendiaries increased in 1941, so did the sophistication of the dummy Special Fires mimicking what the Germans called the *Brandbombfeld* or 'firebomb field'. These later Starfish were laid out in patterned groups and ignited in relays. They contained 'basket fires', crates of wood wrapped in wire netting and hessian, containing wood shavings, sawdust and flammable waste, soaked with layers of creosote. These burned impressively for an hour and were interspersed with 'crib fires' of firewood and coal, dramatic boiling oil fires and yellow-flamed 'grid fires'. The Luftwaffe were not always deceived, for they had their own decoys, but sometimes they were, because navigation in those days was not always accurate. Dobinson concludes that Colonel Turner's decoys probably wasted about 5 per cent of the German bombing effort, and may have spared nearly 2,600 civilians from death and over 3,100 from serious injury.

# 20

# Radio Propaganda

'There is no question of propaganda,' Sir Samuel Hoare told the House of Commons in his capacity as Lord Privy Seal on 11 October 1939. 'It will be publicity and by that I mean straight news.' To British ears, the word 'propaganda' is unpleasant. In 1928, Arthur Ponsonby's *Falsehood in War-Time* exposed many myths of WW1, showing how in that war 'propaganda' came to mean misrepresentation and manipulation. The connotations have remained since mostly negative; except, of course, when you truly believe in what is being propagated or put forward.

The documentary film-makers were one such band of believers. John Grierson first used the word 'documentary' in a 1926 newspaper review he wrote of Robert Flaherty's anthropological film about Western Samoa, *Moana*, saying that it had 'documentary value'. From 1929 'documentary' became the self-defining term for an important group of British film-makers associated with Grierson who were interested in 'the creative treatment of actuality'. Grierson worked closely with a public relations man of genius, a remarkable British civil servant, Sir Stephen Tallents, who had been wounded in the trenches with the Irish Guards, and worked on social reforms with William Beveridge. In 1926, Tallents became the secretary of the Empire Marketing Board. Playing on Tallents's internationalist vision, John Grierson persuaded him that cinema could help make the British Empire 'come alive'. Accordingly, after getting some ideas from Rudyard Kipling at Burwash, Tallents commissioned a film from Walter Creighton called *One Family*, in which a small boy falls asleep over his geography lesson and dreams a dream of the British Empire. A 1930 review found it 'the most extraordinary picture yet made by a British firm':

The portions of the film dealing with men at work express that work with a force and honesty that has never been seen in British films on a large scale, and has rarely been equalled, even in Soviet productions.

What Tallents encouraged in British documentary film-makers was public service propaganda. These non-commercial films looked at the social utilities that linked everybody – electricity, gas, post, railways, shipping, telephones, wireless, and so on – and were the first that allowed ordinary people to speak to the camera. The documentary film-makers were not embarrassed by the word 'propaganda'. John Grierson's epigraph to Paul Rotha's *Documentary Film*, published by Faber in January 1936, 'I look upon cinema as a pulpit, and use it as a propagandist', is confirmed in his introduction to the book:

Our own relation to propaganda has been simple enough. We have found our finances in the propaganda service of Government Department and national organisation . . . Documentary gave to propaganda an instrument it needed and propaganda gave to documentary a perspective it needed. There was therefore virtue in the word 'propaganda', and even pride; and so it would continue for just as long as the service is really public and the reference really social. If however, propaganda takes on its other more political meaning, the sooner documentary is done with it the better.

Most British journalists recoil from the word 'propaganda' as though from a poisonous snake, yet it is really a pet which sits on their desk. All journalism is propaganda when it presents a case or seeks to persuade, because the estimation of 'news value' and the ordering of an argument is intimately linked to a belief system. The greatest journalists understand this. 'I was a professional recorder of events, a propagandist, not a soldier,' wrote one of WW2's finest reporters, Alan Moorehead, of himself. Purge 'propaganda' of negative associations and see it as a branch of rhetoric, or as information directed to public service, and we may get nearer to the way the British came to see it in WW2. In 1936, Sir Stephen Tallents became controller of public relations at the BBC, where a parallel process to Grierson's 'imaginative interpretation of everyday life' was going on among the first radio documentary feature-makers, like John Pudney and Stephen Potter.

In May 1940, the telephone rang at Sissinghurst Castle in Kent. After a long pause, the Hon. Harold Nicolson, the epicene National Labour

MP for West Leicester, was told to hold on for the new Prime Minister. Nicolson had written a Penguin Special in three weeks in October 1939, *Why Britain is at War*, which sold 100,000 copies, and Churchill was inviting him to join the Ministry of Information. As parliamentary secretary to the minister, Duff Cooper, Nicolson found himself in 'the most unpopular department in the whole British Commonwealth of Nations'. (Men from the MoI were known as 'Cooper's Snoopers'.) But Nicolson applauded the British public's 'healthy dislike of all forms of Government propaganda' and sympathised with the 'unconquerable minds' of citizens and journalists frustrated by wartime limitations. Britain was not the Third Reich and nor was he himself 'imitating the technique of Doctor Joseph Goebbels'. In the article he wrote at the end of 1940 for the *BBC Handbook 1941*, Nicolson explained 'the essential difference between the theory and practice of German, or totalitarian, propaganda, and British, or democratic, propaganda'. The first was a 'smash-and-grab raid on the emotions of the uneducated' and the second an appeal to the intelligent, free mind. 'Totalitarian propaganda is akin to revivalism; democratic propaganda is akin to education.'

Nicolson was sometimes encouraged by the British people's spirit, but sometimes was in despair:

I am feeling very depressed by the attacks upon the Ministry of Information . . . And it may be true that if our propaganda is to be as effective as that of the enemy, we must have at the top people who will . . . be caddish and ignorant enough to tell dynamic lies. At present the Ministry is too decent . . . We need crooks.

Duff Cooper had already found one: Sefton Delmer. 'Don't drop your reporting for the *Daily Express*,' the latest Minister of Information said to the journalist, newly escaped from the fall of France with all the other journalists, 'but if you could fit in the occasional German broadcast on the BBC we shall all be most grateful.'

Radio was Sefton Delmer's destiny. When he came to write the first volume of his autobiography, he remembered the summer of 1914 when he was with his mother and sister in the German spa of Bad Sachsa and they watched a cinematograph of the assassination of the Austrian Archduke. A few days later troops in field-grey uniforms camped in the water meadows and a signals unit mounted a brand-new field wireless station, erecting a huge mast, fixing the antennae and cranking the motor of their electric generator. Its roaring splutter was 'the first echo of twentieth century war' for the ten-year-old boy.

Delmer's most celebrated broadcast on the BBC German Service was on 19 July 1940, an hour after Adolf Hitler had spoken at a specially convened session of the Reichstag, praising the German armed forces for their magnificent victories across Europe, and boasting of his strategic skill, before offering peace to England. Hitler had said:

Mr Churchill ought for once to believe me when I prophesy that a great Empire will be destroyed – an Empire which it was never my intention to destroy or harm. If this struggle continues it can only end in the annihilation of one of us. Mr Churchill thinks it will be Germany. I know it will be Britain. In this hour I feel it is my duty before my conscience to appeal once more to reason and common sense in Britain . . . I can see no reason why this war must go on. We should like to avert the sacrifice of millions.

Without time for elaborate prior consultation between the propagandists of EH and the diplomats of the Foreign Office, large Tom Delmer settled under the BBC microphone to make a momentous broadcast. In his autobiography, Delmer says it was his first ever in any language, which is not strictly true as he had already done two or three for the BBC. But as a piece of bare-faced cheek, it counts as one of the great debuts in wireless history. Delmer addressed himself directly to the Führer in smooth and deferential German: 'Herr Hitler, you have on occasion in the past consulted me as to the mood of the British public. So permit me to render your excellency this little service once again tonight. Let me tell you what we here in Britain think of this appeal of yours to what you are pleased to call our reason and common sense. Herr Führer and Reichskanzler, we hurl it right back at you, right in your evil-smelling teeth . . .' This shocked the Germans. They could not conceive that such rudeness would be allowed without the highest possible sanction. Confirmation did come on the Monday when Lord Halifax, former appeaser though he was, spoke for the British government and formally rejected parley with Germany.

Delmer was clearly not a man who could fit into the 'spinsterish' civil service ethos of the BBC full time, with its intolerable 'dreariness and pious unrealism', but he continued to broadcast on occasion for the

BBC because he was both good and quick. In 1941–2, when Hugh Greene's German Service had to respond within hours to the chief domestic radio commentator of Goebbels's Propaganda Ministry, Hans Fritzsche, they naturally asked Sefton Delmer to refute his broadcasts point by point. But Delmer's genius for the medium was finally fulfilled not in the 'white' BBC but in the rougher game of 'black' radio.

Before becoming *Meister im Rundfunk*, Delmer entered into an apprenticeship and alliance with Leonard Ingrams. They had known each other from before the war when Ingrams was 'the flying banker' who piloted his own Puss Moth plane around Europe. Delmer figured that Ingrams (the father of Richard, founding editor of *Private Eye* magazine) was somebody in the cloak-and-dagger world, and found later that he was 'a star operative on the British side of the Secret War'. Ingrams was, in fact, an undersecretary at the Ministry of Economic Warfare who liaised with Electra House, the Secret Intelligence Service, SOE and later PWE.

Arthur Christiansen, editor of the *Daily Express*, sent Delmer to Lisbon in November 1940 where over the next three months the reporter re-immersed himself in the world of Nazi Germany by interviewing hundreds of refugees who had escaped or bribed their way out and were either settling in Salazar's Jew-tolerating dictatorship, or preparing to sail on to the New World. In February 1941, Ingrams obtained security clearance from both SIS and MI5 for Delmer to join the Political Intelligence Department of the Foreign Office to start work on a new kind of broadcasting, in German, to Germany.

The BBC employed fewer than 5,000 people at the start of WW2, and they were soon thrown into upheaval. On Friday 1 September 1939, Val Gielgud (brother of the famous actor John), head of the BBC's Drama and Features department, was in a rehearsal room off Marylebone High Street. He was preparing to direct his first ever play for the new medium of television, *The Circle* by W. Somerset Maugham, when the BBC entered its 'emergency period'. There was a pre-arranged code: when the announcers said 'This is London' instead of 'This is the National Programme', television from Alexandra Palace closed down for the duration. All departments at Broadcasting House except News followed their evacuation procedures and made their way to the regions to escape the imminently expected carpet-bombing of London. Because these bombers could have used British medium-wave transmitters as a navigational aid, the Regional Programmes and the National Programme were merged into one 'Home Service'. Output was restricted to eight news bulletins a day, government edicts, and hours of Sandy Macpherson playing the BBC Theatre Organ.

Since all theatres, cinemas, dance halls and places of public entertainment were also closed by government order for fear of mass deaths from said bombing, it is little wonder that bored listeners scrolling across the radio dial for livelier fare in the blackout came upon German propaganda broadcasting to Britain from Hamburg. On 18 September 1939, Jonah Barrington, the radio critic of the *Daily Express*, heard a voice over the airwaves that he christened 'Lord Haw-Haw of Zeesen':

From his accent and personality I imagine him with a receding chin, a questing nose, thin yellow hair brushed back, a monocle, a vacant eye, a gardenia in his button-hole. Rather like P.G. Wodehouse's Bertie Wooster.

It is almost certain that the particular voice he heard actually belonged to an improbable Polish-German, MG-driving playboy called Wolff Mittler, but the nickname 'Lord Haw-Haw' became wrongly attached to another broadcaster of German propaganda, the razor-scarred Irish fascist William Joyce. Joyce had once been Oswald Mosley's deputy in the British Union of Fascists, and he sounded less a silly-ass-toff than a sarcastic schoolmaster. (BBC Monitoring called him 'Sinister Sam'.) Most British households had a wireless licence in 1939, and the BBC audience was about twenty-eight million people. More than half of them heard Haw-Haw drawling 'Jairmany calling . . . Jairmany calling . . .' at some time or other in the Phoney War, and the keenest BBC listeners heard him most. The government worried about him because he was saying things that you did not usually hear on the BBC, offering criticism of poverty and slums and unemployment and making sneering attacks on the rich and powerful who ignored them.

Haw-Haw's snide commentary during the Phoney War pushed the BBC towards finding new voices who spoke more freely. In early July 1940 Harold Nicolson of the Ministry of Information had a conversation with the Parliamentary Private Secretary to Clement Attlee, the Labour leader who, as Lord Privy Seal, was effectively

Churchill's Deputy Prime Minister. Nicolson reported that

Attlee is worried about the BBC retaining its class voice and personnel and would like to see a far greater infiltration of working-class speakers . . . The Germans are fighting a revolutionary war for very definite objectives. We are fighting a conservative war and our objects are purely negative. We must put forward a positive and revolutionary aim admitting that the old order has collapsed and asking people to fight for the new order.

This is essentially what the novelist and playwright J. B. Priestley started doing. His *Postscript* talks after the 9 o'clock news on Sunday nights between June and October 1940 had enormous popular impact in that crucial year (but were not liked by the establishment). Priestley did not sound like an upper-class chap from Oxford or Cambridge. He was a solid Yorkshire bloke who had seen a thing or two, and he had definite ideas about what kind of broadcasting worked.

In a similar way, everyone who heard him felt that Tommy Handley (1894–1949) was real and true. The first great star of British radio was someone who could play the straight man or be the comic and yet always remain himself. Born in Liverpool, he acquired the scouser's gift for backchat and daft surrealism. From his schooldays, Handley spent all his pocket money on wigs, masks and false moustaches, and loved conjurors, drama, pantomimes and pierrot shows. While serving in the Royal Navy Air Service Kite Balloon Section at the end of WWI he toured with a concert party giving three shows a night to troops and wounded. Tommy Handley had a good singing voice and after the war travelled the country doing musical comedy. In 1924, Handley made his first broadcast from the BBC studios at Savoy Hill, so he was in at the birth of radio, the perfect medium for his kind of quick-fire patter. Ted Kavanagh (father of the poet P. J. Kavanagh) was a stout, balding, red-haired New Zealander who sold his first radio sketch to the BBC for 3 guineas and had the pleasure of hearing Tommy Handley bring it to life. Handley asked Kavanagh if he could write some more; they worked together for the next twenty-three years.

In August 1937, a corduroy-jacketed BBC producer in Bristol called Francis Worsley (ex-schoolmaster, ex-colonial service) had to put together a sound picture called 'Evening in Cheddar' from the famous caves in Cheddar Gorge. Worsley livened things up by getting Tommy

Handley to join the party of tourists being shown around the stalactites. Equipped with a few gags scripted by Kavanagh, Handley got some exuberant repartee going with the guide, but the BBC drew criticism from scientists for allowing 'a red-nosed comedian' to contaminate knowledge with humour.

In June 1939, Francis Worsley was working in BBC Variety in London, looking for a new radio comedy show. Tommy Handley suggested Ted Kavanagh as the writer, and over pink gins at the Langham Hotel, the trio came up with an imaginary venue for the show: a pirate radio station on a cruise ship. It would be a floating mad hatter's party with Handley in charge of broadcasting.

This was a crucial moment in the run-up to WW2. The Spanish Civil War had just ended and Adolf Hitler was all over the news. Whenever the Führer demanded further concessions from the democracies, the *Daily Express* used to run the headline 'It's That Man Again!' Good title for the show, the trio thought. They hired Jack Harris's band from the London Casino and the first programme went out live from the big BBC studio at Maida Vale from 8.15–9.00 p.m. on Wednesday, 12 July 1939.

In the ensuing 'emergency period', BBC Variety and its Repertory Company were at first evacuated to Bristol. Worsley the public-school *New Statesman* reader, Handley the nonconformist, animal-loving Conservative, and Kavanagh the Roman Catholic follower of G. K. Chesterton all got together to think about the new *It's That Man Again!*, which was due for live broadcast in a fortnight's time. What were they going to do now war had broken out? Their cruise ship idea was *kaput* with the peace. All around was evidence of heightened security procedures, new Ministries issuing orders, government bumf, urgent acronyms. Handley was, as usual, doodling on a pad, when he ringed the first letters of *It's That Man Again!* Thus *ITMA* was ready to join ARP, FANY, MEW, MoI, RAF, WVS and the rest of the baffling initials of wartime. Kavanagh's idea of radio-writing was 'to use sound for all it was worth, the sound of different voices and accents, the use of catch-phrases, the impact of funny sounds in words, of grotesque effects to give atmosphere – every device to create the illusion of rather crazy or inverted reality'. The surreal half-hour show contained over eighteen minutes of scripted dialogue and they aimed for a hundred laughs in that time, with a gag, pun or tongue-twister every eight seconds.

TOMMY: Heil folks – it's *Mein Kampf* again – sorry, I should say hello folks, it's that man again. That was a Goebbeled version, a bit doctored. I usually go all goosey when I can't follow my proper-gander.

The madcap world of *ITMA* was like a series of bright cartoons, strung together by the cheery voice of Tommy Handley, playing a busy-body in the government, dealing with bizarre situations. The first *ITMA*, censored, security-vetted, and broadcast from Clifton Parish Hall three weeks into the war on Tuesday, 19 September 1939, did not have most of the characters and the catchphrases that it would develop over three hundred shows in the next ten years, but already ridiculed the officious nonsense that everybody was having to live through.

TOMMY: What's this? Order for the prohibition of peanuts in public places? I'll sign everything that prohibits anything. Get fifty million pamphlets printed.
FUSSPOT: Fifty million pamphlets, sir?
TOMMY: Yes, cancelling all the pamphlets issued already. And phone the BBC and tell them I've got so thin I'm coming along to join the skeleton staff. Finally, if anyone is doing anything, tell 'em to stop it at once.

The next show introduced Funf, the serio-comic German spy, a caricature of Adolf Hitler. 'This is Funf speaking' became a telephone catchphrase of the Phoney War and helped reduce the Abwehr and the Führer to a laughing stock. The feature 'Interned Tonight' had harmless characters like Herr Cut the barber. By November the show was running smoothly. The fictional camouflage unit made everyone invisible – so invisible that Tommy Handley, Minister of Aggravation and Mysteries in charge of the Office of Twerps, could not find his own desk and he was evacuated to 'somewhere in England' to set up a secret broadcasting station.

The nation listened to *ITMA* right through the war and loved its succession of fantastical characters and topical jokes. You could not interrupt the King and Queen of England between 8.30 and 9 p.m. on Thursdays when *ITMA* was on. Princess Elizabeth's sixteenth birthday present in April 1942 was a two-hour command performance of *ITMA* at Windsor Castle.

*ITMA* comforted people by making light of dark times and showing the British that they could still laugh at themselves. The spirit was very similar to that of Heath Robinson's drawings, satirising official pomposity and secrecy with surreal inventiveness; in 1940 his series of cartoons of Winston Churchill included a drawing of the First Lord of the Admiralty 'disguised as a swan' laying magnetic mines in the Thames to discredit the Nazis. Heath Robinson's absurdity, like *ITMA*'s, left you with no other option but laughter. Lord Haw-Haw's aim was the opposite, gloatingly telling people how bad things were in Britain, and how divided the country was, how inevitable their defeat. Poor Mrs Bellamy of Sheffield, who killed herself after listening to Haw-Haw, must count as a small triumph of Nazi demoralisation. *ITMA*'s well-known catchphrases played an important bonding role in wartime. The diver's lugubrious 'I'm going down now' was heard in every lift and even over RAF frequencies as pilots swooped to attack ground targets. When a rescue party located a schoolboy buried in a bombed house, he managed to refer to Mrs Mopp from under the rubble – 'Can you do me now, sir?'

When the German airship *Hindenburg* burst into flames on 6 May 1937, the reporter Herbert Morrison broke down in tears on radio – 'Oh the humanity!' – at the sight, although he did not stop recording sounds and interviews from the scene. That same summer, Edward R. Murrow moved to London as the European Director of the US radio network CBS (Columbia Broadcasting System). He hired William L. Shirer in Berlin, and made his own first broadcast from Vienna during the *Anschluss* of March 1938. Murrow soon became an authoritative news reporter who could speak directly and naturally to the American people in vivid vignettes. During the Munich and the Czechoslovak crises, US audiences became used to CBS news reports from the likes of Bill Shirer and Ed Murrow dramatically interrupting regularly scheduled programmes, because all news was live.

It was on CBS radio that Orson Welles pulled off a Hallowe'en sensation in October 1938, using apparently live radio reports from New Jersey to dramatise H. G. Wells's *The War of the Worlds*. The faked authenticity of the actor playing the radio reporter – Welles made him listen repeatedly to Morrison's *Hindenburg* broadcast and mimic his jerky speech and emotional excitement – was so effective that listeners called the police, and people fled their homes before the Martian invaders.

By the time war broke out in September 1939, Murrow had put together a good team of correspondents. But unlike Goebbels's Propaganda Ministry, who were quick to grant neutral broadcasters the facilities and interviews they wanted, Britain initially bungled its relations with the American radio people. For the first year of WW2, the War Office and Admiralty were unhelpful, the Ministry of Information strangled them in red tape, and Home Security blocked their movements. This all changed with the Blitz.

At 11.30 p.m. on Saturday, 24 August 1940, during the first general night attack on the London region, Edward R. Murrow was standing on the steps of St-Martin-in-the-Fields looking towards Nelson's Column and holding a microphone: 'The noise that you hear at this moment is the sound of the air-raid siren . . . People are walking along very quietly. We're just at the entrance of an air-raid shelter here, and I must move the cable over just a bit so people can walk in.' He crouched down to record the sound of footsteps, the shelter door, the background noise of red double-decker buses and sirens in the background. Even though Murrow was using BBC equipment, British listeners were not permitted to hear such live material in 1940. Everything was scripted, censored, and then checked against the script as it was being read.

The bombing heard faintly on Murrow's broadcast got worse in September and October with air raids every night. He made his first ad-libbing broadcast from the roof of Broadcasting House during an air raid on 21 September 1940, and was inside it during the great raid of 15 October when a 500-lb bomb hit the building while Bruce Belfrage was reading the nine o'clock news. This delayed-action device finally exploded in the music library, killing four men and three women, injuring many others and blowing a hole in the starboard or west flank, five storeys up. Murrow saw the pub on his street corner, the Devonshire, demolished, with thirty dead. His friends and neighbours, Claire and Alan Wells, were killed by bomb shrapnel in Portland Place. Murrow went out at night and talked to people, the poor in the tube shelters, the rich in hotels, the wardens and the first aiders, the fire and rescue services, civil defence on duty, and reported it all in his restrained graphic way, letting neutral America see the dead of London.

This was important at a time when the 'America First' movement, led by the isolationist aviator Charles Lindbergh and supported by the US Ambassador in London, Joseph Kennedy, thought that the USA should remain well isolated from European entanglements and foreign wars. Britain urgently needed America's help, but when Churchill inspected civil defence in Glasgow with President Roosevelt's envoy, Harry Hopkins, on 17 January 1941, he did not actually ask for direct intervention: 'We do not require in 1941 large armies from overseas. What we do require are weapons, ships and aeroplanes. All that we can pay for we will pay for, but we require far more than we shall be able to pay for.'

In the propaganda job of persuading the Americans to help, the British knew that Americans would make a better job of wooing the USA than Britons. The connection between the two countries is caught beautifully in Alice Duer Miller's narrative verse book *The White Cliffs*, a superb piece of emotional propaganda that went through eleven American editions in three months in late 1940, and eight British editions in five months early in 1941.

*The White Cliffs* is the story of an Anglophile New England girl called Susan Dunne who falls in love with and marries a pink Englishman from Devon in the summer of 1914, and slowly becomes 'almost an English woman'. Her husband and his older brother are both killed in WW1, and Susan sees it as her duty to bring up her son in the expected English tradition in their inherited leaky manor. But her Yankee father sees the English as 'the redcoat bully – the ancient foe', and as WW2 looms, Susan is afraid that her only son will die like his father and uncle. She wonders, is contemporary England worth it?

> I thought of these years, these last dark terrible years
> When the leaders of England bade the English believe
> Lies as the price of peace, lies and fears,
> Lies that corrupt, and fears that sap and deceive.

But then she thinks back to Elizabeth I and Cromwell and 'the sullen might/Of the English, standing upon a right' and she comes to understand that American quests for freedom were also rooted in English liberty.

> I am American bred,
> I have seen much to hate here – much to forgive,
> But in a world where England is finished and dead,
> I do not wish to live.

Other Americans helped to get the message across. Quentin Reynolds was probably the best-known American print journalist in Britain. He was due to return to the United States not long after the bombing of London started in September 1940, and the Crown Film Unit wanted him to take a film about the Blitz back under his arm. Humphrey Jennings, Harry Watt and editor Stuart McAllister made *London Can Take It*, a strong, simple film about British stoicism amid wreckage and destruction, in ten days. Quentin Reynolds wrote and narrated its tough-guy script: 'I am a neutral reporter. . . I can assure you, there is no panic, no fear, no despair in London town . . . . [A] bomb has its limitations: it can only destroy buildings and kill people. It cannot kill the unconquerable spirit and courage of the people of London . . . London can take it.' Reynolds screened the film for President Roosevelt.

While American entertainers cheered the British in radio shows like *Hi Gang!*, the BBC set up its own North American radio service in July 1940, staffed by Canadians. Its half-hour *Radio Newsreel* mixed actuality, eyewitness reports and short talks by members of the armed forces or civilian services. Charles Gardner's lively description of a German air attack on a British convoy passing through the Straits of Dover on 14 July 1940, deplored by some in England for its resemblance to a sports commentary on men in mortal danger, was liked in North America because its disjointed spontaneity felt authentic. In 1940–1, the correspondents were still called 'BBC observers', but by 1944 they were no longer coolly detached but warmly embedded. On D-Day, 6 June 1944, the programme *War Report* first went on air, and vivid front-line pieces from the BBC War Reporting Unit became the norm.

By the time the Blitz started in September 1940, the BBC was making much more of its non-British human resources and broadcasting in seventeen European languages. Sir Stephen Tallents was controller (Overseas) with John Salt as the director of European Services. Salt came from the BBC Overseas Intelligence Department, where one of his colleagues, Emile Delavenay (father of the biographer Claire Tomalin), had written an incisive study of how the German propaganda campaign sapped French morale in the first half of 1940. Another colleague, the future poet and translator Jonathan Griffin, wrote in the *Monthly Intelligence Report, Europe* that a campaign was needed 'to lead people to desire an anti-Nazi revolution', using 'concrete facts, slogans, symbols, allusions, martyrs'. Their ideas had a strong impact on Victor de Laveleye, the programme organiser of the BBC Belgian Service, which first went on air on 28 September. On 14

January 1941, de Laveleye proposed the letter 'V' as an emblem to rally his listeners against the Nazis. V had universal appeal because it was the first letter of Victory in English, *Victoire* in French and *Vrijheid* (Freedom) in Flemish. Within weeks, chalked V's were appearing on walls in Belgium, Holland and northern France. But Delavenay wanted more.

On 22 March, the lively French Service dedicated a special edition of their popular show *Les Français parlent aux Français* to the V. The Dutch Service did the same on 9 April. The BBC's Assistant News Editor, Douglas Ritchie, wrote a paper on 4 May, 'Broadcasting as a New Weapon of War', which asserted baldly that 'When the British Government gives the word, the BBC will cause riots and destruction in every city in Europe'. On 26 May, Ritchie chaired the first meeting of the V Committee. He knew nothing of the work of SOE, but he was running with a good idea, backed by his boss, Noel Newsome, another former *Daily Telegraph* colleague, now the BBC European News Editor. Other countries were picking it up: V stood for *vitezstvi* (victory) in Czechoslovakia, *vitestvo* (heroism) in Serbia, and *ve vil vinne* in Norway. Even in faraway Bolivia, whose Andean Indian army had been trained by Germans, defiant V's appeared overnight on pro-Nazi buildings in La Paz.

On 6 June 1941, 'Colonel Britton' made his first broadcast on the English network in *London Calling Europe*. This was in fact Ritchie, speaking in the quiet, conspiratorial tones of an urbane agitator, encouraging disruption in occupied Europe. This smooth, mysterious figure (who got lots of fan-mail) also spoke in polished French, German, Dutch, Polish, Czech and Norwegian. Colonel Britton said there were countless small ways of making things difficult for the Germans. You could spit in their beer or put sand in their gearboxes. You could switch labels on trains or go slow in factories.

According to one account, at the third meeting of the V Committee in June 1941 the ingenious Oxford classicist Tom Stevens remarked that the letter V in Morse code was three dots and a dash. When he rapped it on the table, Jonathan Griffin recognised the rhythm of the opening motif of Beethoven's Fifth Symphony. Whoever had the idea, the morse and the music were first broadcast together on 27 June 1941. From the next day onwards, the same four notes, throbbed on an African membrane drum by the percussionist James Blades in a lower-ground-floor studio in Bush House, became the BBC station identifier in Europe, beating out the pulse of resistance. The Allies were knocking at the German door.

The Prime Minister gave the BBC campaign his stamp of approval by starting to give the 'V for Victory' sign with two fingers. It had to be explained to the aristocratic Churchill that it very much mattered which way his knuckles faced, to avoid giving the UK's most vulgar gesture.*

On 19 July 1941, Churchill broadcast a message to Europe: 'The V sign is the symbol of the unconquerable will of the people of the occupied territories and a portent of the fate awaiting the Nazi tyranny.' Goebbels reacted swiftly and ingeniously by trying to co-opt the symbol. The V was claimed as the German sign for '*Viktoria*', and German stations started playing Beethoven's Fifth. (Actually the German for 'victory' was *Sieg* – hence the salute *Sieg heil!*) The Germans hung a huge V from the Eiffel Tower. In occupied Holland, the German-controlled Radio Hilversum broadcast a Morse V, but on the Dutch streets pro-British groups wore a white V, and pro-German groups an orange one. Chalkers on walls clarified their loyalty by added RAF to the V, or in Norway H7 for King Haakon VII.

The tide began to turn against the V campaign after Brendan Bracken replaced Duff Cooper as Minister of Information in July 1941. There were worries that the BBC campaign might be doing more harm than good. The 'professionals' of the PWE, which came into being that August, did not like the 'amateur' V campaign because despite its popular appeal it was not coordinated with real political and military objectives. It was all mouth and trousers, impotent to deliver real resistance and sabotage at the local level as SOE was beginning to do. The last V committee met in October and the valiant V campaign was finally quashed in 1942. But it must be credited as an imaginative triumph, uplifting hearts and giving hope to the occupied.

---

* British urban legend holds that the digital insult was first made by the long-bow archers of Agincourt in 1415, defying the French who had threatened to cut off the bow-string fingers of any shooters they captured. There is no medieval evidence for this practice. Longbows were drawn with three fingers and archers were killed rather than mutilated. This urban myth probably dates from Mrs Thatcher's premiership in the 1980s.

*TIME* magazine called it 'the first antidote prescribed for the apathy of Europe'. Colonel Britton told his many listeners that 'the V is your sign, the night is your friend'. The V army was invisible, but powerfully symbolic. 'It is a strange army, but one to which it is an honour to belong. It is an army which the Germans fear.'

The Secret Intelligence Service (SIS) had begun to use wireless effectively in 1938 when 'C', Admiral Sinclair, hired Richard Gambier-Parry to organise a secret two-way radio communications network, independent of the Foreign Office's existing system. (The two were eventually combined into the Diplomatic Wireless Service in 1946.) Gambier-Parry had been a public relations officer of the BBC and the UK general sales manager of the American radio manufacturer Philco. He poached the best technical people from among his old wireless contacts and from the Royal Navy to set up Special Communication Units (SCUs) whose job was to handle the traffic to and from embassies and officers abroad, and also to pass on information from the GC&CS at Bletchley Park to military commanders. Eventually, in 1941, Gambier-Parry's unit, Section VIII, took over the Radio Security Service. They were quite distinct from the Royal Corps of Signals who usually did military 'sparks', and for additional mobility they put two-way wireless sets into Packard cars and converted Dodge ambulances.

During the war, Gambier-Parry's section also installed and operated the first covert short-wave radio stations in Britain. In all, there were forty-eight of these clandestine stations, broadcasting in fourteen foreign languages The radio stations were officially referred to as RUs, or 'Research Units', and sometimes as 'freedom' stations. The first of these, code-named G1, went on air in May 1940 just before British troops began to be evacuated from Dunkirk harbour. G1 was a rather desperate response to the secret stations or *Geheimsender* run by Goebbels's 'Büro Concordia' in Berlin. These German outfits included the *New British Broadcasting Station* which first began hectoring Britain in February 1940, to be followed in June and July by other ersatz German stations, *Radio Caledonia, Workers' Challenge, Christian Peace Movement* and *Welsh National Radio*.

The German broadcaster on the British-run secret 'freedom' station *Das wahre Deutschland* – on air from 26 May 1940 until 15 March 1941 – was Dr Carl Spiecker. Spiecker had a history in subversive radio, having run a secret *Freiheitsender* from the outskirts of Paris, broadcasting an essentially conservative appeal to the opposition in Nazi Germany on behalf of the anti-Nazi social democratic *Deutsche Freiheitspartei*. Spiecker's British programmes were now recorded at Wavendon, one of several large country houses on the borders of Buckinghamshire and Bedfordshire that had been taken over by government agencies. All were within a ten-mile radius of Bletchley Park where the Government Code and Cipher School was now based. Sir Campbell Stuart's enemy propaganda organisation, EH, had moved to the riding stables at Woburn Abbey, and after the death of the Duke of Bedford in 1940 took over the whole house as its Country Headquarters (CHQ). The huge grounds, the menagerie of animals, and the gallons of cheap drink available for the isolated staff added what David Garnett called 'more than a touch of madness' to the atmosphere.

The second German-speaking 'Research Unit' or secret station started by the British, code-named G2, was *Sender der Europäischen Revolution* (Radio of the European Revolution), a left-wing intellectual station that broadcast from November 1940 to June 1942. It was soon followed by Rumanian, French, Italian, Norwegian, Danish and Czech clandestine stations. G2 was staffed by exiled German Marxist political scientists and took a revolutionary socialist, left-of-*New-Statesman* line: the workers should throw off the yoke of Nazidom in the spirit of European community good will, and so on. It was run by the clever socialist Richard Crossman, who, although brilliant and attractive (he married three times) was also a contumacious maverick. Even the bullying wartime Minister of Economic Warfare, Hugh Dalton, who first hired Crossman for propaganda work, was afraid of his capacity to quarrel. Sefton Delmer thought that *Sender der Europäischen Revolution* was poor radio, because Crossman did not exercise proper political and editorial control. In his genial way, however, Delmer never quarrelled with Crossman, but just made sure they kept to their different spheres. Crossman eventually rose to become director of Political Warfare against the Enemy and Satellites, while Delmer became director of Special Operations against the Enemy and Satellites. Crossman did 'white' propaganda, and Delmer 'black'.

It was Delmer, when setting up the eleventh secret station in April

1941, who first called the 'Research Units' 'black' stations. The simplest distinction between 'white' and 'black' propaganda is one of origin, between the labels on the tins, as it were, rather than between their contents. The BBC label is 'white' and well known; its propaganda is clearly marked as 'British Broadcasting'. The strategy of 'white' propaganda is to tell the truth consistently over time. In the long run, if you are frank about reporting your defeats, your listeners are more likely to believe in your victories. The tactics of 'black' radio, on the other hand, are short-term, rumour-filled, and deceptive. If you were to hear a station calling itself, let us say, *Gustav Siegfried Eins*, broadcasting in German, purportedly from German territory, and you were not sure of its origin, its agenda or the personnel behind it, you might well be listening to a 'black' station, actually run by the British.

The distinction between 'white' and 'black' propaganda was not at first simply between truth and lies. On one of Delmer's German 'black' stations, *Christus der König* (Christ the King), the Austrian Roman Catholic priest Father Andreas truthfully informed German listeners about what the Nazis were doing to the Jews and the Slavs in extermination camps like Auschwitz. It was factual and true, and only 'black' because no one knew exactly where the broadcasts were coming from. To give *Christus der König* more force with the faithful, Delmer got SOE to spread the rumour in Europe that it was actually a 'black' station of Vatican Radio.

But under the aegis of Sefton Delmer, the contents of the propaganda tins became very different indeed. 'Black' broadcasting began to diverge sharply from 'white' because the camouflage of secrecy gave 'black' a licence to deceive that essentially truthful 'white' did not have. (The deception could be a simple and confusing mixed message, said Delmer, like spitting in a German's soup before crying out '*Heil Hitler!*') Although they were driven by the same aim of defeating the enemy, 'white' propaganda became the open right hand presenting whatever HM Government was prepared to acknowledge publicly, 'black' the closed left fist concealing whatever it could disavow.

G3, the third German station, which began broadcasting on 23 May 1941, was Delmer's first 'black' baby. *Gustav Siegfried Eins*, or *GS1*, was a right-wing station with freedom to use the same kind of bad language with which *Workers' Challenge* from Germany was turning the English airwaves blue. *GS1* was different from the other secret stations because it pretended to be absolutely all for the Führer. It aimed 'to get across subversive rumour stories under a cover of nationalist patriotic clichés'. Delmer's mentor, Leonard Ingrams, favoured 'operational propaganda', or actually getting people to do things, and in a memo to Ingrams a fortnight after the secret station began broadcasting, Delmer said: 'We want to spread disruptive and disturbing news among the Germans which will induce them to distrust their government and disobey it.' He added: 'We are making no attempt to build up a political following in Germany. We are not catering for idealists. Our politics are a stunt . . .'

*Gustav Siegfried Eins* came on air on the eve of Delmer's thirty-seventh birthday. A fortnight earlier, there had been a bizarre episode when Hitler's beetle-browed Deputy Führer and right-hand man, Rudolf Hess, was captured after parachuting into Scotland to try and talk peace to the unwitting Duke of Hamilton. Naturally, Hess's defection became the subject of *Gustav Siegfried Eins*'s first seven-minute broadcast. In it, the ranting character known as *der Chef*, 'the Chief' or 'the Boss' (Hitler's nickname on the election tour that Delmer covered), loudly denied the rumour that Hitler could have had anything to do with Hess's mission to England, thus covertly spreading it. There were nearly 700 more of these short broadcasts before the station's life ended dramatically with the burst of gunfire that cut off the speaker on 18 November 1943. The idea was to make people think that the Gestapo had finally caught up with *der Chef*. Actually, of course, it was another deception in the British 'black' operation.

If 'black' broadcasting is 'pretence' broadcasting, then the principal speaker must be totally convincing. *Der Chef* fitted the bill. The man had to sound like a right-wing, patriotic German, probably a tough but frugal Prussian landowner, a *Junker* on his uppers, outraged by the ostentatious corruption and incompetence of scheming perverts and bigwigs in the Nazi and SS hierarchy who were profiting at home from the sacrifices of the decent and honourable Wehrmacht abroad. *Der Chef* used vigorous and obscene soldier's language to castigate his many enemies. In his very first talk, the British Prime Minister was called 'that flat-footed bastard of a drunken old Jew Churchill', which all added to the effect of a genuine German speaking his mind.

'Black' radio worked precisely because it did not sound authorised. A bored German radio operator, more likely to be military than civilian because more soldiers had short-wave receivers, might be scrolling through the dial one night when he chances to pick up a German voice speaking the salty language of the barracks. Delmer had learned from hearing genuine ships' captains talking to each other, and the 'black' radio listening experience was intended to feel, Delmer said, like 'eavesdropping on the private wireless of a secret organisation . . .' If some of what *der Chef* said rang bells with the listener or if he found the dirty stories funny, he probably kept listening and told his mates about what he heard. In this way, slowly, by word of mouth, the reputation of *Gustav Siegfried Eins* spread.

Sefton Delmer concocted *der Chef* from the angry, barking military types he had known in Berlin before the war, and was able to write his scripts because he was an excellent mimic with a real grip on the language. But to bring the man to life before the microphone he needed the right voice, and found it in Peter Seckelmann, a former Berlin journalist and writer of detective stories who had been living in Britain since 1937, who had volunteered for the British army and ended up a corporal in a Pioneer Corps bomb-disposal unit. After vetting by MI5, Seckelmann came to live with Delmer and his wife at their house, the Rookery, in the village of Aspley Guise near Bletchley. Delmer also found a German journalist, Johannes Reinholz, to play Seckelmann's arrogant, strutting, heel-clicking adjutant.

Sex is always a good way to grab attention. The tabloid newspaper trick of deploring vice while describing it in extensive detail also worked to glue listeners to the radio. There was much interesting material to be found in the works of the German sexologist and advocate of gay rights, Dr Magnus Hirschfeld, 'the Einstein of sex', but the pornographic descriptions that *der Chef* gave in his diatribes against the corrupt elite caused Delmer some trouble. Crossman's 'revolutionary' broadcasters, jealous of Delmer's success, shopped him by making a lurid translation of one of *der Chef*'s racier talks about a voyeuristic German admiral at an orgy. When the script reached the left-wing puritan lawyer Sir Stafford Cripps, he sent a handwritten letter to the Foreign Secretary, Anthony Eden, on 12 June 1942, objecting 'most strongly to such filth being allowed to go out of this country'. According to Delmer, Cripps also said to Eden in person, 'If this is the sort of thing needed to win the war, I'd rather lose it!'

Robert Bruce Lockhart, the director general of the PWE from

March 1942, had to defend Delmer and placate Cripps over lunch by explaining why the depraved German admiral had been targeted: his failures in ensuring supplies to U-boats had already made him unpopular, so whipping up more indignation against the admiral helped to sow dissension among enemy submariners. Delmer wrote, 'We are of course not trying to win Germans to our side by this method; we are trying to turn Germans against Germans and to weaken the German war-machine.' He also justified the violent imagery and excessive language of *der Chef*, deftly finessing the argument into a claim of British moral superiority: 'There is a sadism in the German nature quite alien to the British nature and German listeners are very far from being revolted by the sadistic nature of some of these broadcasts.'

Delmer and his small team on GS1 soon discovered that invention could take you only so far. Rumour and falsehood were best grounded in genuine intelligence. Churchill was to say to Stalin, 'In wartime, truth is so precious that she should always be attended by a bodyguard of lies,' but Sefton Delmer and Dudley Clarke came to realise that deception work turned that maxim inside out. The big lie was so valuable that it needed a bodyguard of truth. When Max Braun, formerly interned as an 'enemy alien' on the Isle of Man, joined the team, he began a huge card index to log personal details of all Nazi Party functionaries and prominent Germans, combing through local German newspapers for authentic stories and genuine background detail to underpin British fantasy and misinformation. For example, Delmer turned a newspaper story about the successes of the German blood-transfusion service on its head by suggesting that the donors had not been tested for venereal disease, so wounded German soldiers might contract syphilis from the plasma.

Delmer gathered intelligence from many sources, including Foreign Office dispatches and SIS reports. The transcripts of 'bugged' conversations between German prisoners of war, derived from the Combined Services Detailed Interrogation Centre (CSDIC), yielded new German technical slang, jokes, rumours, dirt, and insights into everyday German worries. Through the offices of Leonard Ingrams, Delmer was able to get hold of personal letters sent by people inside Germany to family and friends in neutral North, Central and South America. The mail was intercepted by postal censorship in Liverpool, steamed open and copied before being sent on. Gossipy details could be used, altered and inflated to flesh out pointed remarks on air about the inequalities of German wartime conditions, so German listeners heard how the privileged were supposedly gorging themselves on 'Diplomat Rations' and moving safely away from bombing zones.

Delmer found it easy to charm things out of people. In a lecture on Psychological Warfare that Sir Hugh Greene gave to the NATO Defence College in Paris just before he became director general of the BBC, he noted the appeal of Delmer's kind of work:

'Black' propaganda seems to have an irresistible attraction to those in authority and the mere mention of the magic word 'black' will sometimes open up sources of valuable intelligence which might otherwise be withheld. It seems so much more fascinating and romantic than the slowly grinding mills of orthodox propaganda. It appeals to the small boy's heart which still beats under the black jacket or the beribboned tunic.

People trusted Delmer with secrets because he never used intelligence 'raw', but always cooked it so as not to betray its source. The first people to discover this and feed Delmer information were Britain's Naval Intelligence Division, followed by Air Ministry Intelligence, with the War Office last. When Ian Fleming introduced Delmer to his boss in Naval Intelligence, Admiral John Godfrey, Godfrey had straight away grasped what radio could do and set up a new propaganda section called NID 17 Z, under the former *Times* journalist Donald McLachlan, liaising with Fleming.

Just before Christmas 1942, over a champagne lunch, McLachlan talked to Delmer about the long-drawn-out Battle of the Atlantic. One reason to celebrate was that – at last – the decrypt drought was over. From February to November 1942, British Naval Intelligence had not been able to read one single message sent to the U-boats from their command and control centre, the *Befehlshaber der Unterseeboote* (BdU), or any of their replies, because the Germans were using the four-wheel Enigma key, not the three-wheel they had earlier. But on 13 December 1942 the 'Shark' key was cracked and from then on all traffic could be read. The astonishing thing is that German naval intelligence had no idea that their communications were so vulnerable. Most U-boats did not leave harbour with their operational orders, but received them, by radio, at sea; and if the British and Americans could read those instructions, so much the better for them. The Admiralty were keen to step up psychological warfare on the people who were manning them. They proposed that Delmer launch a new, live, short-wave station specially beamed at German submarine crews, featuring a demoralising 'black' news bulletin. This excited Delmer. His previous attempt to do 'black' news with a station called *Wehrmacht-sender Nord* had failed because the broadcasts always had to be pre-recorded and so lost the cutting edge news needed. Delmer was convinced he had the best way of broadcasting live news every night: he could unleash the most powerful transmitter in the world.

To gain advantage in the wireless wars, Churchill had approved the purchase in May 1941 of a giant new radio transmitter from the USA. The Nazi conquest of Europe meant that the Germans had many more transmitters than Britain, some of which were being used to jam UK stations. The 500 kilowatt medium-wave transmitter, which Gambier-Parry called a 'raiding Dreadnought of the Ether', was monstrously powerful and adapted to broadcast on different frequencies, between which it could switch almost instantaneously. (It was nicknamed 'Aspidistra' for its size, after the Gracie Fields song, 'The Biggest Aspidistra in the World'.) Aspidistra was originally destined for the American station WJZ, but the Federal Communications Commission refused it a licence because it massively exceeded the 50 kW limit for commercial radio stations. Section VIII of the British Secret Service managed to get in its bid for the transmitter before the Chinese government did, paying around £165,000 in all.

Aspidistra was initially supposed to fill an old gravel pit in Bedfordshire, but Section VIII's Chief Engineer, Harold Robin, insisted it should be positioned nearer Europe. Eventually somewhere was found near Crowborough in Sussex, the highest spot in the 6,400-acre Ashdown Forest, the largest area of open moorland in the south-east of England. A battalion of Canadian engineers with road-building equipment helped excavate a hole 50 feet deep, and 600 men worked by day and by night in the summer of 1942 to erect the reinforced concrete shell. Robin also supervised the installation of a 3,000-horsepower diesel generator, a cooling tower, workshops and offices. Aspidistra made its first broadcast early in November 1942, supporting operation TORCH, the landings in North Africa. Its output was shared between PWE, the BBC and the RAF.

Aspidistra's power and reach were extraordinary. On 17 November 1943, for example, during an RAF bombing raid on Ludwigshafen, an RAF linguist transmitting via Aspidistra counterfeited the voice of the controller of the German nightfighters, warning them all to land because of the dangers of fog. According to Professor R. V. Jones, when the Germans found out, they employed women controllers. So the RAF found German-speaking WAAFs; and when the Germans used a man and a woman for all orders, the British did likewise. Eventually, the Germans had to supplement verbal orders with music: a waltz meant Munich, jazz Berlin.

Delmer called his new live station *Deutsche Kurzwellensender Atlantik* (German short-wave Radio Atlantic), and it was later familiarly known as *Atlantiksender*. His team moved into a fenced and guarded five-acre compound at Milton Bryan in Buckinghamshire, and got ready to start transmissions in March 1943. It was a strange community that Delmer ran, some of whose members were still enemy prisoners of war. His close assistants included Tom Stevens, the classicist who had been in the 'V' campaign and who had an ingenious, detective-puzzling mind. As Hugh Greene noted, the best psychological warriors were often journalists or university dons.

Delmer envisaged *Atlantiksender* as a really entertaining programme that would lure people into listening with lots of dance music; as chief disc jockey he used Alexander Maass, whom he had known in Berlin and met again in Madrid. They got the latest records of German dance music flown over by fast Mosquito plane from Stockholm, and the American Office of Strategic Services (OSS, the American SOE) helped them obtain new American music. Marlene Dietrich was even persuaded to sing in German for what she was told was a special Voice of America broadcast to Germany. *Atlantiksender*'s German in-house band, led by Henry Zeisel, had been captured by the British Eighth Army when they were touring to entertain Rommel's *Afrika Korps*. Delmer dug other helpers out of PoW camps, finding anti-Nazis or deserters from the German forces, looking particularly for naval men who were up to date on technical terms, correct procedures, and the authentic argot and complaints of the petty officers' messes and the lower decks. Frank Lyndner, a former bookseller who thought Delmer was 'a god', kept the filing system on the German Navy, logging personal details culled from letters to and from captured U-boat crewmembers in PoW camps in Britain and Canada, and sifting German local newspapers, just as Max Braun had for Delmer's earlier station GS1, but this time for dates of births, marriages, deaths, transfers, promotions, awarding of medals, etc., in order to send personalised congratulations over the air.

Most of the listeners in the German Navy guessed that *Atlantiksender* was an enemy station, but they carried on listening despite that, because it was so good. When *Atlantiksender* cheerily reported the results of football matches between U-boat crews in St Nazaire, including the nicknames of goal-scorers with little personal details about them, or broadcast 'special request' music to a specific U-boat that thought its position was completely secret, it made the German Navy feel under constant surveillance. 'The British know everything anyway,' shrugged captured U-boat men, and no longer bothered to resist interrogation. The more the prisoners talked, the more information came back that the 'black' propagandists could use. One of the interrogators on attachment to CSDIC was the veteran BBC broadcaster Charles Wheeler, then a young captain in the Royal Marines who spoke German because he had been born in Bremen. Wheeler was not after technical information. In friendly talks where the prisoners were offered cigarettes, he got stories about the bars and brothels of Brest, Lorient and St Nazaire that helped garnish the chatty broadcasts and to reinforce the German Navy's gnawing *angst* about ever-present spies, phone taps and radar.

The *Atlantiksender* programme of smoochy music, made more seductive by the breathily erotic voice of 'Vicky', the announcer Agnes Bernelle, was interspersed with news bulletins and human interest stories. 'Gallant doctors battle diphtheria in German children's camps' was the kind of apparently upbeat take on a disaster that was actually designed to worry a German parent. Delmer stressed that strict accuracy in professional naval and military matters would make listeners more likely to accept the invented or half-true stories about the economic, political or social situation at home in Germany. The credibility of the news was helped by close attention to BBC Monitoring's *Daily Digest* of German broadcasting, and also by Delmer's acquisition of a Hellschreiber teleprinter left behind by the correspondent of the official German news agency, *Deutsche Nachrichtenbüro*, DNB, when he fled at the beginning of the war. It was still receiving directly from Goebbels's centralised news system, so now the 'black' team could swiftly broadcast his official Nazi news communiqués and speeches, either as impeccable cover or bent to their own disruptive purposes. (In July 1943, Delmer devised an ingenious plan, called *Helga*, to reverse the Hellschreiber process. Instead of receiving news from the centre, he wanted to feed false news, indistinguishable from the real thing, back into the German system while it was idling. Delmer's proposal for what we would now call 'hacking into' the German news system and adding counterfeit stories to the brew was discussed at the very top, but never actually tried.)

When Air Ministry Intelligence became involved, Delmer got detailed reports of the RAF and USAAF bombing raids on Germany, including aerial photographs. Accurate damage reports from specific streets and neighbourhoods were mixed on air with heart-rending descriptions of women and children suffering the incendiaries and high explosive of Allied 'Terror Raids', and horror stories about disease, mutilation, rats, necrophilia. Demoralising news from the civilian home front was followed up by seductive fantasies of a better life for German soldiers who surrendered or deserted. Stories from the International Committee of the Red Cross about escapees from Germany earning good money in Sweden, Switzerland or Spain featured regularly, as did rumours that deserters could get grants of land in Canada, the USA and Brazil. Reprisals from the German authorities were unlikely, the radio added, because they had no way of knowing whether missing people had deserted or simply died.

On 24 October 1943, a new 'black' British radio station came through loud and clear on the same medium wavelengths as the authentic German Radio Deutschland, located in Munich. *Soldaten-sender Calais* (Soldiers' Radio Calais, later called Soldiers' Radio West) followed the same tested formula of music and news. Although it was beamed at German troops occupying France, Belgium and Holland, *Soldatensender Calais* became a big hit with German civilians who liked its hard-hitting style and believed it was a genuine German station dedicated to the military, on the grounds that 'our boys at the front' would have to be told harsher truths than the usual pap civilians were fed. Even Goebbels admitted that *Soldatensender* did a very clever job of propaganda. *Der Chef* was cut down in a fictional hail of bullets three weeks after the new station started broadcasting.

*Soldatensender* went out so strongly on medium wave because Delmer had finally managed to get the underused Aspidistra transmitter back from the clutches of the BBC. Sir Ivone Kirkpatrick, controller of the BBC European Service, could not stand up to the combined weight of the PWE, the Service Ministries and the chiefs of staff. Their case was unequivocal: 'black' radio propaganda was needed to soften up the Atlantic Wall of Fortress Europe for the forthcoming invasion, the great storm of D-Day in June 1944 that would blast the way to Allied victory.

# 21

# 'A' Force: North Africa

———————◆———————

'The Western Desert is a place fit only for war,' begins the script by the *Sunday Times* journalist James Lansdale Hodson for the 1943 propaganda documentary film *Desert Victory*:

Thousands of square miles are nothing but sand and stone. A compass is as necessary, once off the road, as it is to a sailor at sea. Water doesn't exist until you bore deep into the earth. You bath in your shaving-mug. Flies have the tenacity of bulldogs. Bruises turn rapidly to desert sores. Days that are very hot are followed by nights of bitter cold. When the hot *khamsin* wind brings its sandstorms, life can be intolerable. The Arabs say that after five days of it, murder can be excused.

In this testing environment, Major General Archibald Wavell, the soldier who lost an eye near Ypres and walked through the Jaffa Gate into Jerusalem with Lawrence in 1917, had been reviving Lawrence's guerrilla tactics, using cunning, deception, mobility and tiny 'mosquito columns' against elephantine Italian forces. Wavell had been appointed British commander-in-chief of the Middle East Command a month before WW2 broke out. He was now in the same post once held by General Allenby, whose biography he was writing. Just like Allenby confronting the Ottoman Turks in WW1, he was facing a numerically superior enemy.

Wavell had a gift for picking good people. One was Major Ralph Bagnold, an officer in the Royal Signals, one of a select band who knew and respected the great desert that lay behind the cultivated coasts of North Africa. Bagnold had been exploring the Sahara since 1926. He had improved the sun compass for desert navigation, discovered the best way of driving up dunes (full-speed, head-on), invented rope ladders and steel channels for getting unstuck in soft going, and even written a treatise on 'The Physics of Blown Sand' which got him elected to the Royal Society. Bagnold foresaw that the

Italians might send reconnaissance and raiding parties out of the enormous Libyan desert to sever British military communications between Cairo and Khartoum and, when Fascist Italy finally declared war on the Allies in June 1940, got *carte blanche* from Wavell to set up, equip, supply and prepare a new Long Range Desert Group (LRDG).

Bagnold located old companions from pre-war desert explorations, plucking Pat Clayton from Tanganyika and Bill Kennedy Shaw from Palestine, and put them in charge of young men from the backcountry of New Zealand who had lost all their guns and kit in a torpedo attack at sea. Their commander, Major General Bernard Freyberg, VC (the man who swam ashore before the Gallipoli landings), was reluctant to let them go, but the New Zealanders – as always the best troops in the Dominions – took to the desert as though born to it. They became mainstays of the LRDG, doing what Wavell called 'inconspicuous but invaluable service'.

From the beginning, Wavell had been creating the illusion for the Italians that he was stronger and better equipped than he actually was. In June 1940, the *Daily Mail* correspondent Alexander Clifford was able to deduce the 'routine of bluff' by British forces on the Libyan–Egyptian frontier, 300 miles west of Cairo, and made one of the earliest references to the use of dummy tanks:

I saw, gradually, what was happening. Subtly and systematically Wavell was doing his sums and faking his figures. These tiny British patrols were staging big demonstrations. Continually they were making nuisances of themselves, moving rapidly from place to place, shooting up convoys, flinging ambushes across roads, attacking forts and positions, always pretending they were much bigger than they were. Dummy tanks were toted about to give the idea that we had strong armoured units . . . In every way that tiny army set itself to gain time by frightening the enemy.

The dummy tanks were the responsibility of the fake 10th Battalion Royal Tank Regiment (10 RTR) under a Major Johnston. The unit, actually formed from men of the 1st Battalion Durham Light Infantry, deployed and operated primitive dummy tanks and lorries made out of wood and canvas which they carried folded in the unit transport.

In late October 1940, Wavell was in Khartoum with Sir Anthony Eden, the British Secretary of State for War, and other generals, co-ordinating his attack on the Italian Empire in East Africa, *Africa*

*Orientale Italiana.* The British were keeping close tabs on the Italian forces in Libya by deciphering their Air Force communications, reading captured documents and mail, questioning prisoners of war, doing photographic air reconnaissance and above all sending armoured Rolls-Royce patrols to probe the gaps between the scattered outposts and minefields that made up the Italian front line. From November 1940, Wavell initiated a deception operation encouraging the Italians to think that he was really preparing an Expeditionary Force to Greece rather than an attack on them; Wavell's Middle East Intelligence Centre set in motion a paper trail to this effect, spreading rumours and planting false information on a Japanese source in Egypt. (Japan had joined the Axis with Germany and Italy in late 1937.) Compared to later deception operations it was fairly basic, but it prepared the ground. Wavell had further deception operations in mind, but he needed a really good man to oversee them. It was now that he summoned Dudley Clarke to Cairo.

On 18 December 1940, Tony Simonds of British military intelligence was instructed to go in plain clothes to meet an old friend off a civil aeroplane landing just before midday at Cairo airport, and to greet him without surprise. This became a challenge when Clarke arrived looking like a golfer from Chicago, wearing loud black and white plus-fours, a check cap and dark glasses, claiming to be an American journalist called Wrangel. This seems more like showing off than disguise, a trait which would later get Clarke into trouble. He had been travelling for ten days, his journey from England to Egypt made complex by the need to avoid enemy territory.

At 9 a.m. on 9 December 1940, just as Clarke was leaving Lisbon for West Africa, the seven or eight British war correspondents in Cairo were summoned to General Wavell's office at the end of an upstairs corridor at 'Grey Pillars' which housed GHQ Middle East in Garden City. As ever, the commander-in-chief's desk in front of the ten-foot-high map on the wall was bare of papers. 'The Chief' was smiling slightly that morning. He announced 'an important raid', code-named COMPASS, by the British, Indian and Anzac soldiers of General Richard O'Connor's Western Desert Force on the Italian Tenth Army. Taciturn Wavell, described by Alan Moorehead of the *Daily Express* as 'an island in a sea of garrulousness', asked the journalists if they had known an operation was imminent; none had heard a thing.

The little band of reporters, honorary officers dressed in khaki with a shoulder flash that read 'British War Correspondent' in gold letters on green, scrambled to get to the front a day and a half away to the west. No travel arrangements had been made by the Public Relations Unit, so when their cars broke down, they hitch-hiked; they ate what they scrounged and slept when they could. It took them days to catch up to the front because the British Empire troops were going too fast for them, with the infantry division acting as the assault force and the tanks of the armoured division slipping round behind, a method that Richard Dimbleby of the BBC likened to hauling a man up with one hand and punching him in the jaw with the other, again and again. On 16 December they took Sollum and Helfaya Pass and the Libyan escarpment, where British troops could put aside briny tea, biscuit and bully beef to gorge on luxurious Italian rations – ham, cheese, bread, fresh fruit and vegetables, washed down with wine and sweet bottled water. They were amazed to find that every Italian soldier was issued with his own little *espresso* coffee pot. Their 38,000 prisoners included five Italian generals.

When Dudley Clarke presented himself, in uniform, to his old chief on the morning of Thursday, 19 December 1940, Graziani's army was being bundled out of Egypt, and for the first time in more than a year of war the British were not retreating, as they had from France, Norway and Somaliland, but driving forward. Clarke's life as a 'freelance' roving staff officer was ended, as Wavell gave him the secret and 'most gratifying' eighth assignment that would last the next five years. Clarke camouflages his role as just 'being a working part in the smooth-running engine of a General Headquarters at war', because he was obviously not allowed to talk about it. In fact Wavell put Clarke in charge of all bluffs, cover plans and deceptions for his military operations. He remained chief deceiver for all the Mediterranean commanders – Wavell, Auchinleck, Alexander, Wilson – throughout the North African advances and retreats of 1941–2, and did the same job at Eisenhower's Allied Force Headquarters in Algiers from 1943 onwards, ending his war service in northern Italy. Clarke's ideas about strategic and tactical deception would help drive the Axis out of Africa and aid the seaborne landings that led the Allies back into southern and then north-western Europe. Few men of his rank wielded such influence behind the scenes in Cairo, Algiers, London, Washington DC and New Delhi, and when WW2 ended in 1945, Field Marshal Harold Alexander reckoned that Dudley Clarke had done as much as any single officer to win it.

Wavell and Clarke stood before the map on the wall in Cairo. Wavell's 4th Indian Infantry Division had just retaken Sidi Barrani in Egypt. Wavell now planned to pull them back southward and ship them, together with the 5th Indian Division, to Gedaref and Port Sudan for an attack on Italian East Africa. Wavell's other plans included the use of Orde Wingate to lead the Ethiopian Patriot guerrillas of Gideon Force back into Ethiopia from exile in the Sudan.

Wavell's forces were outnumbered by the Italians on paper, but he knew attack was the best form of defence. Clarke's new mission was continually and systematically hoodwinking the enemy about British aims, intentions and capabilities. As 'Personal Intelligence Officer (Special Duties) to the Commander-in-Chief' he reported directly to Wavell and got clerical help from his private secretary, but he had no staff and no 'establishment'. The work was 'Most Secret', so his official cover story and additional duty from 5 January 1941 was a role in MI9, Escape and Evasion by Allied servicemen. He was taken to his office: the door opened on a very small converted bathroom. *Action This Day*.

---

Wavell's attack on *Africa Orientale Italiana* was now the occasion of Clarke's very first deception operation, code-named CAMILLA, focused on British Somaliland. This British protectorate at the top of the Horn of Africa had been evacuated in August by the small British Empire garrison in the face of overwhelming Italian forces. Though British Somaliland was strategically unimportant, Wavell wanted the Italians to believe that Allied troops from Egypt were going to retake it. The Indian Divisions were indeed moving south, but their target was Eritrea on the Red Sea.

Clarke sat down and worked out the staff-work, logistics and communications if British forces really were going to try and take back British Somaliland. Then he constructed a model of the operation for the benefit of enemy intelligence. This elaborate deception involved bogus administration in offices at Aden (which he knew well from his time there in 1935), air and sea raids across the Gulf of Aden apparently designed to 'soften up' Italian naval and military targets around Berbera, the issuing of campaign maps and pamphlets on British Somaliland's climate, culture, clans and customs, sibs spread in civilian Egypt and among the armed forces, more false information planted on the Japanese consul at Port Said and 'indiscreet' fake telegrams and wireless telegraphy traffic. Beginning on 19 December, the plan was intended to peak in early January 1941.

In one sense it succeeded brilliantly. The Italian commander swallowed the bait hook, line and sinker, and started evacuating British Somaliland. Unfortunately this was the exact opposite of what the strategic deception plan intended him to do. He was meant to reinforce his east: instead he moved more troops north to Eritrea, which was of course precisely where the real British attack was targeted. Clarke learned his first lesson the hard way: the point of deception was not getting the enemy to *think* the way you would like, but getting them to *do* what you wanted.

Other parts of Clarke's deception worked much better. He used bogus wireless traffic to make the Italians think there were two Australian divisions in Kenya. This successfully held Italian troops in the wrong place while three coordinated attacks by the multi-racial armies of the British Empire began liberating Ethiopia from five years of Fascist occupation. Emperor Haile Selassie returned to his people in Addis Ababa on 5 May 1941.

Another of Clarke's ideas bore long-term fruit. On 14 January 1941, he met Colonel William J. 'Wild Bill' Donovan, President Roosevelt's personal military emissary, future founder of the Office of Strategic Services (OSS, forerunner of the Central Intelligence Agency), on his six-week strategic tour of the Mediterranean. Clarke told him about the commandos, and wrote a seminal paper for Donovan suggesting the formation of an American commando force. Clarke's love of film served him well. Having recently seen the western, *North-West Passage*, in which Spencer Tracy led a force of buckskinned frontier fighters called 'Rogers' Rangers', he suggested 'Rangers' as an appropriate name for the American commandos. The US Rangers were duly founded in May 1942.

Years later, in 1953, when Clarke wrote a proposal for the sequel to *Seven Assignments*, a book which he wanted to call *The Secret War*, he described his wartime deception work as 'a war of wits – of fantasy and imagination – fought out on an almost private basis between the supreme heads of Hitler's Intelligence (and Mussolini's) and a small band of men and women – British, American and French – operating from the opposite shores of the Mediterranean Sea'. In the event, he was never granted official permission to write that book, but it is clear that Dudley Clarke's deception was aimed high, at the minds of the few, in contrast to Delmer's psychological warfare, which was directed lower, at the guts of the many.

The first four professional British *camoufleurs* to arrive in the Middle East in WW2, led by Captain Geoffrey Barkas, disembarked at Port Said on New Year's Day 1941. Leaning over the rail of the *Andes* in the Suez Canal they surveyed their task with sinking hearts: 'This was the Army we were supposed to know how to camouflage; these huge workshops, depots and store dumps; hutted and tented camps, anti-aircraft batteries and defence works; men and vehicles by the thousands . . . with all the shy unobtrusiveness of a red vest on a fat man.' One answer to their problem was berthed nearby. The huge battleship HMS *Centurion* looked formidable, but was actually a dummy hulk, whose 13.5-inch guns were made of painted wood.

There is a well-known army saying, 'Time spent in reconnaissance is seldom wasted', and the *camoufleurs* soon arranged a flight over the Western Desert and found, looking down, that the terrain was not wholly monotonous. Regions of sand had distinctive patterns which they began to name: the Wadi, the Polka Dot, the Figured Velvet and so on. They noticed, too, how tanks and lorries left distinctive scars and tracks across the desert. Then Barkas sent John Hutton off to the Sudan and Patrick Phillips to camouflage the defences in Palestine, reserving Egypt and Libya for Blair Hughes-Stanton and himself. On 24 January 1941 the two officers and their two drivers set off from Cairo in a Chevrolet staff car with desert tyres to try and find what Churchill liked to call 'the Army of the Nile' somewhere along the coastal plain. Their other vehicle was a fifteen-hundredweight Fordson truck crammed with water, petrol, rations, tents, stoves, tools, canvas, rope, wire, camouflage nets, rolls of garnish, paints and kit for camouflage experiments.

The *camoufleurs* were following the same campaign westward that Alan Moorehead and the other correspondents had pursued. They had some catching up to do because General O'Connor's astonishingly successful attacks were now moving westward into the yellow dust of Libya, having taken the town of Bardia and 40,000 more prisoners in early January. At last they caught up with the Australian infantry (who had replaced the Indian divisions) and Brigadier Horace Robertson, a gingery old tough from Melbourne who had been with the light horse on Gallipoli and in Palestine during the Great War, and who was nicknamed 'the Ball of Fire' by the *camoufleurs*. With the permission of some gunners in a *wadi*, they continued their experiments with camouflage nets.

Things went well until their staff car was smashed up in a violent head-on collision which left their faces cut and bruised. They continued in their truck to the charming town of Derna, which had been abandoned by the Italians and looted by local Arabs in long robes. Also in Derna, relaxing in greater luxury, were the journalists: Alan Moorehead of the *Daily Express*, Alexander Clifford of the *Daily Mail* and their conducting officer, Geoffrey Keating of the King's Royal Rifles. Naked after a long bath, Moorehead was disconcerted to notice the telephone had the owner's name set in the base: 'His Excellency Marshal Graziani'. Two days later the three journalists had the unexpected pleasure of the entire fortress of Ain Mara surrendering to them, thanks to their uniforms and British car. From that glorious height they soon fell over what Moorehead called the 'moving precipice' of the front. There was an ambush by the Italians near Giovanni Berta in the Jebel El Akhdar; three British soldiers were killed and Clifford, Keating and the driver were wounded. The *camoufleurs* saw the point of their craft as trying to save the lives of British soldiers, and they stuck to the experimental work they had been doing all along, carefully garnishing a net to camouflage vehicles and then finding high ground to take photographs of it so they could see if the disguise would work from the air. They witnessed the surrender of Benghazi on 7 February 1941.

Fifty miles south General O'Connor had trapped the remains of the Italian Tenth Army at Beda Fomm, where the 7th Armoured Division annihilated it. The Italian General Bergonzoli (nicknamed 'Electric Whiskers' for his startled beard) was captured, and then the campaign effectively ended. Field Marshal Rodolfo Graziani, known as the 'Butcher' for the ferocity of his tactics in Libya and Abyssinia, was now fleeing westward towards Sirte. O'Connor's British forces could have chased him all the way to Tripoli, driving on in cannibalised or commandeered vehicles to knock the Italians right out of Libya. But it was not to be. Churchill thought he saw the bigger strategic picture, and he was worried about the Balkans. On 12 February he ordered Wavell to halt, change course, and send his best troops across the Mediterranean to assist the beleaguered Greeks.

The *camoufleurs* now headed 900 miles back to Cairo along with the 7th Armoured Division. Though they had captured no flags, Barkas and Hughes-Stanton had acquired practical knowledge about how concealment worked best in the field from all branches of a fighting army. Barkas also remembered his own experience as an ordinary soldier in WW1, and how vulnerable he had felt moving up to the trenches knowing the enemy was watching. Even in an 'unco-operative' environment like the desert, he thought it should be possible 'to blur the picture and confuse the judgment of the most alert enemy'. Slowly, a doctrine of the trade was emerging. Barkas wrote:

The dream of every commander has always been to achieve complete surprise over his enemy . . . Camouflage is merely one factor of surprise. It means deceiving the *eyes* of the enemy.

Alan Moorehead thought the war correspondents did well in the desert because the issues were simpler. 'There were no distractions, no cities, no railroads, shops, cinemas, markets, farms, children or women . . . We never saw money or crowds or animals or hills and valleys. We saw the arching sky and the flat desert stretching away on every side.' The small incident achieved a significance here that it would not elsewhere because they saw it clearly and in isolation. For the *camoufleurs* too, the uncluttered empty desert behind the coastal strip was a bare stage where everything could be seen, which also made it 'an ideal place for . . . the cunning substitution of real for false and vice versa'. This was true, of course, for both sides.

When the actorish German Colonel General Erwin Rommel arrived in Libya to rescue the Italians from total collapse, the first thing he did on 12 February 1941 was march his German troops three times round the same palace at Tripoli so his limited forces could make a bigger show for the *Propaganda Kompanie* newsreel cameras. Then the man later admiringly nicknamed 'the Desert Fox' by the British got Italian workmen to start building and fitting 200 dummy tanks on the chassis of old cars to augment his Panzers. Rommel was an aficionado of feints and deception, and knew display and presentation mattered in this theatre where men could make mirages. Basil Liddell Hart, editor of *The Rommel Papers*, compares him to Lawrence of Arabia in his thinking about unorthodox warfare. Rommel was also acutely visual: he liked to sketch out his battle plans with coloured crayons and used to fly up in a plane to survey the terrain ahead and to take photographs.

Barkas was sure that German visual intelligence was as highly organised and scientifically equipped as the British, employing Luftwaffe aerial reconnaissance, photographic interpretation and all the usual means of observation. Training at Farnham, the British *camoufleurs* had looked at infrared photos taken from above Stonehenge that still showed evidence of stones being moved thousands of years ago. They knew the lessons of WW1: armies left distinctive messes wherever they went. The patterns of their organisation and behaviour scrawled signatures on the ground which photoreconnaissance interpreters read like spoor. Barkas knew that few of these signs could be concealed, but some could be disguised if only units would alter their behaviour. What appealed to him now was 'an aggressive, ambushing use of camouflage as part of the plan of battle'. On the long drive back along the coast, he began to think about using camouflage not just as a passive technique of hiding but as an active performance of misleading display:

The greatest and most respected of military commanders have usually been masters of fraud . . . All generals do their best to mystify the enemy by false threats or movements. But unless they have at their disposal something in the nature of a travelling circus, these deceptive manoeuvres must be carried out by real units at a considerable cost.

Barkas saw his camouflage unit as just that 'travelling circus', a sort of '5% army' ready to construct whatever impression the Commander might wish to present. Engineers realistically simulating the kind of mess that real armies make would be doing 'film production on the grand scale'. In a place of sharp light that threw long shadows, what would work best were 3-D dummies and decoys.

Geoffrey Barkas began with a half-share in a trestle table in the despatch rider's room, a space even smaller than Dudley Clarke's tiny bathroom. But from small beginnings, camouflage and deception were soon achieving great things in the Middle East, and Clarke's and Barkas's people made a team.

'If hopes were dupes, fears may be liars,' wrote the poet Arthur Hugh Clough, and Clarke was swift to play upon the enemy's fear of parachutists, which he read about in a captured Italian officer's diary. In January 1941 Clarke started operation ABEAM, whose aim was to persuade the Italians that the British had parachutists in the Middle East ready to drop behind their lines. They had no such thing, but Wavell, one of the first British soldiers to appreciate airborne tactics, having watched 1,200 paratroops jump during the Soviet manoeuvres of 1936, encouraged Clarke's new ploy.

Two years before any real British paratroopers appeared in the Middle East, Clarke coined the name of a wholly imaginary unit, the Special Air Service Brigade (SAS), and invented their story. The 1st battalion of the Special Air Service Brigade were parachutists, and the 2nd and 3rd battalions deployed in gliders. They were supposedly training south of Amman in Jordan at Bayir Wells camp (a real place Dudley Clarke knew from his time in the Trans-Jordan Frontier Force), guarded by Glubb's (real) Arab Legion. Clarke's '1 SAS' comprised 500 (non-existent) paratroopers in ten platoons (A–K), armed with carbines and grenades. The Italians were led to believe that these men would drop with special containers packed with Bren guns, mortars, explosives, mines, and small-arms ammunition, just like German *Fallschirmtruppen*.

On 2 February 1941, *Parade* picture magazine in Cairo carried a photograph of a parachute-swaddled Abyssinian grinning in front of a large Bristol Bombay transport aircraft. (He was, in fact, just an Egyptian laundryman, dressed up.) Two days later, RAF HQ issued secret instructions to RAF units across the whole of the Middle East to report the stories that Clarke was spreading about the presence of British parachutists and gliders. More documents were planted and further sibs spread in Egypt and Palestine. In early April, Clarke carefully briefed two yeomanry gunners and set them loose in Alexandria, Cairo and Port Said wearing uniforms and badges of 1 SAS, which drew curious questions from other British servicemen. The gunners pretended to be toughly tight-lipped but dropped little hints that they were off to Crete, or raiding enemy lines of communication in Libya. Such information, trickling back through Axis intelligence, was unnerving for the enemy.

In April 1941, Clarke at last got his War Establishment (ME/1941/10/1) and his unit for the first time got its name, institutional recognition and its own offices. Clarke moved north from his bathroom at 'Grey Pillars' to a block of flats which now became the Advanced HQ of 'A' Force.

The name 'A' Force was deliberately vague. It could stand for anything, but Clarke pretended that it stood for 'Airborne', since he was currently fabricating non-existent British airborne forces in the Middle East. By the end of April, realistic dummy 'gliders' (the fictional 'K detachment, SAS') had appeared at Helwan air base near Cairo and were attracting attention locally. Fifty-seven sappers under Lieutenant Robertson, one of Clarke's ten officers, skilfully engineered the fakes. In early May, Robertson's sappers were converting these dummy gliders into the dummy bombers of the 'Desert Air Force' at Fuka near Mersa Matruh. In June, for the benefit of any spies and the captured Italians in the nearby PoW cages, Clarke arranged for the RAF to fly from Heliopolis and drop dummies by parachute over Helwan. They were then collected, repacked and driven back to base for another go. Some genuine parachutes were also diverted to Jock Lewes of B Battalion, No. 8 Commando, to practise with at Fuka.

What began as part of a deception scheme had major (and quite unexpected) consequences in the real world. Among Jock Lewes's men at Fuka was one mad-keen commando who injured himself on a parachute jump. This was David Stirling, 6' 5" tall, whose father was a general and whose mother was a Lovat of the Lovat Scouts tradition. Recovering in hospital, David Stirling re-thought the commando concept. He decided it had grown unwieldy, and instead devised a scheme to reduce the commandos to smaller, mobile four-man teams. Stirling was only a lieutenant and he had to sell this idea to the top brass at Middle East HQ; he later described the bureaucracy he had to get through as 'layer upon layer of fossilised shit', but he persevered. His idea eventually gave birth to Britain's most famous raiding regiment, whose badge is a winged dagger above the motto 'Who Dares Wins'. In 1985, Colonel David Stirling told the TV producer Gordon Stevens that

The name SAS came mainly from the fact that I was anxious to get full cooperation of a very ingenious individual called Dudley Clark[e], who was responsible for running a deception outfit in Cairo . . . Clark[e] was quite an influential chap. He promised to give me all the help he could if I would use the name of his bogus brigade of parachutists, which is the Special Air Service, the SAS.

The Originals in their Own Words

Stirling said that when he 'settled for L Detachment SAS . . . Dudley Clark[e] was delighted to have some flesh and blood parachutists instead of totally bogus ones'. This completed a unique hat-trick. Dudley Clarke helped to found and name three famous striking forces, the British Commandos, the US Rangers, and the SAS.

The principle behind what Dudley Clarke had done with the original, fictional, SAS was also very important in itself. Perhaps the essence of military intelligence is working out the opposition's 'order of battle' and then analysing it. This is the military version of taxonomy, assembling all known information into a coherent and ordered picture. All the pieces – spies' gossip, radio intercepts, newspaper stories, stolen or seized documents, front-line reports, aerial reconnaissance, the badges of captured personnel, phrases from PoW interrogations, vehicle identification marks, camp signposts, guidons, etc. – are part of the huge puzzle being put together so that 'I' or Intelligence can tell 'Ops' or Operations exactly which enemy units are where and what they are up to. Dudley Clarke knew how meticulous and orderly enemy staff were: he had that kind of mind himself and understood the satisfactions of things clicking into place. And so he fed the enemy's hunger for precise information with perhaps his greatest creation, the 'notional' or bogus order of battle.

Inventing phantom forces was a long game that required a good memory, an efficient filing system, military realism and consistency over months and years. Between 1941 and 1945, Clarke invented seven more 'notional' brigades, thirty-two 'notional' divisions, ten corps and three entire armies. The Special Air Service Brigade was a small jewel in what became a large crown. But that initial imaginary force remained a particular success. First, Stirling turned Clarke's dream into a famously effective reality, raiding behind enemy lines. Second, the discovery of the SAS's actual existence in the field confirmed what German and Italian intelligence believed they already knew, and the small pleasure of that confirmation helped blind them to the fact that for much of the time they were being sold dummies.

It was in early 1942 that Clarke started drawing together all his fictitious units into a comprehensive deception plan called CASCADE. A loose-leaf 'Book of Reference' was drawn up which had all real and 'notional' units in it and was circulated to everyone who needed to know so there was no contradiction. Everything bogus had to have a history, a purpose, and physical evidence like identifiable signs which could appear on other (real) units' vehicles for spies to spot. There had to be wireless traffic where necessary, and paperwork. Getting false information on to genuine documents was an administrative headache because once started it could never be neglected. But the grinding details paid off handsomely in the big picture. Clarke's false or 'notional' order of battle, which made the enemy massively overestimate and miscalculate opposition and thus spread their own forces against all possible threats, is completely in the spirit of Sun Tzu's *Art of War*:

For if [the enemy] prepares to the front his rear will be weak, and if to the rear his front will be fragile. If he prepares to the left, his right will be vulnerable and if to the right, there will be few on his left. And when he prepares everywhere, he will be weak everywhere.

By the end of WW2 Clarke and his imaginative disciples had conjured up phantom forces not just for the British but also for their Allies. One of the most important, the 'First United States Army Group' or FUSAG, played a decisive role around D-Day in 1944.

Back in February 1941, when he still had no staff or office and was ill with jaundice, Clarke already had a useful friend, colleague and daily visitor to his hospital bed in Brigadier Raymund Maunsell, the head of Security Intelligence Middle East (SIME), Cairo's equivalent of MI5. This friendship became an enduring asset for 'A' Force. No one had a wider range of contacts than SIME for spreading sibs and planting information useful for strategic deception. Police and security people have their own mafia of professional association and Maunsell, always called by his initials, 'RJ', was in touch with Egyptian, Indian, Persian and Turkish officials, as well as British colonial counter-espionage in Aden, Sudan and Palestine. He watched the Spanish, Rumanian,

Bulgarian and Japanese consulates closely, employed both Sephardi Jewish and Muslim Brotherhood agents in Cairo, and bribed Egyptian policemen and concierges for useful information. Through the Field Security branch of the Military Police (later part of the Intelligence Corps), Maunsell also had access to captured Axis spies who could spread disinformation.

One of the first of these 'double agents' was a Bedouin called Ahmed Sayef, arrested on the Libyan–Egyptian frontier by Lieutenant A. W. Sansom, a fleshy little man, born in Cairo to a British father himself born in Baghdad, who was familiar with the Egyptian *demimonde* and fluent in Arabic, French, Greek and Italian. The man he had captured, Sayef, was working for Sheikh Mustapha ben Haroun, who managed the Arabic-speaking spies for Italian intelligence.

Sansom informed Maunsell of Sayef's existence, and duly let the Bedouin know that he would get double pay if he took back certain information to the Italians. But quite soon Sayef started coming back over the frontier from Italian-held Libya with what was found to be false information. Clearly, the Italians had discovered he was a double agent and were playing the British at their own game. But Maunsell of Intelligence told Sansom of Field Security not to say anything to Sayef because even false information was valuable, once it was known to be false. Finding what the enemy wants you to think may be as useful as truffling up something he does not want you to know. Sansom thought that the British were winning as long as (a) the enemy did not know our information was false, or (b) did not know *we* knew that *theirs* was.

Sansom was appointed chief Field Security officer for the Cairo area and recruited informers from all sorts of communities: Palestinian Jews, Greek Cypriots, Lebanese Christians, Sudanese and so on. He also kept an eye on Axis civilians and Arab nationalists and looked out for spies and security leaks among the usual big city lowlife of crooks, deserters, extortionists, fences, gunrunners and hashish-dealers. Sansom's man at the central telephone exchange tipped him off about any interesting phone calls. 'Mac' (or Mahmoud), the barman at the Kit Kat cabaret, was on the payroll, whereas Joe the Swiss in the Long Bar at Shepheard's was believed (but never quite proved) to work for the enemy.

Madames of brothels told Sansom bedroom secrets; the infidelities of the officer class entered his security fiefdom, but Sansom was never a man to confuse morals with morale. He understood that plenty of available and reasonable prostitutes kept the warriors happy, whatever puritan authorities thought. 'Cairo at this time was one big knocking shop,' he wrote in his entertaining memoir *I Spied Spies*. Many houses and blocks of flats had been converted into furnished accommodation that could be rented by the month, week, day, night or even hour. Privacy was guaranteed and Sansom himself maintained three such flats around the city for meeting informants and changing disguises. Dudley Clarke later used two on the floor below a brothel at 6, Kasr-el-Nil, for his new offices in April 1941. They were a perfect cover: no one paid much attention to the comings and goings of different officers.

On 10 March 1941, the first twelve camouflage captains, trained by Colonel Buckley in England, arrived in Cairo after a two-month voyage round the Cape of Good Hope. They included John Codner, Edwin Galligan, Robert Medley, Peter Proud, Steven Sykes and Jasper Maskelyne. They were all flat broke. Typically, the one who made the telephone call to Barkas asking for money was Maskelyne, the charming but feckless stage magician. Maskelyne's theatrical charisma has cadged him more credit than perhaps he deserves among the *camoufleurs*. Julian Trevelyan, who knew Maskelyne, wrote more matter-of-factly that he was 'at once innocent and urbane, and . . . ended up as an Entertainments officer in the Middle East'.

And yet illusionists and stage magicians do catch people's imaginations. This is one reason why Dudley Clarke employed Maskelyne in 'A' Force. His entertainments were lectures on escape and evasion given to over 200,000 aircrew across the Middle East, and he also helped MI9 in inventing parachute packaging and small devices and tools that prisoners could hide. Maskelyne's presence embodied the power of deception for older heads among the military who remembered his grandfather Maskelyne's Hall of Magic in Egyptian Hall in Piccadilly. Even the dimmest staff officer understood that conjurors made things appear and disappear, so the spell of Maskelyne sprinkled a bit of stardust over the dull business of camouflage, and strengthened Clarke's hand at GHQ.

---

After Wavell, at Churchill's instructions, had pulled back and disbanded General O'Connor's Western Desert Force in order to send 50,000 troops and 8,000 vehicles to Greece and Crete, Rommel launched a *blitzkrieg* attack on the skeleton force left behind in Cyrenaica Command, Eastern Libya. Wavell did not expect this. He missed the warning signs from ULTRA because he did not think Rommel was ready. Wavell simply had too much on his plate: there was fighting in the Balkans to the north and East Africa to the south, and when Rommel attacked from the west, there was also trouble to Wavell's east.

Exactly coinciding with Rommel's advance, a *coup d'état* in Iraq by the Arab nationalist Rashid Ali el-Gailani and a group of colonels threatened the militarily vital oil pipeline to British Palestine. The trouble in Iraq was partly an extension of the Arab rebellion that Clarke had helped to suppress in Palestine in 1936, after which the Grand Mufti and the top guerrilla leader had moved on to Baghdad to foment trouble. El-Gailani's coup succeeded in toppling the pro-British monarchy, but the importance of Iraqi oil to the British meant that it could not be allowed to last long. For Wavell, it was just one more headache at a time when the Germans under Rommel were enjoying great successes.

Through April 1941, the German Panzers swept the British back to the Egyptian frontier, wiping out all the gains of COMPASS and capturing British Generals Gambier-Parry, Neame and O'Connor into the bargain. Only the port of Tobruk in all Libya held out against regular German assault by land and continuous dive-bombing from nearby El Adem airfield. Peter Proud, one of the *camoufleurs* who came out with Maskelyne, found himself trapped inside besieged Tobruk with a lot of Australian infantry and British gunners. Like Geoffrey Barkas, Proud had worked in films, and now put his skills as an art director to good use, improvising camouflage and screening on poles to help defend Tobruk, using scrounged fabric. Real artillery was hidden; meanwhile dummy weapons, lorries and tanks made of scrap formed decoy targets to draw enemy fire. Irregular patches sewn on draped nets effectively imitated broken ground. Nothing was quite what it seemed in the dusty white broken town. Around the thirty-five-mile perimeter mines lay buried, unsleeping camouflaged sentries. Troops with their footfalls muffled by crepe-rubber soles patrolled aggressively right into the German lines. There was much driving of vehicles to make dust for disguise and diversion. The garrison destroyed or altered landmarks to deflect German artillery range-finding, and mixed real

Observation Posts with dummy OPs up poles. When some army vehicles had to be repainted to appear different and thus more numerous, Proud created a starchy coating from condemned foodstuffs mixed with seawater. Life during the 246-day siege was half-troglodytic, half-holiday. Men slept in caves underneath rubble, breathing air through transplanted ship's ventilators, but by day they leapt up from swimming and sunbathing in shorts and boots to man Bofors, Bren, or Lewis guns during frequent air raids. A daily newspaper, *The Tobruk Truth,* was compiled, cyclostyled and circulated.

The harbour was the back door for supply and relief. The entire garrison of Tobruk was changed in the eight-month siege. 27,000 men were shipped out and 29,000 shipped in, including Indians, Poles and South Africans. Much of the movement happened at night, and Peter Proud helped hide the navy lighters and gunboats by day, concealing them among the wrecked shipping in the harbour or in especially netted-over coves. The last three Hurricane fighters were skilfully hidden underground while decoy hangars and model aircraft drew away German bombing and strafing. Camouflage also protected the vital water-distilling plant. The distillery was far too prominent to conceal but it could be made to look wrecked. After a stick of Axis bombs struck nearby, a British camouflage party rushed out to dig bigger bomb holes and to scatter debris prepared beforehand, a cement and paint team created a black, ragged 'hole' in the roof and side of the building, and the demolition team blew up an unused cooling tower. It made a pretty picture for the high-altitude Italian reconnaissance plane, and Rome claimed a direct hit on Tobruk's distillery – which in fact continued to produce its peculiar but potable water.

The next British push-back was operation CRUSADER, in which Clarke's 'A' force also played a role. The *camoufleur* Steven Sykes was sent into the Western desert in 1941 to make nine miles of dummy railhead which ran west of Misheifa to a fake 'Depot no 2', complete with ramps and sidings. The idea was to draw enemy bombing away from the real railway that CRUSADER would use, which ran to Depot no. 1, and also to convince the enemy that the British still had not completed their preparations to attack.

They started laying real rails but used only a few sleepers. Eventually they ran out of rails and used tracks made of flattened petrol tins, bashed into shape and blackened. The eighteen flat cars and thirty-three box wagons were constructed from local hurdles of split palm-branch known as *gerida*, covered with canvas. Sykes was rather proud of his locomotive which made real smoke from an old army cookhouse Soyer stove inside. When materials ran short they had to scale everything down to two-thirds size. Sykes also used objects known as 'net gun pits' that Peter Proud had invented, which could be carried six to a truck. They were made of canvas and covered with camouflage netting, and when raised on poles, each net gun pit looked just like an artillery gun dug into a pit. 'Depot no. 2' was bombed by the enemy, the ultimate badge of honour to a *camoufleur*. (Barkas claimed that 100 bombs fell on the dummies but Sykes modestly said this was 'perhaps more than I would have claimed myself'.) Sykes spent a lot of time driving through the desert too, picking up abandoned 'Sunshields', canvas and wood folding covers that clipped over a tank and made it look like an innocuous 3-ton lorry. The Sunshield could easily be cast aside when going into action. These special hoods were designed to Wavell's specification, and built by Victor Jones of 'A' Force.

Clarke's men were by now doing effective work on many fronts. In the summer of 1941, dummy tanks deployed by Captain Ogilvie-Grant of 'A' Force were a crucial part of the defence of Cyprus. The island was never invaded by the Germans or Italians because it was believed to be strongly reinforced. Before the real 50th Division arrived, the entirely notional 7th Division appeared to have three infantry brigades plus divisional troops, four squadrons of tanks, a 'Special Services' battalion and lots of anti-aircraft guns. There were, in fact, just a few real men who kept busy creating and re-creating a mirage.

Meanwhile, Clarke himself had set off for neutral Turkey on 26 April 1941, carrying a personal letter from Wavell to Sir Hughe Knatchbull-Hugessen, the ill-fated British Ambassador to Turkey. (Only a few months later, Knatchbull-Hugesson's Albanian butler, code-named CICERO by the Germans, started regularly photographing the confidential documents in the diplomat's safe.) Churchill was anxious to get Turkey on the Allied side and sent his Foreign Secretary Anthony Eden to offer 'stimulus and guidance', but the Turks resolutely refused to enter the war until February 1945, when they were sure Nazi Germany was on its last legs.

Clarke's real mission was in Istanbul. He discreetly met up with the British assistant naval attaché, Commander Vladimir Wolfson, RNVR, one of the excellent people picked by Admiral John Godfrey, the director of Naval Intelligence. Working together over the next three weeks, Wolfson and Clarke began what Clarke called a 'long and profitable partnership' that lasted until the end of the war. They began by planting stories about the British SAS and the Free French attacking Rommel from behind, and set up ten new channels for getting deceptive material to German agents, to be triggered by code messages from Clarke to Wolfson sent via the British embassy. These channels included Greeks, Hungarians, Iraqis, Russians, Swedes and Turks who worked in banking, carpetselling, diplomacy, journalism and stenography. The two men also organised a skeleton MI9 system for the area to help Allied servicemen escape from Greece, Bulgaria and Rumania and travel via Turkey back to the Middle East.

Clarke travelled by express train through the night of 17 May south-east from Ankara towards Adana and the Syrian frontier with Turkey. It was a delicate moment of history. Two days earlier, German aeroplanes had begun landing in Syria on their way to support Rashid Ali's anti-British rebellion in Iraq. Wavell had confided to Clarke before he left that he would attack Vichy Syria in just that contingency, and the RAF had promptly bombed Syrian aerodromes at Aleppo, Damascus, Rayak and Palmyra. Oil pipelines that supplied northern Iraq's 2.5 million tons of oil every year from Mosul and Kirkuk ran to Haditha on the Euphrates and then split in two. The northern section led to the port of Tripoli in Lebanon, then part of Vichy Syria, and the southern ran through Transjordan to the port of Haifa in British Palestine. The British had initially closed down the flow to Syria, but in April 1941 Iraqi troops supporting Rashid Ali seized the Anglo-Iraqi Petroleum Company oil fields, re-opened the flow to Vichy Syria and shut down the pipeline to British Palestine. The new commander-in-chief of India, General Claude Auchinleck, diverted to Basra a brigade of Indian troops, Sikhs and Gurkhas, who had been destined for Malaya, and Wavell sent an expeditionary force of 6,000 men across the desert from Palestine in May. The short, sharp Allied campaign was all over by early June. The coup was crushed, and Rashid Ali, the Mufti and the Arab Nationalist guerrillas were sent packing over the border into Iran.

Against this background Clarke evaded the British consular officials at Adana on the Turkish border who were trying to prevent British citizens from entering Vichy Syria. He reached Tripoli on 18 May then drove to Beirut to get the latest information on the military situation for Wavell; the day after that he was in a car with a Jewish refugee family travelling south down the coast road from Beirut to Palestine. Between Sidon and Tyre, Clarke managed to stop the car by what he called 'a lucky stratagem' at 'the key point of the Litani river crossing'. He covertly surveyed the area that the British 7th Division would have to take when invading Lebanon from Palestine. The road followed the flat ground at the edge of the foothills about 1,000 yards from the sea, and crossed the river over the stone arches of the vital Quâsmiyeh bridge. The Litani river was about 40 yards wide and flowed between steep banks lined with poplars. North of the river was a 500-foot hill from which a Vichy redoubt bristling with guns overlooked the bridge amid orchards and cornfields. Clarke saw the fort would have to be dealt with before any invasion could be successful.

Three weeks later, nearly 400 men of No. 11 (Scottish) Commando landed from the sea to attack the redoubt and seize the bridge. The Vichy French, however, promptly blew up the bridge and, because many of the commandos were put ashore by the Royal Navy on the wrong side of the river, they suffered fifty wounded and fifty-four killed, including their colonel, before they secured the crossing.

By 21 May Clarke was back in Cairo, briefing Wavell on all he had done and seen, and preparing a high-level deception to assist the invasion of Vichy Syria, scheduled for 7 June. Clarke intended to spread the story that this attack had been called off because of a flaming row between the Allies. ('There is only one thing worse than fighting with allies,' Churchill had remarked in 1940, 'and that is fighting without them.') The story was that GHQ Middle East had quarrelled with Free French HQ, who had failed to persuade Wavell of the need to invade Vichy Syria. Harsh words had apparently been spoken. General de Gaulle was said to have flown up to Cairo to mediate, been rebuffed, and promptly packed his bags and left. On 4 June, Dudley Clarke got an Arab agent across the frontier into Syria with the news that General de Gaulle had flown off to Khartoum in a dudgeon. It is not clear if the story worked, but the campaign was a success. By Bastille Day 1941, Vichy Syria had fallen to the Allies.

By then, Wavell himself had been sacked by Churchill. After fighting valiantly on five fronts, Wavell was replaced on 21 June, the day that Germany invaded Russia. The Prime Minister who had goaded and harassed him from afar now swapped Wavell with Auchinleck from India. Auchinleck, known as 'the Auk', kept the same team going at GHQ Cairo, and Dudley Clarke still ran 'A' Force.

'The real home of successful deception was the Middle East,' conceded J. C. Masterman in his book, *The Double-cross System of WW2*. However, *The Official History of the Security Service 1908–1945*, completed by John Curry in 1946, points out that Security Intelligence Middle East was pretty much on its own, cut off from MI5. As Curry puts it: 'While the Security organisations in the Middle East had expanded enormously they had to a great extent lost touch with developments in London and during 1940 and 1941 . . . received little benefit from London's experience and knowledge of the Abwehr and its ramifications.' Raymond Maunsell of SIME used a 'turned' German agent called Durrent in spring 1940, very soon after MI5 in London began the practice, but the effective use of double agents thereafter seems to have evolved independently in the Middle East.

The first proper 'play back' or double agent that Dudley Clarke used in Cairo was Renato Levi, a handsome Italian Jew in his mid-thirties. He had been a reliable source of information for the SIS in France before being recruited by Mussolini's military intelligence service, *Servizio di Informazione Militare* (SIM). Levi stayed in touch with British SIS while persuading the Italians that he could set up an espionage network for them in Cairo, where he arrived early in 1941. His first British contact in Cairo was Kenyon Jones, a burly ex-Rugby Blue from SIME hired by Maunsell because he spoke German.

Levi had picked up reasonable English in Australia. He was an easy-going man who lived on his wits and had an eye for the girls and the good life. Levi cheerily assured Kenyon Jones that he was going to be sent a wireless transmitter set via the diplomatic bag of a neutral embassy in the Balkans. Meanwhile he asked for a place to stay and an innocuous job. While they waited for the radio, which never came, Renato chased women. Kenyon Jones suggested using a civilian wireless transmitter (W/T) set instead.

An amateur radio enthusiast built them the right kind of set in a fortnight. Jones used a simple but effective code based on an alphabetic grid-square with a changing keyword on the top line, and Levi took it back to Italian Military Intelligence in Rome, telling them that he had recruited an (imaginary) agent called Paul Nicossof who would be sending messages in Morse from Cairo. After muddles over frequencies, the first successful transmission from 'Nicossof' was made one afternoon in July 1941 from a British radio station at Abbassia near Cairo. Jones had encoded a short message from 'Nicossof' claiming to have made some useful contacts. The signaller tapped out the jumbled letters in Morse code, was rewarded with *grazie* in clear, and then a coded message sent back from Rome.

Jones said later it was 'undoubtedly the biggest thrill I had in the war'. When he told Maunsell about it the next morning, his boss said, 'We must get hold of Dudley at once and you had also better bring in SIS.' In his autobiography, *The Road Uphill*, Jones described Clarke as 'small in stature, humorous, highly intelligent and quick on the uptake', and said he came to like and admire him. What most impressed Jones was Clarke's amazing talent for getting things done. He had access at any time to the Chief of Staff, Arthur Smith, and to Wavell, the Commander-in-Chief, and 'the whole apparatus of GHQ seemed to be at his disposal!'

Clarke provided Kenyon Jones with the messages that the channel – now code-named CHEESE – would send to the Italians. As he became more trusted as a source, his messages went straight to the Abwehr and to Rommel's HQ. Creating the rich CHEESE board of 'notional' contacts in Cairo and across the Middle East was a task handed on to Evan John, an eccentric older writer and artist who had been in the Commandos, then joined the Intelligence Corps, and was posted to SIME in 1941 because, he drily said, he happened to mention 'in some portly presence' that he once talked with T. E. Lawrence in Oxford.

In his autobiographical memoir *Time in the East*, Evan John described Clarke (whom he only called 'The Colonel') as 'a professional soldier with a great love of good English . . . He had read widely in literature, especially spiteful literature. His combined love for malice and good style naturally led him to the eighteenth century, and he knew his Junius better than – as a thorough-going atheist – he knew his Bible.' John's view of Clarke's character is interesting. You need a sharp edge to be good at deception. Clarke's awareness of human folly helped him to take advantage of it; a degree of malice may have spurred him on in his attempts to mislead and deceive, but he also knew his stratagems had a useful purpose. When a trick worked, he must have felt delight at many levels, intellectual, creative and patriotic, something more complex than mere spite.

# 22

# Impersonations

When Clarke arrived in Lisbon on 22 August 1941 to set up ways of distributing deceptive material from the Middle East, he was once again impersonating a bogus journalist, this time in colourful summer shirts. The place was an entrepôt of spooks, heaving with agents and double-dealers from both sides and a natural hub for gossip because flights between England and the Middle East used it as an overnight stopping place, and there were also connections to the USA. Ten days before Clarke arrived, the double agent Dusko Popov had left on a Pan Am flying boat for New York City, carrying in a microdot an Abwehr questionnaire with two dozen questions about the defences of Pearl Harbor in Hawaii. It might have been read as a warning to the Americans that the Axis were considering a strike, but even though the Federal Bureau of Investigation passed on Popov's information to the army and navy, America was still unprepared when, four months later, on 7 December 1941, hundreds of Japanese war planes attacked US ships and aircraft in Pearl Harbor.

Clarke managed to find sixteen new channels for his misinformation. Some were Germans or Portuguese who could pass documents or information directly, others were Axis-sympathisers or just gossips picked from a floating foreign population that included Americans, French, Spanish and Swiss, 'of both sexes and mostly of doubtful occupation'. These people were transient, though, and Clarke missed Vladimir Wolfson who had supported him when they were doing the same job in Istanbul a few months earlier, and who could have kept the supply of contacts going.

As Clarke moved around Lisbon and Estoril for almost a month that summer he must at some point have crossed paths with a young man who was in the same city and absolutely desperate to get into the great game. Perhaps they sat in the same cinema, looking up at the

336

same black and white newsreels, but reading them differently. Had they met, it would have been a most interesting encounter.

Juan Pujol García, a 29-year-old from Barcelona, was the opposite of a professional soldier. His Catalan father was a kindly, ethical, apolitical man who had brought his son up to loathe oppression and war and to believe the pen was mightier than the sword. Pujol was proud that he never fired a shot in the Spanish Civil War: he hid from recruitment, he deserted, he was imprisoned, he improvised, he survived. But when WW2 came, he determined to do the right thing for humanity through practical action that did not involve fighting. He was not a mercenary (unlike Dusko Popov, always a natural businessman doing deals on the side) but an idealist, a bookish and bespectacled person who had passed through occupations from chicken farming to hotel management on his way towards an as yet unknown destiny. He worked hard and could apply himself, but he was better at making things up than making money, and he could talk the hind legs off a donkey.

His father's library had fed Pujol's liberal sympathies. In common with other Spaniards, despite censorship, constant Axis propaganda and the Madrid press baying approval of Hitler's march across Europe, he sometimes managed to hear appalling stories of what the Nazis were doing to people in the extermination camps of Germany and Poland. Like everyone else in the Spanish-speaking world, he was caught up in WW2, willy-nilly.

In January 1941 Pujol approached the British embassy in Madrid and offered to work for them as a secret agent in Germany or Italy, but was turned away. He then went to the German embassy and, boasting of his right-wing credentials, offered to work for them as a secret agent in Lisbon or Great Britain. He was turned away again. But he persisted, and the Germans said they might be interested if he could get himself to Britain. Pujol went to Lisbon to try to get accreditation as a journalist, but in the end managed to get a forged Spanish diplomatic pass by fooling a Portuguese printer that he worked for the Spanish embassy. Back in Madrid he flashed the document and conned two German intelligence officers that he was working with the Spanish security police and would soon be off to England to track down some missing money. Pujol's fast talking and authentic-seeming papers persuaded the Abwehr men that he was genuine, so they gave him US$3000 and a crash course in writing in invisible ink. A genuine German intelligence officer, Fritz Knappe-Ratey, told him to get in touch with the correspondent of the right-wing newspaper *ABC* in London, Luís Calvo, an established German agent there, and said he could use the Spanish diplomatic bag for messages. Pujol bridled, saying that he hoped the Abwehr would not be so indiscreet with *his* name as they had been with Calvo's, and insisted he preferred to work alone.

Pujol wrapped the roll of dollar notes in a condom, hid it in a tube of toothpaste, and crossed the frontier to Portugal in July 1941. There he went to the British embassy and showed them the secret ink and the list of questions the Abwehr wanted him to find answers to. No interest. He tried again. Someone said they would rendezvous with him, but never turned up. Pujol had already sent a letter to Madrid (who had the largest Abwehr *Kriegsorganisation* outside Germany) falsely announcing his arrival in England. Now he had the mind-boggling task of proving to the Germans that he was a secret agent spying for them in a country he had never visited, in a language he barely spoke. Not yet 'Our Man in London', just a man on his own in Lisbon, he had to make everything up from scratch.

And so, like any fiction writer doing research, by August Pujol was haunting Lisbon libraries and bookshops, ransacking reference books like the deep red 1937 Baedeker *Great Britain: Handbook for Travellers*, which had maps and street plans, and *Bradshaw's Guide to the British Railways*. He pored over British journals and newspapers, copied down the names of likely firms from advertisements, and stared at any cinema newsreels which had footage from Britain. He bought a big map of Great Britain, a *Blue Guide to England*, an Anglo-French lexicon of military terms, a Portuguese guide to the British Fleet, and sat down in his rented rooms, in Cascais, and then in Estoril, to compose detailed fantasies for his German masters in Madrid.

Is it so incredible that professionally trained Abwehr intelligence officers treated these amateur inventions as true? There is an almost infinite human capacity for self-deception. We like to hear what agrees with us, and what is most agreeable is a confirmation of the prejudices we already hold. We see what we think we are seeing – a mirage in Iraq, a chimera in Afghanistan, a vital secret agent in London.

In September 1941 Clarke, responding to a summons from the Imperial General Staff, the very top of the British defence machinery, came back from Lisbon to a bleaker London. Four months earlier, his elegant little flat in Stratton Street had been destroyed by a parachute mine in the last great Luftwaffe raid of the Blitz. Back in North Africa, what Clarke had once known as 'Western Desert Force' or 'the Army of the Nile' was being redesignated the Eighth Army, and they were getting ready for operation CRUSADER, Auchinleck's attack on Rommel in Cyrenaica. The Prime Minister hungered to smash Rommel's Afrika Korps, and at this critical moment Clarke was asked to write and circulate a paper on his experience of strategic deception in Middle East Command.

On 29 September, Clarke was introduced to Guy Liddell, MI5's director of counter-espionage, who recorded in his diary that Clarke 'has many double-cross schemes and controls both rumours and purveying false information through Maunsell's channels. Evidently his operations have been very successful'. Clarke told Liddell about the lessons he had learned from the failure of his stratagems in Ethiopia, but also described the effective 'notional' reinforcement of Cyprus, where 'A' Force's cod wireless traffic and dummy tanks had helped to keep the island safe from attack. Liddell saw that things got done quicker in the Middle East because Wavell had ensured deception was part of operations instead of getting stuck further down the food chain in planning or intelligence.

On 1 October, Clarke attended the weekly Wednesday MI5 XX (Double Cross) Committee meeting. The next day he met the people supporting the chiefs of staff at the War Cabinet Office, who had been shown his new paper. Both the Joint Intelligence Sub-Committee (chaired by Bill Cavendish-Bentinck, and including the three directors of naval, military and air intelligence) and the Joint Planning Staff (whose directors ran strategic, executive and future operational planning sections) were impressed by what he had written. Clarke talked with Guy Liddell again and arranged to inform MI5 of all his deception plans and to tell them exactly what support he needed. From now on, London and the Middle East were to share full knowledge, using 'some sort of code within a cipher in order to ensure secrecy' that they had discussed before. (For example, whenever Clarke said 'counter' he would mean 'encourage'.) On the subject of deception, Liddell rather priggishly warned that MI5 'could not embark upon a policy of downright lying. We could, however, put forward tentative half-truths.'

On Tuesday, 7 October, Clarke met the joint chiefs of staff themselves. The chairman, the Chief of the Imperial General Staff, was his old boss at the War Office, John Dill, now a happy widower due to get married again the next day. Clarke already knew Air Chief Marshal Sir Charles Portal from Aden, and Admiral Sir Roger Keyes, the chief of Combined Operations, from his commando days. The next day, the Joint Planning Staff, who had also read Clarke's paper, recommended to the chiefs of staff that a section like 'A' Force should be set up in the heart of the War Cabinet Office, its mission to control deception worldwide. It was a stunning vindication of Clarke's and Wavell's ideas. The section was to be intrinsic to operations, directly commanded by someone with the staff to plan and execute schemes, and able to draw on the resources of all other branches. The chiefs of staff accepted the plan, and Dudley Pound, the First Sea Lord, asked Clarke if he would like to become the new section's first controlling officer.

This was the plum job and Clarke could not have been better qualified to do it. But it was also a sweetly poisoned chalice. Clarke was experienced enough to know it was better to be the big fish in a smaller pond that he understood intimately. Further from the centre in the Middle East, he had more direct power and greater operational freedom. Perhaps he enjoyed greater personal freedom too. As would soon emerge, he was sometimes a risk-taker. Out there, he was treated indulgently when he was driven by Corporal Payne from the front of Shepheard's Hotel to GHQ to arrive exactly half an hour late for the commander-in-chief's morning meeting. Here in London, everything he did would be known, and he could be easily ignored, outclassed, outranked and outmanoeuvred. He declined the job on the grounds of loyalty, saying that he was a staff officer serving General Auchinleck and they were currently mid-operation. He added that it was not quite fair to pinch a man's butler when he had been lent you for the night, which made Pound laugh.

Instead the chiefs offered the job to Colonel Oliver Stanley, director of the Future Operations Planning Section. He was not really right for the role, but among the three General Staff officers that Stanley employed to help him was Dennis Wheatley, whom we last met in 1940 writing a paper for the defence of Britain against Nazi invasion. He had taken up fiction after he lost his wine business in the 1930 slump and came to widespread public attention in 1934 with *The Devil Rides Out*, a novel which blended a John Buchanesque 'shocker' of contemporary politics with black magic and gossip. Wheatley now had to do a three-week square-bashing course at Uxbridge in December 1941 to qualify him as a pilot officer in the RAF. Then he lined his blue greatcoat with red silk and had a swagger stick made that concealed a 15-inch blade. His excellent book *The Deception Planners*, his seventy-fifth, published posthumously in 1980, describes how the new deception system started out 'near impotence'. It was so secret that they were 'kept absolutely incommunicado and not even allowed to tell other members of the Joint Planning Staff what we were up to, though actually for several weeks we were not up to anything at all'.

The new deception system, although approved at such a high level, did not really start becoming effective until Wavell once again took a hand from the Far East, sending a telegram to Churchill on 21 May 1942, saying that if deception was not to remain 'local and ephemeral', it had to be planned and operated from the very centre of strategy. The current approach was too defensive; Wavell suggested 'a policy of bold imaginative deception' to be worked between London, Washington DC and all the commanders in the field. On that very same day, the chiefs of staff approved Lieutenant Colonel John Bevan as Stanley's replacement.

Why did Wavell show a renewed interest in deception in May 1942? After the Japanese attacked the Americans at Pearl Harbor and launched their onslaught on the British and Dutch Empires in the Far East, Wavell the commander-in-chief of Allied Forces decided that he needed another Dudley Clarke to help him. Early in January 1942 he sent for Peter Fleming who travelled out to India via Cairo where he went through all the 'A' Force files and talked to Clarke. By late March, Fleming was staying with Wavell in New Delhi, and setting up General Staff Intelligence (Deception) (GSI (d)). 'It is a one-horse show,' Fleming wrote to Dennis Wheatley, 'and I am the horse.' The first outing for this steed began over dinner in Delhi one evening in late April 1942, when Wavell told Fleming and Bernard Fergusson about Allenby's deceptions in Palestine in WWI and in particular about Richard Meinertzhagen's haversack ruse.

Fleming and Wavell stayed up till one a.m. planning a ruse of their own in Burma, which was later code-named ERROR. The Burmese military situation at that moment was critical: the ferociously determined Japanese Imperial Army was advancing from the south and Generals Harold Alexander and Bill Slim were trying to manage the successful retreat of British, Chinese and Indian forces northwards from Rangoon to Mandalay and then north-east towards the Indian frontier. Operation ERROR intended to convince Japanese Intelligence that the British commander-in-chief, Wavell himself, had had a car accident during a hurried last minute visit to the front, leaving behind a briefcase of his important secret papers.

Peter Fleming assembled the contents of the briefcase with care. He persuaded Wavell to part with some genuine personal letters as well as a favourite photograph of his daughter Pamela. Among the papers were Wavell's 'Notes to Alexander', which alluded to 'two armies' in Burma, growing air strength and a new secret weapon. Altogether the documents made the Allies seem far stronger in India than the Japanese had estimated. Fleming also forged an apparently indiscreet letter by Joan Bright Astley, who was running General Ismay's 'Secret Intelligence Centre' for commanders-in-chief in Winston Churchill's Defence Office in London, and he borrowed one of Wavell's impressively ribboned coats.

Fleming flew to Burma early on 29 April 1942, accompanied by Wavell's ADC, Captain Sandy Reid-Scott. As a Second Lieutenant in the 11th Hussars, Reid-Scott had won the MC in December 1940 and then lost an eye in an air attack on the Bardia to Tobruk road. At Shwebo, on the plain in central Burma between the Chindwin and the Irrawaddy rivers, Fleming and Reid-Scott conferred with General Harold Alexander. His chief of staff suggested that a good place for the ruse was about sixty miles south: the Ava bridge across the Irrawaddy river at Sagaing, just downstream from Mandalay, where the last of the Burma Corps would be retreating north. With his usual unerring luck, Fleming now ran into Mike Calvert who had helped him with Auxiliary Units in Kent in 1940, and was here commanding a battalion. They obtained an almost brand-new, green, Ford Mercury V8 staff car and drove it south to the Ava bridge and 17th Divisional

HQ at Sagaing. Panicking Indian refugees and retreating Chinese soldiers were coming the other way, and seemed 'a ragged and sorry sight' to Reid-Scott.

Major General Bill Slim, the Burma Corps Commander (and future Field Marshal) was marshalling the retreat, trying to see that the Chinese Fifth Army and 7th Armoured Brigade got north across the Irrawaddy before the iron road-and-rail bridge was blown up by midnight. Around 7.30pm, on a curve 400 yards south of the bridgehead, Fleming and Calvert made some skidmarks and then ran their empty sedan staff car off the road. 'The results were not spectacular', Fleming wrote. 'The car flounced down the embankment without overturning, crossed a cart track, and plunged into a small *nullah*, at the bottom of which it came to rest with its engine still ticking over self-righteously.' They let the air out of one tyre, punctured the other, banged the bodywork about a bit, threw stones to break the windscreen and headlights and took the ignition keys. In the boot of the car they left the briefcase together with Wavell's service dress jacket with its ribbons and commander-in-chief's medal on the breast, two blankets and three novels nicked from the Shwebo Club library. The vehicle was in good condition, visible from the road, and likely to attract enemy attention. By eight o'clock the conspirators were back north of the bridge in the jeep.

Slim saw the last Stuart tanks in the retreat – each one weighing 13 tons – singly and gingerly crossing the Irrawaddy on a roadway that was meant to have a capacity of only 6 tons, and applauded the British engineering of the Ava bridge in its last hours. 'With a resounding thump it was blown at 2359 hours on 30th April, and its centre spans fell neatly into the river – a sad sight, and a signal that we had lost Burma,' wrote Slim at the end of the chapter 'Disaster' in *Defeat into Victory.*

As with many deceptions, it is not known exactly what ERROR achieved. The Japanese never invaded India. But Peter Fleming came to think that the Intelligence branch of the Japanese Imperial Army was not worthy of the name. Stolidly dim, they could not be relied upon to make the right deductions from obvious facts, and they were either ignored or despised by people in Japanese Operations. Fleming surmised that the briefcase plant was probably far too subtle a diversion. 'What we want', he cabled London Controlling Section, 'is not red herrings, but purple whales.' PURPLE WHALES in time became the code-name of a double agent among Chiang Kai-Shek's Chinese in Chungking, used to pass (or even better, *sell*) false high-level documents, all imaginatively concocted by Peter Fleming, to the Japanese.

When 'Johnny' Bevan took charge in June 1942, he wrote his own energising directive for 'London Controlling Section' (LCS). Reporting directly to the chiefs of staff, the LCS was to prepare and coordinate strategic deception and cover plans, support their execution, and do all and anything 'calculated to mystify or mislead the enemy whenever military advantage may be so gained'.

Bevan was an Old Etonian stockbroker with good social connections who had impressed Churchill in the First World War. Both Bevan and Wheatley were gregarious and clubbable, which was important when they needed to make important friends and pull strings. *Bon viveur* Wheatley kept his cellar well stocked through his old mates in the wine trade, and was a great believer in lunch, as well as a good dinner. Getting tight was part of the job; sociability lubricated vital contacts. A typical wartime lunch at Rules started with two or three Pimms at the bar followed by a snorter of absinthe (known as 'Chanel No. 5'), red or white wine with the smoked salmon or potted shrimps, Dover sole, jugged hare, salmon or game, and then port or Kümmel with a Welsh rarebit. Little wonder Wheatley took secret afternoon naps in bedrooms reserved for cabinet ministers. Bevan was much fussier, more nervous, and driven, so the two men complemented each other well.

Others who came to join LCS included Major Ronald Wingate, son of Sir Reginald Wingate, Sirdar of the Sudan in T. E. Lawrence's day. Wingate was a former civil servant who relished the Byzantine hierarchies of British bureaucracy. Because he understood perfectly what he called 'the working of the protocol' he achieved results that no outsider could ever hope to match. Whereas Bevan could get irritable and rude, Wingate was always imperturbable and urbane. Wheatley considered him 'as cunning as seven serpents'.

It was Bevan who moved the LCS down into the fortified under-ground basement of the No. 10 Downing Street Annexe. Upstairs they had been physically isolated, but now they were right in the thick of things. Wheatley compared the Cabinet War Rooms to the lower deck of a battleship. Steel support girders held up five-foot-thick concrete bombproofing, and pipes and cables ran overhead through the whitewashed narrow passages of a warren of over a hundred rooms, with chemical lavatories and stale air, guarded by armed Royal Marines. This was where strategic decisions were taken and where planning happened. Bringing deception right into the heart of war operations worked well: only in the place where the most truths were known could the best lies be formulated. The Prime Minister took a keen interest in LCS's work. Sefton Delmer, the psychological warrior of black propaganda, would also become a frequent visitor. From the summer of 1942 until victory the whole machinery of British deception began to synchronise and work together to keep the Axis guessing about where and when the next blow would fall.

When Dudley Clarke left London for Portugal on 12 October 1941, after his meeting with the chiefs of staff, he had every reason to feel both proud and successful; he had been listened to with great respect; he had been offered the top job, and turned it down. Now he was going to Lisbon to continue misinforming the Abwehr that Auchinleck would not be ready to attack in North Africa until after Christmas 1941. Clarke was preparing the ground for the story that agent CHEESE would be sending over the radio in late October: to wit, the fiction that Auchinleck's CRUSADER offensive was off because three British divisions – one armoured, two infantry – had to go to the Caucasus to help the Russians, who were now battling the huge German invasion. Wavell was said to be leaving India to lead the three divisions. The whole thing seemed to make sense. The British had quashed the revolt in Iraq, dealt with Syria and then, at the end of August, moved into Persia, thus securing the Iranian oilfields. Logically, their next primary military focus might well have been to focus on the Caucasus. Clarke's job was to persuade the Axis that the Allies were about to do just that, rather than poised to make a move in North Africa.

But when the head of 'A' Force went to Madrid, presumably to spread the same plausible rumours through Spanish channels, he came a mighty cropper. The secretive activities that had always worked for him now came unstuck in a most ridiculous manner, and Clarke almost lost everything. The incident remains to a degree mysterious even today, partly because it could only be referred to at the time in an oblique manner, but the bones of what happened are clear.

Some time in the middle of October 1941, Dudley Clarke was arrested in Madrid dressed in women's clothes, and detained by the Spanish police. Where exactly, and in what circumstances, is still not known. But the Churchill Archives Centre in Cambridge holds copies of the four photographs taken of Clarke, which are not at all the usual police mugshots. There are two seated poses, and two full-length where Clarke stands by a table-chair in front of a sheet pinned on the wall of a room with a bare wooden floor. The photos show two different incarnations of the head of 'A' Force, Middle East. One is a perfectly conventional fair-haired, middle-aged man in a pinstriped suit, a slightly floppy-collared shirt and a checked bow tie, frowning slightly. The other shows a lipsticked, kohl-eyed 'woman', her expression hard to read but possibly a touch defiant and even amused, wearing an elegant, slim-fitting day dress printed with passion flowers and three strands of pearls, perfectly accessorised by dark stockings and high-heeled court shoes, a chic small white handbag and a pale close-fitting turban, the only incongruous note overlarge hands, which are only half-concealed by dark elbow-length ruched satin gloves.

Guy Liddell of MI5, the Director of Counter Espionage who had last met Clarke riding high in London the month before, rather acidly recorded Clarke's 'difficulties' in his diary for 21 October:

He has been imprisoned by the Spanish authorities, presumably on his way to Switzerland. I am afraid that after his stay in Lisbon as a bogus journalist he has got rather over-confident about his powers as a secret service agent. It would be much better if these people confined themselves to their proper job.

How did Clarke get out of police custody? According to Liddell, a man Clarke had contacted earlier in Lisbon, and whom he believed to be a German agent, was in Madrid at the time and saved Clarke's bacon by telling the Spanish authorities that Clarke was 'an important agent who was ready to assist the Germans'. Liddell thought that Clarke's 'speedy release' could only be explained 'by the Germans having intervened on his behalf'.

The circumstances of his release were to say the least of it peculiar. At the time he was dressed as a woman complete with brassière etc. Why he wore this disguise nobody quite knows. He seems, however, to have played his cards fairly well . . . Dudley Clarke is now on his way home. Nobody can understand why it was necessary for him to go to Spain. Before he is allowed to go back to the Middle East he will have to give a satisfactory account of himself.

Once again a supercilious note creeps in. Liddell comments: 'It may be that [Clarke] is just the type who imagines himself as the super secret service agent.'

Perhaps Clarke's sudden pre-eminence in the London meetings of early October had grated on MI5's Director of Counter Espionage, but Clarke had also crossed a line. The man who was in charge of 'A' Force, who knew many British and Allied secrets (including ULTRA) and who had recently been offered the post of Controlling Officer for deception worldwide in the War Cabinet Office, had let himself be caught in suspicious circumstances in the field dressed up as a woman, like the most amateurish agent. It is easy to speculate on the attractions of such risky behaviour from Clarke's point of view, though he never publicly commented on the incident. We also have to remember the extraordinarily loud black and white plus-fours and checked cap that Clarke had worn when impersonating a journalist on his first arrival in Egypt. His love of theatre since boyhood, his past as a theatrical entrepreneur, performer and stager of pantomimes, could explain his wanting sometimes to dress up and play a part rather than just writing the scripts. This is certainly how Thaddeus Holt sees it, as a kind of comic caper: 'no more to it than *Charley's Aunt*, or d'Artagnan's escape from Milady'.

Was there a sexual element in the charade? Clarke never married, but there is no evidence of homosexuality. As early as 1917, as we have seen, when Dudley was still a teenager, he confessed to overwhelming excitement at the vision of women in uniform on motorbikes. The curious detective thriller that Clarke wrote after the war, *Golden Arrow* (Hodder & Stoughton, 1955), shows persistent and unusually passionate attention to feminine clothing, which the author describes with almost poetic relish and the same obsessive attention to detail that made him a great deception planner. The hero of the novel, Giles Wreford, is a retired colonel rather like Clarke who ran an 'anti-sabotage' unit in the war and now does freelance security work. ('Few ever guessed at the capacity for taking infinite pains which lay well-hidden behind an easy-going exterior.') Giles loves women, but they frequently annoy him by getting something slightly wrong:

With all [Paula's] natural beauty she seemed sadly incapable of acquiring a proper flair for dress . . . superbly endowed with grey-green eyes and Titian hair, with the long slim legs and the gently-curved figure of a model, she could usually be relied upon to ruin the effect of the most exquisite outfit with some shocking misalliance . . . As she swung proudly into the hall of the hotel, his first glance went straight to the revolting little hat in exactly the wrong shade of mustard.

Clarke the novelist never forgets the accessories. As in deception operations, the slightest wrong or missing detail may draw attention to itself and bring the whole illusion crashing down. The novel ends with a paean of love for a present from a rich admirer:

Paula gazed down upon a honey-coloured fur stole nestling in its tissue paper. It was a topaz mink . . . In front of the hall mirror she help[ed] the mink to frame itself in silky folds around her . . .

Perhaps he took fetishistic pleasure in women's clothing; perhaps he also sometimes enjoyed wearing it; perhaps dressing up as a woman really did seem to him the best way of disguising his identity when making a contact: all three things could be true. He was small (5' 7" according to his passport) and slight, and with the gloves on he does make quite a plausible woman in the photographs. Or perhaps it was less complicated, and Clarke just wanted to have fun.

The Germans may have actually sprung Clarke from prison, as Liddell says, but the British naval attaché in Madrid, Alan Hillgarth, who had good relations with the Spanish authorities, certainly helped. Hillgarth worked both for Sir Samuel Hoare, the British Ambassador in Madrid, and for John Godfrey, the director of Naval Intelligence in London, and managed to thwart Italian and German machinations in Spain while maintaining all the diplomatic proprieties of formal Spanish neutrality. Certainly it was Hillgarth who obtained and sent to London the Spanish police photographs of Clarke, both in drag and out of it, where they were circulated as far as Churchill himself (who took a personal interest in Clarke's escapade). They aroused considerable interest and, reading between the lines of the memos, some amusement:

Dear Thompson,

Herewith some photographs of Mr Dudley Wrangel Clarke as he was when arrested and after he had been allowed to change. I promised them to the Prime Minister and thought you might like to see them too.

Yours ever,

Alan Hillgarth

It was probably Alan Hillgarth who conveyed Clarke safely back to British territory in Gibraltar, from where Clarke was recalled to London to explain himself; in other ways, too, his troubles were not yet over. The next convoy's sailing from Gibraltar was delayed. Enemy submarines had just sunk a British tanker and a freighter in the Atlantic west of Gibraltar, and the Royal Navy sent out a dozen warships to conduct anti-U-boat sweeps. The Admiralty knew from decoding German wireless traffic that a group of six U-boats, code-named *Breslau*, were deployed in the approaches to the Straits of Gibraltar. One had been sunk by the British and the others had withdrawn, but only temporarily, to the west where four Italian submarines were also lurking. They had certainly not given up.

Convoy HG-75, comprising eighteen merchant ships, left Gibraltar for Britain at 4 p.m. on 22 October. Dudley Clarke was on the *Ariosto*, a new merchantman out of Hull which was carrying the convoy's commodore and a cargo of cork and ore. The ships were escorted by three British destroyers; nine other Royal Navy warships of the 37th Escort Group had sailed ahead an hour and half earlier to hunt for U-boats. But German agents in Ceuta and Algeciras immediately signalled the departure of the convoy to the commander of the U-boats and within half an hour the U-boats knew too.

As the ships headed into the Atlantic, the British destroyers escorting the convoy were alerted that enemy submarines were around by their own pinging sonar echoes coming off the submarine hulls and by 'Huff-Duff' from the Admiralty. The High Frequency Direction Finder or HF/DF system located enemy submarines via high-frequency, short-burst radio transmissions which they could only make on the surface. When Allied listening stations in Britain, Canada and the Caribbean detected a high-frequency German submarine signal, they drew a line on the map to it. The intersection of several of these 'cuts' gave a rough location. On the second night out, two British corvettes from the 37th Escort Group had just finished chasing a sonar contact when there was a red explosion at the rear of the convoy. The destroyer HMS *Cossack*, astern of the port wing, had been torpedoed by *U-563* about 250 miles west of Gibraltar. The destroyer's bridge was blazing fiercely and its short-range ammunition was exploding in the heat. Cold men on oval Carley Float life rafts sang in a darkness illuminated by their burning ship. HMS *Carnation* picked up forty-nine survivors.

Six hours later, in the darkness before the dawn of 24 October, *U-564*, skippered by 25-year-old Reinhard Suhren, fired five torpedoes at the convoy and then escaped. The torpedoes hit three separate British cargo ships at ten-second intervals, the *Carsbreck*, the *Alhama* and the *Ariosto*, which was carrying Dudley Clarke. *Carsbreck*, loaded with 6,000 tons of iron ore, sank like a stone within a minute. *Alhama* lingered for ten minutes before foundering in a sea made chiaroscuro by the escort ships firing star shells and fierce white flares called 'snowflakes' high into the sky. On the *Ariosto*, six men died in the explosion, but the forty-five other crew and passengers had five minutes to get to the boats and rafts before the ship went down. As he scrambled to escape, Clarke wondered if it were true that, being born with a caul over his head, he was not meant to drown. Most of the *Ariosto* survivors were picked up by a Swedish ship, but Clarke was among the seven rescued by the British destroyer HMS *Lamberton*. He was lucky in more ways than one. The *Lamberton* ran low on fuel chasing the U-boats and was forced to double back to Gibraltar, together with its unexpected passenger. By the end of October, Clarke was restored to the Rock. The breathing space seems to have given the authorities in London time to calm down about the incident in Spain. People in high places may also have realised that there had been a risk of Clarke drowning when the *Ariosto* went down, taking with him the secrets of many plans not yet executed, or of his being captured by the Germans, with even more frightful results. His value to the war effort may therefore have suddenly become clearer. All this allowed Clarke to bargain with fate.

A message from Sir John Dill to Churchill on 31 October records that Clarke had sent a telegram to London the night before telling them about the torpedoing and asking 'whether he is still to come back to U.K.' or whether he should go on to Egypt to return to his duties. It must be remembered that Dill had an old liking for Clarke.

In the first weeks of Dill's appointment as CIGS Clarke had done good service as his military secretary. Now Dill played a central role in finessing Clarke's return to favour. In the missive to Churchill Dill pointed out that Clarke had already been delayed 'about a week' by the 'mischance' with the U-boat, and also said that he was 'wanted in the Middle East', that is, by Auchinleck, who was 'in the best position to take proper disciplinary action with knowledge of all the facts'. Dill suggested to Churchill that Clarke could be dealt with out in Gibraltar, where he could be questioned by Field Marshal Lord Gort, Governor of Gibraltar. If Gort considered Clarke's story 'reasonable' and if he found him to be 'sound in mind and body', Dill says Gort should 'send him on to Middle East by first possible aircraft as he is urgently required there'. Churchill approved Dill's suggestion on 1 November.

Unsurprisingly, the master of deception indeed managed to convince Lord Gort that he was fit to return to duty. The long report Gort wrote is nowhere to be found, but John Dill's account of it is recorded in a minute to Churchill on 18 November. Dill advised Churchill that the report was 'of such length that you certainly should not be bothered to read it'. (Or perhaps Dill thought the report contained damaging information about Clarke.) He told Churchill that 'the Report clearly shows that Col. Clarke showed no signs of insanity but undertook a foolhardy and misjudged action with a definite purpose, for which he had rehearsed his part beforehand'. Dill's partial quotation from the covering letter Lord Gort wrote for the report begs some questions, beginning as it does ' . . . he seems in all other respects to be mentally stable'. In which respects had Lord Gort judged Clarke *not* to be mentally stable? The quotation from Gort's letter continues: 'We can reasonably expect that this escapade and its consequences will have given him a sufficient shock to make him more prudent in the immediate future.' Gort accordingly sent Clarke back to the Middle East, and Dill says, 'We can safely leave it to General Auchinleck to deal with him, both from a disciplinary point of view and as regards further employment in his special role.'

Churchill's curiosity was evidently aroused, and he sent Dill's message straight back with 'a definite purpose' underlined and the handwritten question underneath 'CIGS. What was his purpose? WSC.' Dill replied, 'Colonel Dudley Clarke had worked up contact with certain German or German-controlled elements, with a view, later, to their providing a channel for the dissemination of false information which was designed to provide a "cover" for British operations (in the Middle East).' This seems to have satisfied Churchill; Clarke was sent back to Egypt, and there is no evidence that he was ever disciplined.

By the end of 1941 and the beginning of 1942, the horrors of the World War had spread far beyond the Atlantic Ocean, the deserts of North Africa, and the plains of Western Europe. In June 1941, Nazi Germany had launched BARBAROSSA, attacking the Soviet Union with 150 divisions, and by December German troops were forty miles from Moscow. On 7 December the Japanese launched their crippling air attack on the US Navy at Pearl Harbor and all American bases across the Pacific, before going on to demolish the British and Dutch Empires in the Far East, overrunning Hong Kong and Singapore, and driving the British out of Borneo, Burma and Malaya. This became one of the darkest periods of the whole bloody conflict.

But the great giants were now in the struggle; the USA had joined the World War and both they and the USSR were fighting a common enemy. In the endgame, though British brains did their bit, Russian blood and American treasure would bring down the Axis of Germany, Italy and Japan.

# 23

# The Garden of Forking Paths

In December 1941, the month when the USA joined the war against the Axis, a small book of seven short stories appeared in Buenos Aires. The opening tale's arresting first sentence – 'I owe the discovery of Uqbar to the meeting of a mirror and an encyclopaedia' – sounded like a mysterious cipher. The stories were by Jorge Luis Borges, the 42-year-old poet, essayist and librarian who disliked Nazis in Argentina, and consistently wrote in favour of the Allies during WW2. The seventh story, the one which gave the volume its title, *El jardín de senderos que se bifurcan* or 'The Garden of Forking Paths', is about the mysteries of time, but is also a detective story involving WW1 espionage. Yu Tsun, a Chinese spy in England has to communicate a place name in France to his spymasters in Imperial Germany. He therefore kills a scholar whose surname, Albert, is the same as the town on the Somme, so the newspapers will carry his message to Berlin. 'The Garden of Forking Paths' was republished in December 1944 together with nine more stories in the momentous volume *Ficciones* or *Fictions*. In WW1, camouflage and cubism developed side by side; in WW2, deception grew into an infinitely branching series of Borgesian fictions.

As the war wound towards its climax, nothing was really what it seemed. Take the brutal lot of British prisoners of war: their regime of mind-numbing routine was, for the bulk of the 140,000 PoWs held in Europe, just that: a hard, hungry, tedious life. Only a small percentage of PoWS tried to escape. But escape remained the Holy Grail, particularly among officers, who did not have to work in their camps, unlike the 'other ranks' who were treated like slave labour. These officers felt it was their duty to make things difficult for their captors by trying to get out. So, from time to time the dull prison camp routine was concealing a busy hive of subversive activities. Classic postwar accounts of PoW activities like Eric Williams's *The Wooden Horse* (1949) and Paul Brickhill's *The Great Escape* (1951) focus on dogged British attempts to fool the prison guards or 'goons' with an elaborate pretence of normality. A loitering man is really on the alert; a closing window is a signal; behind the false wall is a hiding place; under the lavatory is a wireless set receiving coded messages via the BBC, and so on. 'Stooges' keep watch for 'ferrets' or 'snoops' while artistic and mechanical work goes on – the painstaking forging by hand of correctly stamped, sealed and embossed German ID and travel passes or the tailoring of fake civilian clothes from blankets or uniforms with the nap shaved off. Tiny compasses were engineered from melted gramophone records and 78 rpm needles, and underground railways were put together from bedsteads and beading. As everyone who has seen John Sturges's film of *The Great Escape* knows, seventy-six air force officers got out of Stalag Luft III in March 1944. Three reached England, but all the rest were recaptured. Fifty were shot by the Gestapo, on Adolf Hitler's orders.

The books and films of the 1940s and 1950s imply that the prisoners of war improvised, stole or scrounged everything themselves. Censorship shaped by official secrecy did not allow the authors to tell the whole truth. Prisoners could receive parcels from voluntary and charitable organisations on the outside. (These were not the Red Cross boxes, which had to be respected under the Geneva Conventions.) Some of these charity parcels had blankets, scarves, underwear, clothing, etc., others contained books, puzzles, paper, pencils, playing cards, sports equipment, musical instruments and other entertainments for bored men confined in the *Stammlager*. Parcels from the Licensed Victuallers' Sports Association or the Welsh Provident Fund, for example, might contain a box of Monopoly, the property board game with paper play-money and little wooden red hotels and green houses. Made by Waddington's, these Monopoly sets looked and felt completely kosher. Yet if you peeled away the London streets from Old Kent Road to Mayfair, inside the folding board you might find several useful maps of your part of Germany, printed on silk squares. These Get-Out-of-Jail-Free cards came courtesy of the ingenious Clayton Hutton, technical officer of MI9, the British secret service dedicated to escape and evasion, whose Middle East section Dudley Clarke also ran.

In his remarkable autobiography, *Official Secret*, Hutton reveals how, as well as designing compact ration packs for airmen who were in danger of being captured by the enemy, he hid compasses in fly-buttons, fretsaws in pencils, and flexible Gigli saws in the bootlaces of flying boots with false heels. He also tells of sending escape aids (from batteries to blades, crystal wireless sets to wire cutters) into the PoW camps hidden inside innocuous items such as cricket bats, skittles or chess sets, and entire maps divided among fifty-two playing cards in a sealed pack. The PoW books and movies of the 1950s and 1960s were never allowed to reveal that one officer on every camp's 'escape committee' would be in regular secret communication with London, so that requests and information could be sent both ways coded in innocent-looking letters from wives, girlfriends, family members.

In particular, these told the prisoners which special parcels to look out for among the innocent ones. German money stitched into book covers, together with the tobacco, cocoa and coffee sent in legitimate food parcels, could be used to bribe guards and for escapers' expenses. After this method was discovered, the Germans started ripping all the covers off all books, so Hutton arranged for the pressing of special 78 rpm records with money hidden in the centre, under the label around the hole. Among Hutton's clothing parcels were woollen blankets specially selected with the help of the Wool Association to be easily converted into suits, apparently new issue RAF and Marine uniforms which, once stripped of British insignia, matched Luftwaffe uniforms. There were also packets of handkerchiefs tied up with black and white ribbons that just happened to match those from which Iron Crosses dangled. The ingenuity the British showed in trying to help their men behind the wire was remarkable, and even when the escapes did not work, it kept the Germans on their toes and helped to raise morale.

But MI9 was not the only secret service hiding things. SOE had an entire unit, largely recruited from film-industry buyers, craftsmen and prop-makers, dedicated to camouflaging anything in everything. Section XV was based at the Thatched Barn roadhouse on the Barnet bypass, not far from Elstree Film Studios, and was run by a large genial man called J. Elder Wills who worked with Paul Robeson on *Song of Freedom* and was involved in the early days of Hammer Films. He made two training films for the Army School of Camouflage, where he also built dummy planes and tanks, before coming to SOE late in 1941 and starting his first workshop in January 1942. Their camouflage section became used to hiding ammunition, stores, weapons and wirelesses or almost anything else in all kinds of boxes. They could hide microfilms and messages virtually anywhere, and they could make sniper suits and hides which exactly matched the foliage of a specific region, or cobble Japanese split-toe boots or sneakers that left an apparently bare footprint, for use in the Far East.

SOE disguised its people with careful copies of authentic refugees' clothes, taken apart and examined by foreign tailors for cut and stitching and made up by specially recruited seamstresses in London's garment district south of Oxford Street. Getting the right labels, collars and collar studs, buttons and buttonholes, even drilling off the name 'Lightning' on a zip-fastener, were all part of the attention to detail. For the shabbier parts of Europe, clothes and luggage were carefully aged, scuffed and distressed. Realistic foreign identity papers required a forgery section, whose staff of fifty, including several craftsmen whose extensive holidays at His Majesty's Pleasure meant their credentials were well known to Scotland Yard, produced over 275,000 authentic-looking documents. The make-up section in Knightsbridge provided agents with wigs, gum pads, nose plugs, hair and skin dyes, spectacles, special dentistry and even plastic surgery.

SOE equipped its agents with deadly devices that could have come from Q's toyshop. There were guns that slid up a sleeve or were hidden in a pen, thumb-knives stitched into a lapel, exploding briefcases and incendiary luggage. They designed a range of bombs and booby traps and tyre bursters that looked like an amazing range of everyday objects, from bicycle pumps and books to tobacco boxes and toothpaste tubes.

The most common disguise for bombs for sabotage use was probably the lump of coal, since this was the usual fuel for train locomotives, boiler houses and power stations. A film-set plasterer called Wally Bull moulded the first fake coal in two plaster halves that were then crimped together around the explosive. Later models had liquid plaster poured around a metal general-purpose explosive charge, so showed no join at all. Section XV produced about 3½ tons of explosive 'coal' between 1941 and 1945, as well as 43,700 incendiary or explosive 'cigarettes' for SOE agents.

Section XV's work manufacturing devices and gadgets for the secret armies was keenly examined by King George VI in SOE's demonstration showroom at the Natural History Museum in London. He thus continued the tradition of his father, who had inspected Wilkinson and Solomon's camouflage at the Royal Academy and in Hyde Park in WW1.

Borges said that whatever is imagined becomes real. He liked to mix the real with the fantastic, to slip invented books on to the library shelves next to genuine ones, or to have a known writer review an imaginary text, shuffling truth and fiction just as deceivers like Dudley Clarke and Sefton Delmer did in the war. In his story 'Tlön, Uqbar, Orbis Tertius', a group of men invent a fantastic ideal world which then begins to intrude into the real one. Peculiar objects called *hrönir* start materialising. Some are better versions of something that was lost, others were previously unknown. Purest and weirdest are the *ur*: things produced by suggestion, created by the power of hope.

The eventual meeting between Juan Pujol and British Intelligence seems like just such an intrusion of the fictional into the real. The situation was both serious and absurd. Pujol was living in Lisbon but pretending to be in London for the benefit of his Abwehr controller in Madrid. He had invented an elaborate charade in which he said he had bribed a KLM pilot to bring his secret messages (written in invisible ink in between the lines of ordinary letters) from London to a bank in Portugal where the Germans could pick them up (in fact he wrote and delivered them himself in Lisbon). He had sent his histrionic wife Araceli Gonzales to Madrid to check whether the Abwehr actually believed him. There she pretended to be a jealous wife, saying she knew Pujol was just away having an affair with some tart, thus forcing the Germans to reassure her that, no, he was doing useful work for them in London, and not to worry. The Abwehr really did think that he was a top agent.

By then Pujol was seriously worrying the British. Since October 1941 they had monitored and decrypted his messages at the point when they were transmitted by the Abwehr from Madrid to Berlin. Apparently this 'Agent ARABEL' had gained sub-agents in Glasgow, Liverpool and the West Country, and was about to get a job at the BBC. A message that happened to be partially true, about a convoy to Malta, rattled MI5. An Abwehr spy at large in England, neither interned nor turned, could threaten the nineteen double agents that Tar Robertson was already running as well as the whole super-structure of the Double Cross committee. Yet who was this ARABEL or ARABAL? MI5 realised there was something funny about him as an agent. He was unable to understand English pounds, shillings and pence, knew nothing about regiments and said, oddly, that people in Glasgow would do anything 'for a litre of wine'. Much of his information was patently untrue. Could they be sure this man was not still in Iberia? Lots of the German agents in Lisbon and Madrid lied to get more pay. Might ARABEL be one of the fantasts?

Meanwhile, Pujol had only heard once from the Abwehr. They wanted many more details about troop movements. Pujol, who barely spoke English, knew nothing about the British military set-up and had no British contacts to help him concoct anything plausible. He was about to give up when either he or his wife had one last try at the American embassy in Lisbon. The USA was two months into the war and the US Naval Attaché, Edward Rousseau, was bright enough to realise he had landed some sort of a fish. Rousseau contacted the British; eventually in February 1942 they put two and two together and realised that Juan Pujol García and ARABEL were the same man, potentially a most valuable double agent. But who was he going to work for? MI6 wanted him in Lisbon, MI5 wanted him in London – but MI5 won.

Pujol was quietly shipped on a British freighter from Estoril to Gibraltar, from where he flew to England, apprehensive, balding, bearded. When he landed at Plymouth on 24 April, there were two MI5 officers waiting to meet him at the foot of the steps. One was an Englishman, Cyril Mills, who said his name was Mr Grey, and the other was Tomás Harris, a lean dark handsome man with swept back hair who spoke fluent *castellano* because he was half Spanish. Pujol shook his hand. It was a firm grip. Then Harris put his arm round Pujol's shoulder in a gesture of protection and friendship. *Bienvenido, hombre.* This Pujol liked; they trusted each other at once.

Tomás 'Tommy' Harris, the wealthy and well-educated son of Lionel Harris, a Jewish picture dealer from Hampstead who knew Solomon J. Solomon, followed his father into picture dealing, became a friend of Anthony Blunt's and later published a classic account of Goya's etchings. Soon Harris and Pujol became a key double act of WW2, partners in the creation of fictions, like Borges and Bioy Casares.

Borges's real-life friend Adolfo Bioy Casares appears on the first page of the fiction 'Tlön, Uqbar, Orbis Tertius' as the man who quotes the (invented) heretical saying that 'mirrors and copulation are abominable because they multiply the numbers of men'. Bioy and Borges first met in 1932 through the Argentine avant-garde magazine *Sur*, and Bioy was impressed by Borges's essays, 'The Postulation of Reality' (1931) and 'Narrative Art and Magic' (1932) which are precisely about how writers get readers to suspend disbelief and accept the illusory truth they have created by skilful use of detail and atmosphere. The pair decided to write parody detective stories together in December 1941.

Meanwhile Pujol and Harris began collaborating in a similar way in London. Working from an office in Jermyn Street and a five-bedroomed MI5 safe house in Crespigny Road, Hendon (a respectable, mostly Jewish neighbourhood), they scripted what was in effect a serial novel or a soap-opera for the Abwehr in which two dozen busy sub-agent characters came and went, rose and fell, busily compiling a farrago of facts, figures and useful information. At the centre of it was the narrator and principal character whom the Germans knew as ARABEL and the British, at first, by the unglamorous code-name BOVRIL. But Pujol was such a good actor that the British soon changed his name from a beef extract to a movie star, GARBO. Pujol grew into his leading role as the top German secret agent in Britain. The character that he and Harris co-wrote was that of a hard-working diva, assiduous, bossy, demanding in the touchy way of 'talent', quick to take offence if not assuaged and flattered, pouring out endless thoughts and suggestions in a pompous and flowery Spanish style that is worthy of Borges and Bioy's creation, H. Bustos Domecq.

Harris's sister Enriqueta Harris Frankfort later became a world authority on Velázquez and Goya. In WW2 she was working in the Ministry of Information, and fed scraps of detail to Harris and Pujol, who used them to invent an unwitting source of intelligence referred to as J (3). J (3) was supposed to be an official high up in the Spanish section of the Ministry of Information with whom ARABEL/GARBO had been able to ingratiate himself by posing as an exiled Republican writing cheerful propaganda for distribution in Spain. J (3) would become a key source of governmental contacts and 'secret' documents for GARBO. Absolutely none of this was real, of course. *La vida es sueño*, life's a dream, as they said in the Spanish Golden Age.

# 24

# The Hinge of Fate

---

In Winston Churchill's huge narrative history of WW2, the turning point comes in the fourth volume, aptly entitled *The Hinge of Fate,* which shows that the 1942 North African campaign was a crucial step on the path to victory. 1942 dawned gloomily with military disasters for the Allies everywhere, but ended on a much more promising note. 'This is not the end,' said Churchill on 10 November 1942. 'It is not even the beginning of the end. But it is the end of the beginning.' It was the year when British *camoufleurs*, British double agents, British military deceivers and British black propagandists fell into step.

In his autobiographical *Deceivers Ever: Memoirs of a Camouflage Officer*, Steven Sykes (who kept a contemporaneous diary of his 'scurrying about' through the Western Desert and European D-Day) noted that things were changing in 1942, but did not quite grasp the importance of the man who came to put them in place.

Early in February a Col Clark[e] appeared – a very spruce senior (and elderly) Staff Officer in an immaculate British camelhair coat. There was an air of mystery about him, and on the 9th I met him for discussions on Wireless Telegraphy for 37 Royal Tank Regiment – also details of Bedouin tent colours. Col Clark[e] wielded deceptive power via wireless messages and agents . . . it would seem that the tanks of 37 RTR were to become Bedouin tents – a further sign that the deceptive side of desert camouflage was being taken seriously.

Sykes is referring to 'A' Force's deception plan BASTION, which attempted to block Rommel's advance into Egypt at the beginning of 1942 by making him think he was running into a trap at the Gazala line. Hiding tanks in Bedouin tents was an old British trick that Rommel was very well aware of and had even copied. Accordingly, on 15 February Victor Jones put up 150 (in fact empty) tents deep in the desert behind the left wing of the British army, and surrounded them with faked tank tracks, people moving about and lots of dummy wireless traffic.

In March the writer Julian Trevelyan came out to Egypt to report on the new techniques of camouflage in the Middle East for the authorities in the UK. He saw Colonel Geoffrey Barkas, the head of camouflage at GHQ, who told him about the complexity of deception in the desert:

You cannot hide anything in the desert; all you can hope to do is to disguise it as something else. Thus tanks become trucks overnight, and of course trucks become tanks, and the enemy is left guessing at our real strength and intentions. All this involves complex staff work, and Barkas can claim credit for having sold the idea to the high-ups since Wavell's first advance.

Trevelyan went west in a truck with a dour driver called Jock Harris. They passed burnt-out aeroplanes and overturned German and Italian lorries on the way to bomb-scarred Tobruk and Eighth Army HQ, a scatter of tents and vehicles in a shallow *wadi*. They visited an armoured brigade in no-man's-land, navigating by compass through a terrain with no landmarks, thumping and rattling over an endless succession of prickly and stony patches. This was not the Foreign Legion desert of the movies, shifting dunes and plodding camels, but 'a remorseless plain of glaring stone and dust'. Past 'Knightsbridge', a lonely crossroads marked by petrol cans, they pressed on to 'some tanks dressed up as trucks, and to some old trucks dressed up as tanks, bumping along over the stones with flapping skirts like old Cockney dowagers'.

The adventurous soldier David Smiley commanded a 'squadron' of eighteen such dummy Crusader tanks, and found the life congenial. Hot days and cold nights were healthy, and co-operating with real tanks was good fun. Once a map error by Divisional HQ sent them too close to an Italian fort. They had been shadowed by a German reconnaissance plane, popularly known as the '*shufti-wallah*', who duly summoned Stuka bombers to attack them. But to Smiley's delight and amazement the German planes accidentally bombed the Italian tanks that had sallied forth from the Rotunda Segnali to see what was going on. When Smiley's unit, 101 Royal Tank Regiment, were equipped with some of the first dummies of the new American Grant tanks, they were so secret that they were wrapped in sacking before they left Cairo, and were unveiled at night, 500 miles away in the desert, so they cropped up like mushrooms in the morning.

On 2 April Trevelyan talked with Sykes, 'the most intelligent and sympathetic camouflage officer that I have yet met out here', and then drove the long and bumpy road to 'one of Camouflage's show-pieces in the desert':

The dummy railhead looks very spectacular in the evening light. No living man is there; but dummy men are grubbing in dummy swill-troughs, and dummy lorries are unloading dummy tanks, while a dummy engine puffs dummy smoke into the eyes of the enemy.

Trevelyan was with another camouflage officer the next day when enemy fighter planes machine-gunned them by their broken-down car. Armour-piecing bullets splintered the stones beside him as he squirmed into the ground. Shaken, the two men returned to find the *camoufleurs'* camp in commotion, a wrecked lorry burning and everything shot up except the dummy railhead. Trevelyan said the Germans 'later paid it the compliment, I believe, of dropping a wooden bomb on it'.

But Tobruk fell on 21 June and a week later two British corps were shattered at Mersa Matruh. There was a disorganised retreat east to the Nile Delta; the Royal Navy left the harbour at Alexandria causing '*panique*' in the city's high society. Privileged and well-connected womenfolk were evacuated from Cairo to Palestine, and everyone else had contingency plans to evade German occupation. So much paperwork was torched in Cairo during what became known as 'The Flap' that 1 July 1942 was dubbed 'Ash Wednesday'. You could buy peanuts in twists of paper headed 'MOST SECRET'. They had gone up unburnt in the hot smoke and then fluttered down all over Cairo.

When all seemed lost, Auchinleck boldly took personal command of the Eighth Army. He reorganised them into battle groups, and with his back to the Nile, halted Rommel's advance at the First Battle of El Alamein in July 1942. Dudley Clarke's deception plan for 1st Alamein was called operation SENTINEL. As usual it drew on his well-stocked chest of 'notional' forces. 'A' Force whirled up a *khamsin* of camouflage and deception to buy the British some time. SENTINEL managed to persuade German Intelligence that there was an army camped in the sandhills before them. Through the dust of bogus activity the Germans seemed to glimpse at least two motorised divisions and a light armoured brigade. Faced by such a force and with his supply lines stretched, Rommel could not press forward. Auchinleck did not win a decisive victory, but he held the pass.

British security had to grow tighter now. The British captured Rommel's radio monitoring station 'Schildkrote' (Tortoise) at Tel al Aysa the same month, and discovered that Rommel's SIGINT unit (621st Signals Battalion) had learned about British plans and the Allied order of battle from careless wireless traffic. The Germans had also broken the US military attaché's code. Two German signallers were both found to possess copies of Daphne du Maurier's best-selling novel Rebecca in English, although they didn't speak a word of the language. The books were in fact being used for coding and decoding messages from two German spies who had been working in Cairo with the Egyptian Army officer Anwar El Sadat. Driven across the desert from Rommel's HQ in May by the explorer Laszlo Almasy (fictionalised in Michael Ondaatje's The English Patient), the spies, code-named kondor, now lived on a sleazy houseboat near Cairo's Zamalek Bridge with a transmitting wireless hidden inside a large radiogram. They had been spending Abwehr-forged English £5 notes in Shepheard's Hotel, Groppi's, the Turf Club and the Kit Kat Club. Sansom of Field Security managed to track them down and in a raid on their houseboat at 2 a.m. on 25 July, the agents failed to throw their matching copy of Rebecca, with an already encoded message, into the Nile. The British then turned this to their own advantage by using the spies' radio to send false messages to Rommel as if from kondor, expressing 'British fears' of an attack on the vulnerable Alam el Halfa Ridge (which in fact was heavily defended). The false messages were accompanied by a classic haversack ruse. A blood-stained British armoured car was left half wrecked and abandoned on the edge of a minefield for the Germans to find. It yielded for the eyes of enemy intelligence a map deceptively marked up for armoured vehicles: hard ground was deemed 'impassable' while the soft sift that drained three times as much precious fuel was indicated as 'good going'.

On 8 August, Churchill's impatience for movement drove him to a controversial decision: he sacked Auchinleck. In some people's view, Auchinleck had saved the entire Middle East by outmanouevring

Rommel at 1st Alamein, but halting the German advance was not, in Churchill's view, enough. He made General Harold Alexander, the man who successfully brought out the rearguard from Dunkirk, the new Commander-in-Chief, Middle East, and on 13 August 1942 the egotistical General Bernard Montgomery took over Eighth Army.

The controversial Montgomery was alert enough to understand that this army – the first real Commonwealth army – was evolving its own characteristics, particularly as regards dress, or the lack of it. The style of the 'Desert Rats' is well caught in 'The Two Types', cartoons by Jon (W. J. Jones) that appeared in various British Army newspapers at the end of the war. Monty came out wearing a conventional red-banded officer's peaked cap, but soon, seeing that other 'Desert Rat' officers found suede boots, silk scarves, and sheepskin jackets more comfortable than service-issue uniforms in the heat and cold of the dusty Western Desert, he swapped his cap for an Anzac bush hat, finally settling for his characteristic double-badged tank commander's black beret, worn with jerseys and corduroys. Appearances in Cairo, though, were deceptive. If you saw casually dressed officers in Cairo, they were probably 'gabardine swine', desk-bound box-wallahs disguising themselves in scruffy camouflage to look authentic, whereas the real fighters from the desert or Special Service were more likely to show up in Shepheard's or the Mohammed Ali Club in immaculately correct uniform.

Montgomery already knew Dudley Clarke, having taught him infantry tactics when Clarke was one of the candidates re-sitting the Staff College exams in 1931. Montgomery was a good teacher, for 'the whole thing became plain and simple' to Clarke, who scored well in the exam. A dozen years on, he told Clarke that 'A' Force now needed to prepare deception plans for the Second Battle of Alamein, which was due to begin on 23 October, the night of the full moon.

Clarke delegated the spadework for the deception that helped to win this battle. He had just had a visitor from LCS, a rather brilliant regular soldier called Lieutenant Colonel David Strangeways who let him into a big secret. In early November, the Americans were going to be landing at the other end of North Africa, on the coasts of Algeria and Morocco. This was operation TORCH, and lots of strategic deceptions would be needed to avert enemy eyes from the area. Clarke was to go to Washington DC and London to help with that.

The *camoufleurs* Geoffrey Barkas and Tony Ayrton had to take over.

On 16 September, two days after Clarke talked to Montgomery, they went to see Montgomery's Chief of Staff, Brigadier Freddie de Guingand. They heard the plan of attack, which was rather like WW1, face to face with no open flanks. The battle would have to be stage-managed so as to blast a hole in the enemy front through which forces could pour. Montgomery needed the help of the *camoufleurs*, and said they would be given full resources and 'Operational Priority'. In the north, Montgomery needed concealment for the real attack: in the south, he needed a big display to suggest the attack was really coming there, delayed until November by problems with the American Sherman tanks. The goal was to hold up half of Rommel's armour in the south.

Operation BERTRAM, the overall deception plan for the Second Battle of Alamein, was made up of seven subsidiary operations which all interlocked in a complex version of the three-card trick, or pea-and-thimble. The deceptive camouflage experts had learned that the objects they were using did not have to stay the same: both appearance and reality could change, especially if the switch was performed at night. The netted coves at Tobruk first hid real ships, then half-concealed fake ones. A supply dump could be made to look like a lorry, the lorry could look like a tank, and a tank could hide itself inside an apparent supply dump. A 'Cannibal' was a device which from the air looked like a lorry in a dispersed park of other lorries. When the poles and canvas of each 'lorry' were pulled down, however, they revealed either a 25-pounder gun-howitzer with its limber, or the Quad gun-tractor that towed them. There were 400 such guns, in batteries increasingly coordinated by wireless.

Ayrton and the illustrator Brian Robb were in charge of what Churchill called 'a number of ingenious deceptive measures and precautions'. In the north, the Royal Army Service Corps had 6,000 tons of stores and supplies to hide near the front, over half of it near El Alamein railway station. On the ground Ayrton and Robb found a hundred sections of slit trench, nicely lined with masonry. An extra facing of War Department petrol tins, stacked three high, hid two thousand tons of fuel undetectable from the air, with good ventilation for the notoriously leaky containers. The food stores were stacked at night in the shapes of three-ton trucks and covered with a standard camouflage net pegged properly so extra stores could fit under the wings. Other stores went into ordinary soldiers' bivouac tents. From the air, the whole dense assemblage looked just like any other congregation of thin-skinned vehicles dotted about the desert.

The placing of 722 'Sunshields' or dummy truck covers were vital to one of the subsidiary operations, MARTELLO, which was to mask the move of real tanks towards the front. First, hundreds of real lorries were parked in the area often enough to get enemy reconnaissance used to their presence. Then the lorries were driven off at night and replaced by dummy Sunshields. Each Sunshield was numbered and earmarked for an individual tank to drive up to and hide inside the next night before dawn. The tanks had come from a rear area code-named MURRAYFIELD, and when they left, they too were replaced by dummy tanks.

This was all part of concealing the real attack in the north, but Ayrton and Robb also had to coordinate the 'distraction' display in the south, supplied with material and devices by Barkas and his deputy and successor as head of camouflage, Major R. J. Southron. The camouflage centre at Helwan went into overdrive to supply 400 dummy Grant tanks, 100 dummy guns and over 2,000 dummy lorries. A *camoufleur* called John Baker designed a prototype truck that could be assembled from *gerida* palm hurdles, stitched into hessian and painted by teams from East Africa, Mauritius and Seychelles, helped on in the final days by the first British Camouflage Company from Palestine.

Operation DIAMOND was another subsidiary deception that started weeks before the battle. It involved the continuation of the genuine water pipeline, buried in a trench which ran from El Imayid into the MARTELLO assembly area, with a twenty-mile dummy pipeline made of beaten and shaped empty petrol tins, heading south and luring enemy eyes towards the dummy dumps code-named BRIAN where 700 tarpaulins had been draped over miscellaneous objects so they looked like 9,000 tons of ammunition, food, oil and ordnance.

From 15 October, three field regiments of dummy artillery were located at a site in the south code-named MUNASSIB. They had some signs of life and the camouflage was deliberately not quite good enough to hide their fakeness from the enemy, who therefore discounted them. But after the Battle of El Alamein began, the dummies were switched at night for real guns. Their crews lay hidden until a tank attack in their sector allowed them to start a surprise shelling.

Overall then, a German or Italian intelligence officer surveying the terrain south from El Alamein and considering his *shufti-wallah* or *Fliegerführer* reports on the morning of 22 October 1942 would have seen not much change. The big British tanks were still at the back, so there was surely two days' grace before they could get into position. Large dumps, a completed pipeline, and radio traffic analysis indicated more activity in the south. Something would be coming there, maybe, but not yet.

In the first stage of the Second Battle of Alamein, 'A' Force's deception plan BERTRAM gained for the British what General Alexander called 'that battle-winning factor': surprise. The attack began with an artillery barrage by nearly 900 guns which blasted two corridors through the enemy minefields and defences for British infantry and tanks to advance. The second and third stages went on for twelve long days, and the battle was won by grim and chaotic fighting.

Overwhelming force carried the day by 4 November. Rommel had 530 tanks, but 300 of these were Italian, whereas Montgomery had 1,200 tanks, 470 of them heavy Shermans and Grants from the USA. Above all, Rommel did not have enough fuel. Through Enigma decrypts the British knew exactly how much petrol he had and which tankers were coming to supply him, and they made sure to sink these individual ships. In September 1942, 33 per cent of Axis military cargo and fuel was sunk before it reached Libya, but in October the figure reached 44 per cent, and, in *The Hinge of Fate*, Churchill claimed the Germans lost 66 per cent of their petrol. In a disorderly rout, the defeated Germans and Italians fled west from Egypt along the coast road.

Historians today tend towards the view that the real pivot of the war against Hitler was not Alamein, where casualties were relatively light, but Stalingrad, where, between August 1942 and January 1943, the Russians lost half a million men, and killed or captured 250,000 Germans. Nevertheless, 'The Battle of Egypt', as Churchill told the House of Commons on 11 November 1942, 'must be regarded as an historic British victory.' Churchill ordered Sunday church bells to ring out across the nation. He added 'a word about surprise and strategy':

By a marvellous system of camouflage, complete tactical surprise was achieved in the desert. The enemy suspected, indeed knew, that an attack was impending, but where and when and how it was coming was hidden from him. The 10th Corps, which he had seen from the air exercising 50 miles in the rear, moved silently away in the night, but leaving an exact simulacrum of its tanks where it had been, and proceeded to its points of attack. The enemy suspected that the attack was impending, but did not know how, when or where, and above all he had no idea of the scale upon which he was to be assaulted.

Not an iota of this camouflage and deception could be described in the official 1943 Ministry of Information book *The Battle of Egypt*, nor could it be shown in *Desert Victory*, the film about El Alamein directed by Roy Boulting from footage shot by British Army and RAF Film Units. But then, like all films, *Desert Victory* was fake too. The grim handsome faces illuminated by the gun flashes of the opening artillery barrage were filmed at Pinewood Studios. Cameraman Peter Hopkinson told film historian Kevin Brownlow that the famous shot of the advance of the Australians through smoke was in fact staged behind the Ninth Divisional cookhouse in the Egyptian desert. British soldiers put on German uniforms to play corpses lying beside captured Panzer tanks. But in his 11 November speech Churchill confessed: 'I must say, quite frankly, that I hold it perfectly justifiable to deceive the enemy even if at the same time your own people are for a while misled.'

---

Within days of the end of 2nd Alamein, an even bigger surprise hit the Axis: operation TORCH, the Allied landings in the Vichy-French-held colonies of North Africa. Apprised of TORCH in September 1942 by Strangeways, Clarke had flown to the USA and the UK in October to coordinate what the Americans and LCS were doing to divert enemy attention. This time there were eight different, overlapping deception plans. LCS and 'A' Force spread false information that the Allied objectives were in places as far apart as Dakar in West Africa and Malta, east of Sicily. When TORCH began, therefore, half a dozen German and Italian Atlantic submarines were lurking south of Dakar, with another forty between Gibraltar, the Azores and Cape Verdes, rather than directly off Casablanca, where they could have wreaked havoc on the American invaders. Nor did the German Focke-Wulf reconnaissance aircraft find the US convoys. Hundreds of Axis bombers and fighters were too far to the east, on Sicily or in Southern

Italy, waiting to bomb the armada on its way, as they believed, to relieve Malta. As a result, the three American convoys were not attacked from the air.

MI5's 'B' Division, who ran the double agents, were beginning to liaise closely with the deceivers. From London, GARBO reported the unfortunate long illness and death of his invented agent two, William Maximilian Gerbers, who was supposedly based near Liverpool, a carefully arranged alibi for the lack of any reports from Gerbers of the TORCH convoys gathering on Liverpool's River Mersey. The death notice was in the *Liverpool Daily Post* on 24 November.

In October 1942, John Bevan hosted a deception conference in London to which representatives came from Washington DC, Peter Fleming travelled from India and the by now legendary Dudley Clarke from Cairo. 'We had all heard so much about Dudley Clarke that we were most intrigued to see the "great deceiver" in the flesh,' wrote Dennis Wheatley. 'He proved to be a small, neat, fair-haired man, with merry blue eyes and a quiet chuckle which used to make his shoulders shake slightly.' From 8 October to 1 November 1942, Clarke also touched base in London with the CIGS Alan Brooke, General Ismay, the Admiralty, MI5, MI9, SOE, the whole secret kingdom. On Wednesday, 14 October, Churchill received Bevan, Clarke and Fleming in his private rooms.

On 8 November 1942, Allied forces under American command began making three separate landings in Morocco and Algeria. US General George S. Patton Jr led the Western Task Force in 100 ships directly across the Atlantic from the USA to land near Casablanca. Centre and Eastern Task Forces, comprising British and US troops who had sailed from the UK and assembled at Gibraltar, landed at Oran and Algiers. A huge Allied propaganda blitz accompanied the landings: a repeated radio broadcast by President Roosevelt, speaking French, and 22 million printed leaflets of his speech dropped by plane. '*Mes amis,*' it began, and Roosevelt spoke of his friendship for France:

The Americans, with the help of the United Nations, are doing all they can to establish a healthy future as well as the restoration of the ideals of freedom and democracy . . . We are coming among you solely to crush and destroy your enemies.

The giant radio transmitter Aspidistra, used for the first time, blasted Roosevelt's speech from the South Downs to the Atlas Mountains so effectively that Moroccans thought it was coming from Rabat. A message from General Eisenhower to the armed forces and people of North Africa was also broadcast and printed in leaflets and the Free French leaders General Henri-Honoré Giraud and General de Gaulle broadcast a request to French commanders, soldiers, sailors, airmen, officials and colonists to rise in the war of liberation: 'Help our Allies. Join them without reserve. The France which fights calls upon you. Despise the cries of traitors who would make you believe our Allies want to seize our Empire. Forward! The great moment has come!'

In the event, the American TORCH landings were unrehearsed, shambolic, and followed by up to three days of bloody fighting. But they could have gone a great deal worse had it not been for British deception work on three continents.

In January 1943, Churchill and Roosevelt met in plenary conference at Casablanca in Morocco. In comfortable villas at Anfa, surrounded by flowers, fruit and sunshine, the British Prime Minister and the US President gathered with their Anglo-American Chiefs of Staff to decide how they could at last bring the war to a successful conclusion. Led by the CIGS, Sir Alan Brooke, the British delegation knew exactly what they wanted and finally got the Americans to agree to it: first clear the Axis out of North Africa, and then jump to Sicily and mainland Italy in order to knock the Italians out of the war. Meanwhile, U-boats were to be sunk, Germany's industrial heartland bombed, and plans made for a subsequent invasion of north-west Europe to ensure Germany's final defeat. At the press conference, Roosevelt surprisingly declared that only 'unconditional surrender' was acceptable.

With regard to North Africa, the Anglo-American summit agreed that the British Eighth Army would come under US General Dwight D. 'Ike' Eisenhower's supreme command as soon as it crossed the Tunisian border from Libya, and that the British General Alexander would be deputy to Eisenhower, with operational command over all Allied forces in Tunisia, including the British First Army, the American II Corps and the Free French. General Alexander's combined forces would be known as 18th Army Group.

Campaigning with allies is rarely easy. Churchill said, 'There is one thing, however, which you must never do, and that is mislead your Allies. You must never make a promise which you do not fulfil. I hope we shall show that we have lived up to that standard.' But the British and the Americans – two people separated by a common language, as the cliché goes – were actually foreigners to each other, no more natural partners than Germans and Italians. The British referred to the Americans as 'our Italians' and many Americans, of all ranks, loathed 'Limeys' as patronising snobs and ancient enemies. Others, however, saw the relationship with America more positively. When the future Conservative Prime Minister Harold Macmillan was sent out to the new Anglo-American Allied Force Headquarters (AFHQ) in Algiers as Churchill's personal political emissary, he told the future Labour minister Richard Crossman, who was running Psychological Warfare, that the British were now Greeks to the Americans' Romans: 'We must run AFHQ as the Greek slaves ran the operations of Emperor Claudius.'

The British Eighth Army advancing west met the British First Army and the American II Corps advancing east, and General Harold Alexander joined them all together. He had a small tactical team from 'A' Force at his headquarters and managed to surprise the enemy totally by a German-style blitzkrieg attack which drove a narrow offensive blow straight through their lines to capture Tunis. German soldiers were caught sitting astonished at café tables, aperitifs undrunk, while others came out of the hairdresser's, mouths agape in shaving foam, or still draped in the barber's sheet, with their hair only half cut. David Strangeways was at the point of the spearhead, leading a small tri-service advance team of intelligence officers and men from 30 AU Commando – called 'S' Force – into Tunis and Bizerta to seize vital Axis intelligence material before it was destroyed; he won the DSO. The journalists Alexander Clifford, Alan Moorehead and Geoffrey Keating got into Tunis on 7 May just behind the first troops of the 11th Hussars and the Derbyshire Yeomanry and found a kind of madness: in one street people were throwing flowers, in the next grenades; here there was sniping, there cheering as hundreds of British captives were freed. Alexander kept up the pressure on the peninsula east of Tunis, the last area to hold out; suddenly it cracked, and thousands and thousands of German and Italian soldiers were throwing down their arms and surrendering. There were no aeroplanes for the generals; no boats for the soldiers; no Dunkirk for the Axis in Africa. At 19.52 hours on 12 May 1943 it was over. The next day, General Alexander sent his famous signal to Winston Churchill: 'Sir, it is my duty to report that the Tunisian campaign is over. All enemy resistance has ceased. We are masters of the North African shores.'

# 25

# Mincemeat

For Churchill, North Africa was 'a springboard, not a sofa'. Now the combined American and British armies would have to get back into Europe to achieve the final defeat of Italy and Germany. But where exactly should the attacks first thrust? In their conference at Casablanca in January 1943, Churchill and Roosevelt had decided on Sicily. Logically, it had to be the big island at the toe of Italy, just at the narrowest point of the middle of the Mediterranean. But Churchill wanted the Axis forces to fear and prepare for the two alternatives to what Alan Moorehead called 'the obvious route'. To the east lay 'the attractive route': through Greece and the Balkans, and to the west lay 'the quick route': via Sardinia and Corsica, stepping stones to France and the north of Italy. The deception planners had six months to distract the Axis. The code-name for the (real) invasion of Sicily was HUSKY; the name of the Mediterranean deception plan was BARCLAY, which would mark the peak of the deception effort in the Mediterranean theatre. What did the Allies want the Axis to do? Reinforce everywhere but Sicily.

The broad outline of the deception story went like this. The British Twelfth Army in Egypt (which, being one of Clarke's 'notional' units, did not exist at all) was going to attack Crete and Greece in May. Turkey would be persuaded to join the war on the Allied side, then major Allied forces would move through Bulgaria and Rumania to attack the Germans in Russia from behind. Meanwhile, the British Eighth Army would land in the south of France in early June, and together with French forces, drive up the Rhône valley. At the same time General Patton would lead the US Seventh Army in the attack on Corsica and Sardinia. These wholly fictional attacks would be postponed first to June and then to July.

John Bevan of LCS and Dudley Clarke of 'A' Force met in Algiers on

15 March to coordinate BARCLAY. For its part, 'A' Force gathered a mighty host of dummies to simulate the Twelfth Army in Cyrenaica, eastern Libya, within easy view of German reconnaissance aircraft flying south from Crete. There were dummy landing craft in the harbours, a division of dummy gliders and eleven squadrons of dummy aircraft on seven airfields (protected by real aircraft that 'scrambled' whenever German 'bandits' appeared and real anti-aircraft guns that blotched the sky with flak), an entire '8th Armoured Division' of dummy tanks, dummy camps, dummy training areas, and lots of dummy wireless traffic along the Mediterranean coast.

'A' Force gave a strong Greek flavour to the assembly. Greek troops were given conspicuous amphibious training; calls went out for Greek-speaking British officers, and maps and pamphlets on Greece were distributed; there was heavy buying of Greek drachmas on the Cairo exchange, and fifty heavy strongboxes labelled as Greek bullion arrived from London and were taken under armed guard to a Cairo bank.

But Greece was, in reality, treacherous terrain to work in. Early in 1943, Dudley Clarke managed with difficulty to get a list of all their secret agents from all six Secret Organisations operating in the western Mediterranean ('A' force, MI9 Middle East, ISLD or MI6, SIME or MI5, SOE and Hellas, the Greek Intelligence Services). He found that 40 per cent of all agents in Greece were working for no fewer than three different Allied secret services, wasting time, money, information and security. Nevertheless, to assist BARCLAY, a six-man SOE team (no Greeks were included) blew up the railway viaduct at Asopas and stirred up the Greek resistance; men landed by submarine on the Greek island of Zante and left evidence of reconnaissance; in Crete, SOE officers like the future writer Patrick Leigh-Fermor were busy laying on guides to help Special Boat Service commandos across the mountains to blow up a few German planes and installations and sink ships with limpet mines in Heraklion harbour. The Germans were worried enough to move a spare Panzer division across Europe from France to the Peloponnese.

In London, John Bevan was providing the Foreign Office with rumours and gossip for diplomats to spread. Also in London, the Double Cross Committee came up with one of the most famous ruses of the war, modelled closely on real life events: in October 1942, a Catalina plane carrying Lieutenant Clamorgan of the Free French to

Gibraltar to liaise with the US Army for operation TORCH had crashed in the sea off La Barrosa, south of Cádiz. Two bodies had been washed ashore in Franco's Spain carrying ID cards, letters and a report including the names of many secret agents in North Africa. It was believed that the Abwehr agent in Cádiz managed to see and photograph the papers before the British could get there from Gibraltar.

Inspired by this, in the Double Cross Committee meeting of the same month Flight Lieutenant Charles Cholmondely of the RAF (on attachment to MI5) suggested deliberately dropping a dead body from an aeroplane into the sea near enemy territory, carrying apparently important letters or secret documents. Cholmondely invited Ewen Montagu of Naval Intelligence to help him, and the two got cracking.

Their first job was finding a dead body. They went to see the pathologist Sir Bernard Spilsbury for advice, and the coroner Bentley Purchase pointed them in the direction of the body of a 34-year-old man, 'a bit of a ne'er-do-well' who had died in January 1943 'from pneumonia after exposure'. The corpse was kept on ice, and Montagu claimed that a relative gave permission for his body to be used on condition that his true identity never be divulged. In return, the family were promised that he would later get a proper burial, though under a false name. Sir Bernard assured them that no pathologist in Spain would be able to detect that the man's pleural effluvia did not come from drowning in an aircraft lost at sea.

The dead body would be packed in dry ice in a 400-lb steel container labelled 'Optical Instruments', wearing the battledress uniform of a major in the Royal Marines, with a trenchcoat and a life-vest over the top. A black leather briefcase would be chained to the trenchcoat belt. Alan Hillgarth, the naval attaché from Madrid, advised that Huelva, west of Seville on the Spanish coast that runs towards Portugal, was the best place to make the drop. Montagu went to see Bill Jewell, captain of the Royal Navy submarine *Seraph*, who was used to special operations. He agreed to deposit the body in the sea north-west of the mouth of the Rio Tinto in late April.

The documents the dead man would be carrying had to be carefully planned. His Royal Navy identity card and his pass to Combined Operations Headquarters named him as Major William Martin, RM. He was carrying three letters with authentic signatures. The first, a covering note introducing 'Martin' to Admiral Sir Andrew

Cunningham, commander-in-chief Mediterranean, was from Lord Louis Mountbatten, chief of Combined Operations. This covering note said that 'Major Martin' was well up on experiments with barges and landing craft and had been 'more accurate than some of us about the probable run of events at Dieppe'. (An admission by Mountbatten of failure in the Allied assault on Dieppe in August 1942 would be an interesting tit-bit for the Germans.) 'Let me have him back, please, as soon as the assault is over.' Mountbatten's note ended, 'He might bring some sardines with him – they are on 'points' here.' This was supposed to be read as a veiled allusion to Sardinia. This covering note also asked Admiral Cunningham to pass on a second letter.

This main letter had to be A1, something to make an enemy intelligence officer really sit up. How about the vice-chief of the Imperial General Staff, General Archibald Nye, writing to General Harold Alexander, the active British army commander under General Eisenhower in Tunisia, about future plans for the Mediterranean? General Nye wrote a magnificent letter of Byzantine duplicity. He used HUSKY, the code-name of the real attack on Sicily, as the name for the fictional attack on Greece, and also referred to operation BRIMSTONE, a fictional assault on Sardinia. Even better, Nye referred to an attack on Sicily as if it were merely the cover plan or deception for the supposed attacks on Greece and Sardinia! So in this Borgesian world, the real attack was offered as the cover plan for the fictional ones. Built in to the letter was the equally fictional reason why it had to be delivered by hand rather than being sent by signal which an American might read at Allied Headquarters: there were references to a British disagreement with the Americans about the awarding of medals to the wounded. This letter went through several drafts coordinated with Dudley Clarke. Chance favoured it, too, as the days went by: General Alexander's defeat of the Germans in Tunis in May added to his status and made it all the more likely to German Intelligence that he would be leading the advance towards Europe, as the letters suggested.

A courier could easily have slipped these two letters into a pocket, so to justify 'Major Martin' carrying a briefcase, Montagu and Cholmondely also gave him two proof copies of the forthcoming Ministry of Information illustrated book by Hilary St George Saunders, *Combined Operations 1940–1942*, together with a letter from Mountbatten to General Eisenhower asking him to write an

introduction for its American edition. This (real) book drove the threat
of Allied invasion home: it recounted the growth of the Commandos
and Special Service troops, the birth of Combined Operations and
their raids and landings from the Lofoten Islands to Madagascar. (A
very careful German intelligence officer could have read a partial
account of a certain Lieutenant Colonel D. W. Clarke's role in all this
on page 11.)

Now Montagu and Cholmondely carefully built up Bill Martin's
identity with fictional letters from his father in Wales, his fiancée, Pam,
in Wiltshire and his Lloyds Bank manager, bills from the tailor Gieves
and the Naval and Military Club, two 10/6 ticket stubs from the
Prince of Wales Theatre (the show was *Strike a New Note*), a pair of
bus tickets, a photo of Pam drying herself on a beach, his bunch of
keys and wristwatch and leather wallet with £8 and a used book
of stamps, his Players cigarettes and box of Masters safety matches.
('Details are always poignant,' as Borges observed.) Major Martin
wore a silver cross on a chain round his neck and carried a St
Christopher medallion in his wallet. The pair of regulation British
identity discs of hard cardboard was attached to the braces of his
trousers (the higher octagonal green one with two holes stays with the
body, and the lower, red, round disc is removed as token of death). A
soldier's name, rank, number and religion – his was RC – were
stamped on the front, and his blood group on the back.

After permission was given by Churchill and Eisenhower, His
Majesty's Submarine *Seraph* left Holy Loch, Scotland on 19 April and
was off Huelva early on the 30th. When they unbolted the metal
container, the body inside was wrapped in a blanket secured by
knotted tapes. They could smell that the dry ice had not been wholly
successful; decomposition had started and the lower face was green.
But the briefcase stamped with the official crown was secure and the
Mae West needed no more air. Captain Jewell said a few words from
the burial service as the other four officers bowed their heads. They
put the body in the water at 04.30 hours and saw it drift inshore.
There were Spanish fishing boats in the distance. They dropped a
rubber dinghy in, upside down, half a mile south of the corpse, and
then riddled the metal container coffin with small arms gunfire till it
sank. Jewell then signalled: 'Operation Mincemeat completed.'

The corpse was duly retrieved, and the British vice-consul in

377

Huelva, Francis Haselden, was summoned to the morgue, where the Spanish followed the formal procedures of examination and identification; the port medical officer performed a swift autopsy on the corpse and pronounced death by drowning. In Andalusia's African heat, a late April afternoon can be much hotter than midsummer England, and Francis Haselden was glad to get out of the small morgue. The body was placed in the coffin the vice-consul had brought along. The Spanish naval officer attending wondered what should be done with the black briefcase and other possessions. The vice-consul, pretending to be a stickler for protocol and a respecter of Spanish neutrality, suggested they should be deposited overnight with the naval commandant of Huelva and collected officially the next morning. When he came back he was told they had been forwarded to the Almirantazgo de Cádiz, local naval headquarters.

Hasleden duly cabled Alan Hillgarth, the naval attaché in Madrid, who was apparently getting urgent signals from London about the briefcase and its documents. The British ambassador in Madrid, Sir Samuel Hoare, pressed for their return. The black briefcase and its contents travelled from Cádiz to Seville and then to Madrid through the coils of Spanish bureaucracy (and, of course, under the clicking shutters of German cameras) until Hillgarth was finally given it back, with the key still in the lock, on 13 May.

Meanwhile, the British consulate had buried Major Martin in Huelva Roman Catholic Cemetery on 2 May. A wreath was laid from 'Pam' and the family, then later a flat marble slab, inscribed

William Martin
Born 29th March 1907
Died 24th April 1943
Beloved son of John
Glyndwr Martin
and the late Antonia Martin of
Cardiff, Wales
*Dulce et Decorum Est Pro Patria Mori*
*RIP*

This gravestone has more recently been amended at the foot:

Glyndwr Michael
served as Major
William Martin, RM

Glyndwr Michael is one of the few civilians in the Roll of Honour on the Commonwealth War Graves Commission website. His Additional Information box reads:

Mr Michael posthumously served his country during the Second World War under the assumed rank and name of Major William Martin, Royal Marines, date of death given as 24th April. These details are recorded on the original ledger which marks the grave. History knows him as 'THE MAN WHO NEVER WAS'.

The day the 'drowned' corpse was buried, the real Allied plan for the invasion of Sicily was at last emerging from the mist. Up until that day, there were still several competing plans for the landings, but General Montgomery cornered Eisenhower's Chief of Staff, Bedell Smith, in a washroom at Allied Forces Headquarters in Algiers and drew a map of Sicily on the steamed-up mirror over a hot-water basin. None of the other plans would work, Monty said. In two months' time he would land in force with Eighth Army here – indicating the south-east corner – and the Americans here – pointing to the south-west. Months of bickering ended with the meeting of two tense and aggressive men in an Algerian lavatory. Basically, that was it: the terrier and the bulldog agreed.

And what happened to the deception plan? We know the false information in the briefcase – suggesting operation HUSKY was aimed at Greece – travelled fast up the food chain and had been signalled to Berlin by 9 May 1943. It reached Admiral Dönitz and General Keitel, and Adolf Hitler himself. Admiral Canaris, in charge of the Abwehr, discussed it with Propaganda Minister Goebbels. MINCEMEAT thus played its role valiantly in the overall BARCLAY deception. And BARCLAY was a great success. Hitler sent 1st Panzer division from France to Kalamata in Greece in May and ordered General Rommel to Salonika to defend Greece, not Sicily, in July. German torpedo boats were ordered from Sicily to the Aegean; three new minefields were laid there and shore batteries installed.

All, naturally, to no avail, for on 10 July the Anglo-American Armies, Monty's Eighth and Patton's Seventh, landed in Sicily and took the island in thirty-eight days, not the ninety they expected. Mussolini was deposed; Italy surrendered in September.

In June 1943, Juan Pujol (GARBO) had just been entrusted with the Abwehr's newest, high-grade cipher. Everything seemed to be going well when an unexpected crisis threatened the work he was doing with what even the sober Official History of MI5 called 'passionate and quixotic zeal'. The crisis was what the British police term 'a domestic'. Pujol's wife Araceli Gonzales and his infant son had joined him in England in the summer of 1942, but a year later she was fed up and missing her mother badly. Tomás Harris, GARBO's case officer and co-author, described Mrs Pujol as 'highly emotional and neurotic' and as 'a hysterical, spoilt and selfish woman [who was] nevertheless, intelligent and astute and probably entered into her husband's work because it was dangerous and exciting'. On 22 June Guy Liddell of MI5 recorded in his contemporaneous diary: 'Mrs GARBO is extremely homesick and jealous of GARBO who is completely absorbed in his work and has consequently to some extent neglected her.' Mrs Garbo was, in fact, desperate to go home.

What had happened was a flaming row between husband and wife on the evening of the 21st. Pujol did not want to go to dinner with some Spanish people at the Spanish Club, thinking their links to the Spanish embassy might be dangerous. Mrs Pujol got very upset, and phoned Harris, threatening to 'spoil everything' by revealing all to the Spanish embassy if she did not get her papers to leave the country. Liddell said: 'She thinks that as the whole of GARBO's network is notional we have no further use for his services.'

One of Liddell's reactions was the ingenious idea of warning the Spanish embassy that a woman of Mrs Pujol's appearance was intent on assassinating the ambassador. 'It would however result in the police being called which would be a bore.' In the event, Pujol/GARBO himself came up with an even more brilliant and theatrical solution, which he got Harris and his colleagues to help him enact.

After her husband had gone to work on the following morning, the 22nd, Mrs Pujol received a telephone call and was told she would have a decision about her papers by 7 p.m. that night. In fact, at about 6 o'clock two CID officers turned up at her house with a note from Pujol saying that he had been arrested and was in a police cell: could she please pack his sponge bag and pyjamas? She became hysterical and rang Harris, in tears, pleading that her husband had always been loyal and was ready to sacrifice himself for the cause, so why was he arrested? Harris gravely explained that 'the British Secret Service' was perfectly prepared to accede to her request to return to Spain, but they had had to ask her husband to write a letter breaking off contact with the Germans, in order to protect British interests against his wife's threatened 'betrayal' to the Spanish. Pujol, Harris said, had protested that he would rather go to prison than sign such a letter. At the mention of betrayal, Pujol had become abusive and violent and had to be arrested on disciplinary grounds. His wife said he was behaving just as she expected him to behave; Pujol loved his secret work, and was only trying to protect her.

Later that night there were more hysterics. The man who operated the wireless that communicated with Madrid was summoned to the house and found Mrs Pujol in the kitchen with the gas taps turned on. Harris thought this was play-acting, but there was a 10 per cent chance of an accident, so Harris's wife Hilda spent the night in the house to keep an eye on her. The next day, at her request, Mrs Pujol was formally interviewed. Nervously weeping, she pleaded that what had happened was all her fault. If her husband was pardoned, she would never interfere again, never again ask to go back to Spain.

At 4 p.m. a car picked her up and drove her to Kew Bridge and then on in a closed Black Maria police van to a Victorian mansion by Ham Common. Latchmere House, a former hospital for 'shell-shock' officers in WWI with at least one padded cell, was now the barbed-wire-enclosed Camp 020, where Nazi spies were held and questioned. The commandant and chief interrogator was the colourful Lieutenant Colonel Robin Stephens, nicknamed 'Tin Eye' for his steely monocle. Mrs Pujol was taken inside, and her blindfold removed. Her husband was brought to her, unshaven, wearing Camp 020 uniform. The first question he asked his wife to answer, on her word of honour, was if she had gone to the Spanish embassy. She swore she had only used it as a threat to get him to pay her more attention, and told him she had signed a confession taking all the blame on herself. Pujol said he was facing a tribunal in the morning, but hoped to convince the judges she had never really intended going to the Spanish embassy. She left Camp 020 more composed, but still weeping. The next day Mrs Pujol was told by a chief of the security service that her husband would be allowed to continue his work, but was warned against any repeat behaviour. She returned home chastened and never gave any trouble thereafter.

That day, 24 June, is the Fiesta de San Juan (St John's Day), and when Juan Pujol came back home on the evening of his saint's day, he gave his wife a copy of a statement he had supposedly made. GARBO had composed a small masterpiece in defence of a poor, weaponless woman and the hot pride of a Spaniard insulted by the taint of treason, ending:

I know that I am appealing to a chivalrous Tribunal and I trust in their decision . . . fortunately this country, innocent of artifice and subterfuge, controls with scrupulous legality the common weal of her people and of those who collaborate with them.

The three-dimensional brilliance of this swiftly improvised deception shows how unnervingly good Pujol had by now become at his job, even when the person to be duped was his own wife. In his 1985 autobiography, *Garbo*, written with Nigel West, Pujol conceals the fact that he was ever married.

It was Winston Churchill who really started off the posthumous celebrity of operation MINCEMEAT or 'The Man Who Never Was'. As Prime Minister he was the one who gave permission for it, and he spellbound dinner parties afterwards with the story. One of those who heard it was Churchill's friend Duff Cooper, the hedonistic former diplomat and Minister of Information who became a liaison link with the Free French and then British Ambassador in Paris.

Cooper turned the story of the deceptive corpse into his first and only novel, the bittersweet *Operation Heartbreak*, published by Rupert Hart-Davis in November 1950. It is a story of accidental heroism, of a decent chap who loves his regiment and longs to serve his country in WW2 but is dogged by ill luck and misses out on the action. His death from pneumonia redeems his failure in life, for providence leads his corpse to be used in a military deception that will save thousands of lives. It was an instantaneous success, selling 30,000 copies by Christmas; the US film rights went for $40,000. 'Faint resistance was offered by certain branches of the secret service,' wrote Cooper in his diary, 'but I was able to overcome them.' He told them he had the story from Churchill himself and would say so if they tried to prosecute him for breaching security. Dennis Wheatley thought the book would have led to jail for anyone but a friend of Churchill's.

Less than three years later, in January 1953, a right-wing journalist called Ian Colvin, who had contacts with former German intelligence officers, started making enquiries in Madrid, Gibraltar and southern Spain about the body of a British officer that had washed up ten years before, in wartime, carrying important papers. Clubland gossip had convinced him that Duff Cooper's novel was based on truth, and indeed the name he found on a grave in Huelva cemetery, William Martin, was not a million miles from Willie Maryngton, Duff Cooper's fictional character. Colvin prepared to publish a book investigating the British deception operation, to be called *The Unknown Courier*. The Joint Intelligence Committee ('very unsportingly' Professor R. V. Jones thought), decided to hold back Colvin's account and gave Ewen Montagu, formerly of Naval Intelligence but now back in the law as a QC and judge, permission to write an officially approved account of his role in the ruse. It was a spoiling operation against what Montagu smugly called 'the possible dangers and disadvantages which might result from publication by partially informed writers'. But as a result, the tiny operation MINCEMEAT became the best-known British deception of the war.

Montagu says that he wrote *The Man Who Never Was* in one weekend. He was helped by Jack Garbutt of the *Sunday Express*, which serialised the story before it came out and helped make the book a big hit in 1953. 'The Goons' spoofed it on their zany radio show, which was about as popular as you could get. 1953 was Coronation year, which gave a sense of renewal, of old matters being turned over afresh. A collection of 'ancient' bones known as the Piltdown Man was revealed to be a forgery, so the scientific intelligentsia were intrigued by deception. When Ronald Neame turned *The Man Who Never Was* into a BAFTA-winning black and white film in 1956 ('the strangest military hoax of World War II' said the Cinemascope poster) with a saturnine Clifton Webb playing Montagu, and Montagu himself playing someone else on the Double Cross Committee, the scriptwriter Nigel Balchin gingered up the plot with two sinister Irishmen (Stephen Boyd and Cyril Cusack) spying for the Germans, and the Goon Peter Sellers imitated the growling voice of Churchill. By this time, fact and fiction and fictionalised fact were inextricably entangled. The British love their gardens of forking paths.

# 26

# The Double

———————————•—————————

In a well-known piece, *Borges y yo*, 'Borges and Myself', Jorge Luis Borges comments that he likes Robert Louis Stevenson's writing, and that the other Borges, the famous one he has to live with, also enjoys it, but in a show-offy way, like an actor. In another piece, Borges reflects on Stevenson's *The Strange Case of Dr Jekyll and Mr Hyde*, pointing out the difference between the book itself and the film versions that followed. Whereas the films concentrate on dramatic transformation scenes in which one man splits into two, the book is really about how two separate men turn out to be one and the same (unstable) person. 'He, I say – I cannot say, I.'

Like Borges, Dudley Clarke also loved the cinema; deceivers appreciate a really good performance and a convincing *mise en scène*. Working late at night, Dudley Clarke spent hours at Cairo's many cinemas, seeing the same films again and again, meeting with associates in the noisy privacy of the auditorium, sometimes dictating letters, other times staring at the screen. Movies seemed to help him to think and to imagine new deceptions.

In January 1944 Clarke watched *Five Graves to Cairo*, a romantic melodrama of the recent desert warfare in North Africa. A British corporal, John Bramble, staggers delirious out of the Western Desert to find shelter in a hotel. When the hotel is commandeered by the Germans as the billet for General Rommel, Bramble impersonates a waiter just killed in an air raid, who turns out to have been an undercover German spy. The plot turns on Bramble's finding out that the five 'graves' between Tobruk and Cairo are actually Axis supply dumps, allowing him to thwart German plans to conquer Egypt.

Clarke was fascinated by the sinister panache of the actor playing Erwin Rommel. 'Count Erich von Stroheim und Nordenwall' he called himself, and he was apparently a scion of Austrian nobility. In fact he was the son of a Jewish hatmaker from Vienna who completely re-invented himself when he came to America in 1909. Von Stroheim made his name playing sneering, cruel German officers in WW1 propaganda films (*The Hun Within*), became a noted silent film director (*Blind Husbands, Greed, The Merry Widow*) and then an actor specialising in arrogant Prussians (as in Jean Renoir's *La Grande Illusion*).

Seeing Erich von Stroheim playing Rommel gave Clarke a brilliant idea. Maybe one man could actually be two men, in two places at once. What if an actor impersonated Rommel's rival, Bernard Montgomery? The month before, December 1943, Eisenhower had been with some fanfare appointed supreme commander of OVERLORD, the proposed 'liberating assault' on Western Europe to destroy Nazi Germany, while Montgomery was quietly made commander of the British 21st Army Group and acting commander of all land forces in the amphibious invasion of Normandy, code-named NEPTUNE. Early in 1944, Monty was back in the UK, but he was not the kind of man to live somewhere peacefully in mufti as he was supposed to. He took a suite at Claridges Hotel and, when spotted in full uniform in a box at London's Palladium Theatre, stood for a five-minute ovation. Dennis Wheatley found such behaviour 'vainglorious', but it did make Montgomery readily recognisable.

What if the well-known figure of Monty were to show up and show off in Gibraltar, Algiers and Cairo? Dudley Clarke knew that would divert Axis attention away from the Channel, where the real Allied attacks would shortly be coming, and back to the Mediterranean.

Operation COPPERHEAD got underway in the spring of 1944. Lieutenant Meyrick Clifton James was a professional actor who had started out with Fred Karno and now occasionally performed with the Pay Corps Drama and Variety Group, but he had one talent he did not need to work on: he had once rescued a patriotic show in Leicester just by donning a black beret and a British warm (overcoat) and walking out on the stage. In this garb Clifton James was indistinguishable from General Montgomery, the victor of El Alamein, and the whole audience clapped and cheered for five minutes. In March 1944, the *News Chronicle* ran a photo and a brief story about this remarkable resemblance; when Dudley Clarke's idea was put into action, someone in MI5 remembered it. James was traced to Leicester, and that was why a middle-aged lieutenant in the Royal Army Pay Corps received a telephone call out of the blue from the famous British film star David Niven, innocently asking him to take part in some army films.

*I Was Monty's Double* by M. E. Clifton James (ghosted with the help of Gerald Langston Day) was published in June 1954, ten years after the events it describes and the year after the publication of *The Man Who Never Was*. It is the study of a self-conscious performance: a rather shy man with no natural authority has to play an incredibly assertive man of magnetic personality. From the first lies that he is forced to tell his wife in the name of Official Secrecy, James sees things in theatrical terms, as a story of stage fright and elusive confidence. 'Only those who have been on this deception work can realize quite how nerve-wracking it can be.' His astute MI5 handlers are like the producer, the stage director, the stage manager, and he is surprised by their sense of humour, the fun they are having and their skill at mimicry. He finds himself among a cast of people who seem to change their rank and their uniforms with bewildering ease. While watching the real Montgomery at a coastal rehearsal for D-Day he is overcome with a sense of unreality, yet Monty seems to be acting too:

On the stage I have seen even rank bad actors and singers get away with it because they had personality, and I have seen really competent artistes without personality who could get nowhere at all. This man was what we should call a 'natural' . . . He would have made a fortune on the stage, I thought. Here in this great war drama he had carefully chosen his cast, appointed the cleverest directors, managers, technicians and property men, and from the leads down to the walk-on people he was making certain that everyone knew his part.

The army on active service reminded James of the theatrical profession and the times when a cast stale from weeks of rehearsing finds itself in the 'dead spot' before opening. The show feels a flop and the run looks short, but then a producer comes in and with a few quiet words switches on cheerful confidence again. What James most admired about Monty was the way he could inspire people, 'us[ing] his showmanship to brilliant effect,' James says. Because Eighth Army distrusted 'Brass-hats', Monty stuck a black beret on his head 'and talked to Pit and Gallery in a way that no General in the field had ever talked before'.

Before playing his part Clifton James needed to study Montgomery's bird-like gestures and mannerisms, the way he pinched his cheek, the way he ate his vegetarian food, his impulsive visits to schools, his boyishness, and the rigid rituals of deference and procedure around him. James finally met the general, in person, on a holiday near Dalwhinnie in the Scottish Highlands, aboard his private train.

As we stood facing each other it was rather like looking at myself in a mirror. The likeness struck me as uncanny . . . On the stage it is something if you can resemble a man after using every artifice of make-up, but in this case there was no need for false eyebrows, padded cheeks, or anything of that kind. I was extraordinarily like the General, and as I afterwards discovered, the two of us were remarkably alike when we were boys.

They were both Antipodean: James grew up in Perth, a son of the chief justice of Western Australia, Montgomery in Hobart, a son of the Bishop of Tasmania. Though James at 46 was eleven years younger than Montgomery, he looked older because he smoked and drank heavily. (Monty was teetotal and detested cigarettes.) Now the younger man studied the older's high-pitched, incisive way of talking, and his voice as parched as the desert. 'Everything will be all right. Don't worry about it,' said the general to the actor.

The RAF took James up for a spin to make sure that, like Monty, he did not get airsick, and then Wing Commander Dennis Wheatley of LCS drove him back from RAF Northolt. There were costume-fittings for the right uniform with five rows of medals. A gold breast-pocket watch chain was bought from Woolworth's and James's missing middle finger was covered up with a dummy finger strapped to the others, ready for the final rehearsal with a man from MI5 who called himself Brigadier Heywood. James trimmed his moustache shorter, greyed his temples with greasepaint and put on Monty's leather flying jacket and black beret at the right angle to pose for the photographs Churchill wanted to see. He was given khaki handkerchiefs monogrammed with the initials B.L.M. to leave around, and a small Bible like the one Monty always carried. Whenever he felt paralysed with fear, the equivalent of Hyde had to master Jekyll: 'With a violent effort I pushed James aside and became Monty.'

James landed at Gibraltar early on 27 May. As Sefton Delmer had observed with Hitler, a famous man is 'on' as soon as there is a public. The curtain went up at the top of the aircraft steps. James thought that the Rock looked like a painted backcloth on a Drury Lane stage set. Other actors – top brass, troops, drivers with cars – were drawn up to greet the leading man. In the background, hidden among real Spanish workmen, were the villains of the piece, Hitler's secret agents, watching and noting as they were supposed to. James went through the meeting and greeting with 'a curious sense of unreality. It was if everything were taking place in a dream.' As they drove through the streets of Gibraltar, troops came running and shouting 'Good old Monty!' The Guard of Honour saluted at Government House: the Governor of Gibraltar was now Lieutenant General Sir Ralph 'Rusty' Eastwood, who had been at Sandhurst with Montgomery and was in on the secret. 'Monty' saluted the Guard, shook hands and, chatting with Eastwood, walked into the Governor's Palace holding him by the arm. In the Governor's study, Eastwood took off his hat and sat down at his desk, staring fixedly at James, then smiled, jumped up, and warmly shook his hand again. 'I wouldn't have believed it possible,' he exclaimed. 'You're simply splendid. I can't get over it. You *are* Monty. I've known him for years.'

'Monty' and the Governor were positioned in the garden, in front of the stone frieze depicting Nelson's victory at the Battle of Trafalgar, when two Spaniards, clean-shaven businessmen in dark suits, came through the garden gates on their way to talk to Lady Eastwood about the ancient Moroccan carpets in the house. Clifton James says they were known to be 'two of Hitler's cleverest agents, Gestapo-trained and quite ruthless'. He was babbling about the War Cabinet and 'Plan 303' when the pair approached and were introduced. They looked at him with awe and respect before going into the house, from which a 'workman' had also been studying him through a telescope. For the last act in Gibraltar, James talked more nonsense about 'Plan 303' and gave audible orders and instructions near the airport canteen where a Norwegian contact of the Germans worked.

By the time he took the final salute and flew off, James was really into the part: 'While actually impersonating Monty I felt calm and sure of myself, but these off-stage intervals between the scenes were nerve-racking.' There was the fear of the plane being shot down, of assassination or kidnap. There was no cast of fellow-thespians to buoy him up backstage, no proper audience response out front, just the man he knew as Brigadier Heywood briefing him about the next act in Algiers, 'a regular hot-bed of intrigue with dozens of enemy agents posing as free Frenchmen and loyal Italian collaborators'.

One of the 'spies' notionally employed at Algiers airport was an 'A' Force channel, a wireless operator recruited by the Gestapo in Paris and parachuted into Algeria in 1943, who had promptly given himself up to the British and been 'turned'. The double agent duly reported the arrival of General Montgomery to his Abwehr controllers in Dijon, excited by Montgomery's reception by British, French and American officers and his high-speed twelve-mile journey with a siren-wailing motor-cycle escort to Allied Force Headquarters. Two Italian spies in the crowd asked what was going on and a Frenchman (sent there especially to follow them) said, 'Monty was coming to North Africa to form a great new army that would strike the soft under-belly of the Germans in the south.'

According to Clifton James's narrative, he spent a week in a sort of recurring dream, flying to various airports, 'landings, official receptions, guards of honour, bogus talks on high strategy; crowds of civilian spectators, no doubt with enemy agents among them; the streets lined with cheering troops'. He met various people believed to be spies and slipped into his role 'so completely that to all intents and purposes I *was* General Montgomery . . . Even when I was alone I found myself playing the part.' James could never escape; asleep, he dreamed of Monty. So the let-down, the fall from celebrity, came hard:

I drove up to General Wilson's headquarters as Monty, in a blaze of glory, but the moment I passed through the door the glory was gone for ever.

Upstairs I changed into the uniform of a Lieutenant in the Pay Corps . . .

When you are *khaleef* for an hour you have at least the borrowed splendour of your position to bear you up, but when you shed the trappings of exalted rank and return to your humble station with all the backwash of the strain through which you have just passed, you certainly need all the courage and stamina you have.

James was smuggled out through the kitchen back door, up a lane and into a small villa which was promptly locked. The impersonation was over. He felt tired and overwrought: 'I could only see again and again the scenes which had just taken place as if I were chained to my seat in the cinema.' A sergeant brought a 'high tea' of sausage, egg and chips, and then James met the real brigadier who had planned his exploit: 'the famous Brigadier Dudley Clarke who founded the Commandos'.

He was the man who early in the war thought of the idea of training a gang of tough young men to strike at the enemy behind the lines: men who would stop at nothing, and who would use every appropriate 'un-English' means to gain their ends. This bold plan did not appeal to the pundits at the War office who told him it was 'not cricket'. But he refused to take no for an answer and was so persistent that at length he was given a hearing. When Mr Churchill heard about it he at once gave orders for Dudley Clarke to go ahead.

James says he was left alone in the villa on his last night, with strict instructions to lie low, and that he talked like Montgomery to two parrots who then embarrassingly shrieked 'Monty! Monty! Monty!' In the morning, Clarke saw James sitting openly on the balcony of the villa in the sunshine and told him off – if he were seen, all his good work would be endangered.

There is also a more malicious account of what happened. In 1979, Jock Haswell alleged that the deception operation was 'abruptly switched off and Monty's Double disappeared' because, according to rumour, Clifton James got drunk, and 'was seen to be drunk while wearing the famous double-badged beret, uniform, insignia and medals of the teetotal Montgomery'. In this version, James was eventually taken back to England and threatened with court martial if he opened his mouth. A similar rumour had appeared in 1946 in *My Three Years with Eisenhower* by General Eisenhower's long-time friend and naval aide Harry C. Butcher, where there is a report of Monty's double seen 'staggering about in Gibraltar, drunk, smoking a large cigar'.

As Dennis Wheatley tells the story in *The Deception Planners* (1980), the ending of the story was 'rather pathetic. James was flown on to Algiers. Dudley met him, had him taken to a small hotel where he exchanged his gorgeous plumage for an ordinary Lieutenant's battledress, gave him a bottle of whiskey and told him not to leave his room until further notice.'

Clifton James later flew to Cairo in an American cargo plane and stayed in the Cairo flat of Terence Kenyon who worked for 'A' Force.

He was kind to the actor, who had become a nervous wreck. 'A' Force's Betty Crichton also looked after him. She told Thaddeus Holt, years later, that Clifton James was 'a very nice man who always got a bad press. He was under terrible pressure and strain, and coming out of that part was very difficult for him.' James was afflicted by toothache, and the heat, flies, smells and squalor of Cairo in June appalled him. He eventually made his way back to Leicester via Gibraltar after a five-week adventure, still having to tell lies about what he had done.

But he was also mysteriously changed by the experience. Timidity and diffidence had been replaced by confidence and a feeling of superiority. Although he saw out the rest of the war without promotion in the Pay Corps and was 'treated shabbily' (Wheatley's phrase), receiving no official recognition for his services, James was eventually rewarded. He was allowed to have a ghostwriter tell his story in a book called *I Was Monty's Double*, which in 1958 was made into a film directed by John Guillermin. In it, James took the roles both of Monty and his Double. The script by Bryan Forbes contained a wholly fictitious action-packed attempt by submarine-borne German commandos to kidnap the fake General Montgomery, heroically foiled by John Mills as his minder Major Harvey, but in other respects it largely followed James's book, and gave him a special celluloid immortality. And he appeals to theorists, too. As Harry Pearson points out in *Achtung Schweinehund!* (2007), by playing himself being himself, as well as playing the man he had been playing at being, M. E. Clifton James became postmodern.

# 27

# Overlord and Fortitude

When an eccentric genius called Geoffrey Pyke proposed constructing unsinkable aircraft carriers or freighters from enormous icebergs, Churchill ordered him to proceed – no idea that could conceivably help the Allies to win the war was too outlandish for this Prime Minister. Pyke's team of scientists invented a kind of super-ice, made by mixing in 4 per cent cotton wool or wood pulp to a slurry of freezing water, making an incredibly tough substance that melted very slowly which was called in Pyke's honour 'pykrete'.

Pykrete became an exhibit at the Quebec conference of August 1943 at which the Allied leadership discussed the plan for the final liberation of Europe, operation OVERLORD. Churchill had crossed the Atlantic on his way to the conference on one of the world's largest liners, the *Queen Mary*, which weighed 86,000 tons; but Pyke was proposing something even bigger, a 600-metre long, self-refrigerating aircraft carrier made from Pykrete, to be called *Habbakuk*, which would weigh more than two million tons and could carry and launch 200 aeroplanes. You could use it to invade Japan! Pyke was already building a prototype on a lake in Ontario.

Admiral Louis Mountbatten, the head of Combined Operations, used showmanship to demonstrate the power of Pykrete to the Americans. Two cold blocks were produced, one of ice, one of Pykrete, and burly General 'Hap' Arnold of the US Army Air Corps was invited to demolish each with an axe. Arnold shattered the brittle ice with a mighty blow. Then it was the Pykrete's turn: but the American general howled with pain as the axe-head jarred off the Pykrete, leaving the block intact. Mountbatten then drew a pistol and finished off the ice, but once again the Pykrete stood firm and a spent bullet ricocheted uselessly off it, narrowly missing a senior RAF officer. Churchill roared with laughter. The demonstration was a propaganda triumph, though in the event Pykrete was never used.

Churchill had come to Quebec to put on a brave show, and he was flanked by two fire-eating British warriors who he hoped would impress the Americans as much as the Pykrete had: the handsome and much-decorated air ace Wing Commander Guy Gibson VC, DSO and bar, DFC and bar, famous for the 'Dambusters' raid, and Brigadier Orde Wingate, ferocious leader of Patriot guerrillas in Abyssinia and now of bearded Chindits in the Burmese jungle.

The main item on the agenda was the forthcoming attack on what the Germans called *Festung Europa*, Fortress Europe. Where was the best place to enter the Continent if you were setting off from the UK? There were several options, but the American and British team led by Lieutenant General Frederick Morgan, called Chief of Staff to the Supreme Allied Commander or COSSAC, charged with planning OVERLORD, had actually decided on Normandy. Yet Normandy's fifty miles of beaches did not seem suitable for a massive invasion. The swirling currents and the daunting difference between low and high tides (up to 21 feet or 6.4 metres) made unloading heavy gear on sandy beaches implausible. Conventional wisdom said you required a proper deep-water harbour with wharves and cranes to disembark the 50-ton tanks, huge guns, and great pallets of stores necessary for an invasion. Hence the raid on Dieppe on 19 August 1942 – a trial run at seizing a port.

But the bold and imaginative answer that so appealed to Churchill was huge floating harbours. He had been thinking about this idea since July 1917, when he imagined a way of seizing two Frisian islands from a moveable atoll of concrete. In May 1942, he had written a note to Mountbatten: 'Piers for use on beaches. They must float up and down with the tide. The anchor problem must be mastered. Let me have the best solution worked out. Don't argue the matter. The difficulties will argue for themselves.' On board the *Queen Mary*, on 6 August 1943, there was a scientific demonstration in a bathroom by Professor J. D. Bernal, one of Mountbatten's physicist boffins, who put a fleet of twenty paper boats at one end of a half-filled bath. At the other end, a naval lieutenant made waves with a loofah. The paper boats were swamped and sank. Then Bernal put more folded newspaper boats into the bath, but surrounded them with an inflated Mae West lifejacket. The lieutenant made vigorous waves, and this time the boats did not sink. 'That, gentlemen,' said Bernal, 'is what would happen if we had an artificial harbour.'

A fortnight later, the Quebec Conference approved the concept of two artificial harbours – one British and one American, code-named 'Mulberries', and said they should be constructed and fully operational two weeks after D-Day. The Quebec Conference also approved the outline OVERLORD plan. The team were told to plan in more detail for an assault by three divisions and three airborne brigades. A section called Ops (B) was set up to prepare 'an elaborate camouflage and deception scheme', but there was only one officer working on it.

At the next Allied Conference, held in Teheran from 28 November to 1 December 1943, Joseph Stalin, Franklin Roosevelt and Winston Churchill concerted their 'plans for the destruction of the German forces'. The American and British Allies promised to leave the Balkans alone but agreed to help relieve the pressure on Russia by opening 'the Second Front' in May 1944, invading northern France in operation OVERLORD and southern France in operation ANVIL (which in the event got delayed). Stalin agreed to coordinate his big push on the Eastern Front with the Allied attack in the west, and all agreed on the need for a deception plan.

By now, the Wavell/Clarke thesis that major operations should have a cover plan, if practical and useful, was taken for granted. The Soviets believed in military deception, which they called *maskirovna*. An American deceiver later sent to Moscow to coordinate OVERLORD deception plans with the Russians was talking to a Russian deceiver when the subject of the media came up. When the American said that in a democracy you could not use the press to fool your own people, the Russian shrugged, 'Oh well, we do it all the time.' It was at Teheran that Churchill said to Stalin, 'In wartime, truth is so precious that she should always be attended by a bodyguard of lies,' and Stalin replied, 'This is what we call military cunning.'

On 6 December John Bevan of LCS was brought in to work up the strategic deception plan for OVERLORD, and gave it a new name, BODYGUARD, in a nod to the Prime Minister's observation. Strategically, it aimed to make the Germans dispose their forces in the wrong places – in the Balkans, in northern Italy, in Norway and Denmark, anywhere but northern France. Later, the operational challenge would be to deceive the Germans about exactly when, where and in what strength the invasion was coming. This part of the deception plan would evolve down an endless series of forking paths as executive control shifted.

Dwight Eisenhower was given command of OVERLORD ('Over Lord and Under Ike' was the joke) and he took up his responsibility in January 1944, when what had been COSSAC became SHAEF, Supreme Headquarters Allied Expeditionary Force. Eisenhower brought his own chief of staff, Walter Bedell Smith, with him, so Frederick Morgan became *his* deputy. The Ops (B) or deception side of COSSAC expanded as bigger fish started to arrive at SHAEF. Dudley Clarke's deputy in 'A' Force, Colonel Noel Wild, arrived from Tunis to take over, and also became the SHAEF member on the Double Cross Committee. Major Roger Hesketh of SHAEF intelligence worked closely with Tar Robertson and the other officers in MI5's Section B1A which controlled the double agents. Hesketh and Wild were also in close touch with John Bevan and others at LCS. Deception was a small club in an old boys' network; the official historian Michael Howard described them as 'a handful of men who knew each other intimately and cut corners'.

The British made sure that they retained executive control over the crucial Channel-crossing and landing part of OVERLORD, the actual D-Day invasion, code-named NEPTUNE. Allied air forces and navies were both under British control. The temporary commander of all the Allied ground forces for NEPTUNE was Montgomery. He and his chief of staff Freddie de Guingand set up their own deception staff, called G (R), modelled on Clarke's 'A' force which had helped Eighth Army so much in the desert. The man in charge of this was David Strangeways, the 'A' Force Tactical HQ commander who had led the successful surprise raid into Tunis to seize German intelligence materials, and who was probably Clarke's best pupil for ingenuity and sharpness.

The first thing Monty did was tear up the NEPTUNE plan that COSSAC had prepared. He thought the Normandy front should be doubled to fifty miles, and preceded by an assault from the sky by three airborne divisions, not three brigades. In the first wave of sea landings, he wanted not three but five divisions on five separate beaches, supported by two more divisions behind. If he did not get this, he said they could find another commander. Eisenhower concurred, but getting what Monty wanted meant a massive increase in ships and equipment, including another thousand landing craft to add to the three thousand-odd already prepared for.

Feldmarschall Erwin Rommel, charged by Hitler with defending the coast of France at the end of 1943, knew that his best chance was to smash the Allied attacks on the beaches, and that the first day would be 'the longest day'. From desert warfare experience, he was a great believer in anti-vehicle and anti-personnel mines. As well as making a 'Devil's Garden' of obstacles at different tidelines along the beaches, he wanted to mine, wire and fortify the entire coastal strip into a 'zone of death' five or six miles deep. To defend the Atlantic Wall, he dreamed of sowing 200 million landmines along the entire coast of France, although he never achieved it. 'Up to the 20th May 1944,' says the War Diary of German Army Group B, '4,193,167 mines were laid on the Channel coast, 2,672,000 of them on Rommel's initiative, and most of them after the end of March.' He also planned to fill all potential landing fields with patterns of ten-foot-high wooden stakes that would rip flimsy gliders apart. Many of the stakes were to be wired to artillery shells whose detonation would cause further carnage.

The Allied intelligence reconnaissance for the D-Day landings was high, wide and deep. Thousands of mapping photos were taken from different angles in the air. Low-level missions along the beaches to photograph the arrays of obstacles the Germans were building were known as 'dicing' missions, as in 'dicing with death'. They were taken so close that you can see individual engineers running for cover, and count their footsteps in the sand. Things seen from the air were sometimes investigated by divers from the sea, and commando raids brought back prisoners and samples of barbed wire and metal defences. Geologists and oceanographers were consulted and recruited. Following an appeal on the BBC wireless in 1942, the great British public had sent in over ten million of their pre-war French beach 'holiday snaps'. These Brownie Box-photos and picture post-cards were sorted, graded, assembled and scrutinised for tiny details of Normandy.

Through General de Gaulle's Free French Intelligence service, *le Deuxième Bureau*, run by 'Colonel Passy' or André Dewavrin, the French resistance was mobilised to report on every detail of the German construction of their defences. The *Centurie* network, radiating out of Caen, eventually had 1,500 agents noting every gun emplacement and mine field, every concrete caisson and fifteen-foot-deep anti-tank trench. One house painter in the resistance managed to purloin a blueprint of the defences from the office of the *Organisation Todt* that were building them.

The Allies agreed to cross the Channel as two armies, one British, one American, fighting side by side, but not mixed together. The British (including the Canadians) would go in on the left, to SWORD, JUNO, and GOLD beaches, preceded by the paratroopers of the 6th British Airborne Division. The US First Army would go in on the right, to OMAHA and UTAH beaches, preceded by the paratroopers of the US 101st and 82nd Airborne Divisions.

Co-ordinating the effort required awesome organisation and logistics. The Allies had to marshal and maintain over 2 million men, 11,000 aircraft, and 7,000 ships in England. The prodigious industrial output to meet their requirements had to be matched by efficient distribution. The engineering work behind the landings was staggering, and thousands of construction workers were recruited to work night and day. The Petroleum Warfare Department pioneered PLUTO (Pipeline Under The Ocean), ready to pump millions of gallons of petrol across to the invaders. To get the astonishing volume of men, equipment and supplies ashore in north-western France, Churchill's pet project, the technologically ingenious 'Mulberry' floating harbours, were essential. Two were to be constructed off Normandy. Over a hundred enormous 6,000-ton reinforced concrete caissons called 'Phoenixes' (each 60 feet high, 60 feet wide and 200 feet long) would be towed across the Channel from Selsey Bill and Dungeness by some of the fleet of 132 tugs and then filled with sand from 'Leviathans' so they sank to form a breakwater in the Bay of the Seine. Outside this artificial reef was a floating line of 'Bombardons' towed from Poole and Southampton to calm the waves, and inside, in shallower water, a line of 'Gooseberries', formed from two dozen redundant merchant navy vessels, Liberty ships and one old dreadnought that were scuttled and sunk where needed. In the calmer waters within the two-square-mile Mulberry harbour, strong Lobnitz or 'Spud' pier heads were sunk deep into the sand which allowed long bridges or floating roadways to the shore, known as 'Whales', to float up and down with the tides. The menagerie of code-names was augmented by power-driven pontoons called 'Rhinos' and amphibious vehicles known as 'Ducks'.

Further amazing engines onshore also sprang from Churchill's 'inflammable fancy': armoured tank bulldozers and ploughs, special fat-cannoned Churchill tanks for blasting blockhouses, other 'Crocodile' Churchill tanks that could squirt petrol and latex flames over a hundred yards, great machines for laying fascines across mud or barbed wire, or for thrashing their way with flailing chains clear through exploding mine fields. These devices came from Churchill's direct encouragement and protection of a brilliant maverick, Major General Sir Percy Hobart of 79th Armoured Brigade, and were collectively known as 'Hobart's Funnies'.

The deception plan for NEPTUNE, the cross-Channel attack, was called Plan FORTITUDE, and its object was 'to induce the enemy to make faulty dispositions in North-West Europe'. FORTITUDE NORTH aimed to keep Hitler worrying about Scandinavia, and the danger to Germany posed by an Allied attack on Norway and Denmark. Dummy wireless traffic and bogus information from double agents indicated that the (notional) British Fourth Army in Scotland, supported by American Rangers from Iceland, was going to attack Stavanger and Narvik and advance on Oslo. British deceivers also worked hard on the neutral Swedes. The commander-in-chief of the Swedish Air Force was asked for 'humanitarian' assistance in the event of an Allied invasion of Norway. As his office was being bugged by the pro-Nazi chief of Swedish police, this information went straight to Berlin. When Hitler read the transcript he ordered two more divisions to reinforce the ten already in Norway. Thus 30,000 more soldiers were diverted away from France.

Operation FORTITUDE SOUTH, developed by David Strangeways, aimed in the first instance to convince the Germans that there was another mighty force in Britain, as well as Montgomery's (real) 21st Army Group: this was the First United States Army Group, or FUSAG, stationed in the south-east of England, opposite the Pas de Calais, the quick route to Germany via Antwerp and Brussels. FUSAG was, of course, notional, a ghost army created and sustained by the deception plan QUICKSILVER. It had its own insignia, a black Roman numeral I on a blue background inside a red and white pentagon, and it was supposed to comprise the Canadian First and the US Third Armies.

Most importantly for the story, it was apparently commanded by the profanely theatrical, ivory-handled-pistol-packing US General George S. Patton Jr, 'Old Blood-and-Guts' himself, from a headquarters at Wentworth, near Ascot. ('One must be an actor,' Patton once wrote about overcoming ever-present fear.) Hitler thought that Patton – who had got into trouble for slapping a shell-shocked soldier in Sicily – was easily the Americans' best man, because he was ruthless. Of course he would be leading the Allied fightback.

From 24 April 1944 onwards, the eleven divisions of FUSAG were brought to life by dummy radio traffic to hoodwink the German 'Y' or wireless-eavesdropping service. The radio deceivers went on genuine army exercises where they recorded all the radio voice traffic, learned accurate technical terms and questioned people about their activities before writing their own scripts, which they got Allied servicemen in Kent to read out. They tried not to make it sound too polished, because in real life people often did not hear and asked for repeats. For Morse work they got American radio operators of 3103 Signals Service Battalion who had been in Sicily and North Africa, and whose 'fist' or way of signalling the German listeners might recognise. There was also physical camouflage work, particularly the planting of scores of dummy landing craft – known as 'big bobs' – at Great Yarmouth, Lowestoft, and on the East Anglian rivers Deben and Orwell, for German aerial reconnaissance planes to spot. The dummy landing craft weighed about six tons, were built out of scaffolding pipes and canvas and floated on 55-gallon oil-drums welded together. They were painted and stained to look old, and were 'serviced' by crews who hung out washing, flew ensigns, sent up smoke signals and moved around in small boats. The idea was to keep the Germans looking eastwards; the Pas de Calais had to remain the invasion site uppermost in their minds, with the Fifteenth Army there ready to repel an invasion, and Seventh Army in Normandy less on its guard.

Great activity, using lighting at night, was simulated at Dover and Folkestone, where the 2nd Canadian Corps and the US VIII Corps were notionally based. The architect Basil Spence oversaw the building of a fake oil terminal with pipelines, storage tanks and jetties. It was solemnly inspected by King George VI and General Montgomery among other notables, and duly reported in the press – because since early March 1944, Royal visits had been coordinated with the deception planners.

It all worked: a German *Oberkommando der Wehrmacht* intelligence map, captured later in Italy, of what they believed to be the British Order of Battle on 15 May 1944, reflected their belief that most Allied forces were stationed in the east of the UK. The (mostly imaginary) units the map showed had been carefully built up over the last fifteen months through a mass of detail sent to the Abwehr by their trusted spies in England, who of course were actually MI5-controlled double agents. The three most important were a Pole, Roman Garby-Czerniawski, code-named BRUTUS, reporting to Paris, a German, Wulf Dietrich Schmidt, code-named TATE, reporting to Hamburg, and our Spanish spy, Juan Pujol García, code-named GARBO, now sending his reports directly by coded wireless message to the Abwehr *Kriegsorganisation* in Madrid.

With everything to play for, the battle for morale became all-important. Sefton Delmer's 'black' radio station, *Soldatensender Calais*, now broadcast loud and clear, and by early 1944 PWE 'black' and BBC 'white' broadcasting were working well together in their different spheres, the distortions of 'black' weaving a smoke of lies around the 'white' buttresses of truth. Delmer was working closely with PWE, the BBC, Naval Intelligence and LCS, and had an office in Bush House. Here men and women from the European resistance, Polish, Danish, Norwegian, French and Dutch, came to see him, and he helped them with forged notices, posters, proclamations and identity papers. *Soldatensender Calais* played its part in operation OVERLORD, helping to soften up the morale of German troops defending the Atlantic Wall, encouraging slacking by saying, 'Units which show themselves smart and efficient are drafted to the Eastern Front. Promotion in France is a sure way to death in Russia.' Delmer's cheery-toned but deeply depressing black radio broadcasts abraded German soldiers' confidence by saying that Russian successes were due to their being supplied with (imaginary) American 'miracle weapons' like the new 'phosphorus shells', which could destroy reinforced concrete and pierce any armour.

In May 1944, the month before D-Day, Delmer launched a daily newspaper for the German troops named *Nachrichten für die Truppen* or 'News for the Troops'. It was a joint British–American venture. SHAEF gave him a team of editors and news writers to command, and the paper ran for 345 editions, using rewritten radio material. Two million copies a day were dropped by American bombers across France, Belgium and Germany, with pieces about German difficulties fighting an air war without fuel, or detailing 'impossible' political interference with the army leaders' decisions. Delmer, the lifelong newspaperman, later said that this was the wartime enterprise of which he was proudest.

Meanwhile, Juan Pujol's role as ARABEL, the Abwehr agent, was also moving steadily towards its climax. He was by now the chief spy in an extensive (and entirely fictional) network code-named ALARIC. He had not only invented, for his Abwehr spymasters, four supposedly important contacts providing him with information, he had also recruited seven equally imaginary sub-agents who in turn got military information from some fifteen notional sources. So Agent THREE in Glasgow, the Venezuelan student called Carlos ('recruited' or invented while Pujol was still in Lisbon), supposedly knew a drunken NCO in the RAF, a British infantry officer and a Communist Greek seaman who had deserted but who wanted to help the Russians open the Second Front. Agent FOUR, a Gibraltarian NAAFI waiter based in Kent, got much information about the arms depot and underground railway in the imaginary 'Chislehurst caves' from a guard stationed there, and many details about FUSAG (including gossip about quarrels between US & UK Commanders) from an American NCO based in London. Agent SEVEN, an ex-seaman in Swansea, was particularly active, with sub-agents in Exeter and Harwich, and also apparently knew a Wren in Ceylon, a soldier in the 9th Armoured Division, an Indian fanatic, and the leader and 'brothers' of the Aryan World Order Movement, a group of extreme Welsh Nationalists.

These colourful imaginary agents and sub-agents were spread across the country, and Pujol sent their 'information' on by radio. From 1 January to 6 June 1944, he sent 500 wireless messages from London to Madrid, putting over the deceptions that SHAEF wanted. From Madrid, ARABEL's reports went to *Oberkommando der Wehrmacht* and to *Fremde Heeres West*, the German Intelligence department dealing with the Allied armies in the west. The false information made them calculate the number of divisions in the UK as seventy-seven, overestimating them by 50 per cent. The whole fantastic spider's web of inventions was not just the work of the writing partners Pujol and Tomás Harris. They were advised by David Strangeways and LCS, and behind them, by the presiding genius of 'A' Force, Dudley Clarke.

Now Pujol's role as GARBO went up a notch. Permission was granted to let ARABEL break the news of the Normandy landings, to give him even greater credibility with the Abwehr. So, just before D-Day, Pujol's imaginary Agent FOUR apparently broke out of a high-security army camp at Hiltingbury together with two American deserters and brought ARABEL the news that the 3rd Canadian Infantry Division, having been issued with 24-hour ration packs and vomit bags, had left the camp. This information was transmitted eight minutes after those very Canadians landed on JUNO beach, just before 8 o'clock on the morning of 6 June 1944. It was too late for the Germans to do any more to prepare for the landings, but Pujol retained his status as the Abwehr's top man in Britain.

With hindsight, the Bailey bridge of history seems solid, with its incidents all bolted together in due order. But before the event, things are very different. On the eve of D-Day, the future was blank, unclear. Nothing was inevitable; everything was at hazard. Eisenhower knew it was a gamble he could lose, and handwrote the gloomy message he would have to give if the landings failed.

Winston Churchill was feeling his responsibilities, and his age. He had been Prime Minister for four long years, actively running a country that was fighting for its life in the greatest conflict the world had ever known. His mind went back to the past, to the 'hecatombs' of WW1 which he had survived but thousands and thousands had not. He worried about the D-Day landings too, telling an American visitor, 'It is not because I can't take casualties, it is because I am afraid what those casualties will be.' At the back of his mind was Gallipoli, the amphibious landing that wrecked his political career nearly thirty years before. Things had also gone wrong in the landings at Narvik, at Dieppe, and at Anzio in Italy where it had taken four months for 125,000 men to break out from the trap of the beachhead. What would happen in Normandy? Six months short of his seventieth birthday, Churchill the warhorse now determined to be there, watching the D-Day landings from a bombarding ship. King George VI said he would do the same. This caused consternation. What if Monarch and Prime Minister were to be killed? Both men were finally dissuaded. On the night of Monday, 5 June, Churchill dined with his wife and then spent time in the Map Room, glaring at the dispositions. Before going to bed, he said to Clementine, 'Do you realise that when you wake up in the morning 20,000 men may have been killed?'

Thirty years before, Philip Gibbs and the other journalists had been barred from the front by Lord Kitchener. But by 1944, media-savvy generals like Montgomery were welcoming news organisations like the BBC. At D-Day, Richard Dimbleby had eighteen reporter colleagues: Guy Byam jumped with the paratroops, Chester Wilmot went in on a glider, Richard North in a landing-craft, Stanley Maxted in a minesweeper, and other BBC correspondents were with different units and at SHAEF HQ with their 40-lb 'midget' recorders, getting actuality and eyewitness accounts from the battlefield. *War Report*, broadcast nightly after the nine o'clock news from 6 June 1944 until 5 May 1945, was a new kind of radio reportage. The war correspondent joined the combatants in the field on behalf of the citizens at home, bringing the front line into the back parlour.

Sefton Delmer's radio scooped the world with its report of the landings at 4.50 a.m. on D-Day, taken almost verbatim from a teleprinter flash on Goebbels's DNB news service, but augmented with extra disinformation. Delmer was also proud of that night's edition of *Nachrichten für die Truppen,* which reported that the Atlantic Wall was breached in several places, and that attacks were taking place at the mouth of the Seine and at Calais. This claim was carefully coordinated with the deception planners to spread maximum confusion.

In operation TITANIC in the darkness before the dawn of D-Day, handfuls of SAS men from Fairford in Gloucestershire were dropped from the sky at four sites behind the German lines, attended by scores of dummy parachutists, the simple sacking ones known as 'Paragons', the more elaborate inflatable rubber ones christened 'Ruperts'. They parachuted down with assorted pyrotechnics that simulated the sound and the chemical smell of battle. The few real SAS men shot off flares and fireworks, stirring up the ants' nest with plenty of noise, and then slipped away to join the French resistance or to make their way back to the British lines. Because the best way to deal with parachutists is to tackle them as soon as they land, thousands of German troops were out scouring woods and fields inland, and so were not ready to fight the forces landing on the beaches.

Electronic and electromagnetic deceptions also played their part. Dr
R. V. Jones, the head of British Scientific Intelligence, had kept a
watchful eye on all German radar developments – the Bruneval Raid
by Commandos in February 1942 was a scientific swoop on a radar
station in Normandy made at his request – and now organised a
massive fraud upon the German system. After RAF and USAAF fighter
planes destroyed 85 per cent of the German radar chain, what
remained was duped in two operations called TAXABLE and GLIMMER.
As the huge invasion fleet pulled out from behind the Isle of Wight, it
split. The bulk of the ships turned south towards Normandy, but a
decoy flotilla continued eastwards. Above them, Leonard Cheshire's
617 Squadron of Lancaster bombers flew back and forth in a moving
grid, eight miles long by two miles wide, continuously dropping
reflective tinfoil to create the radar image of a large fleet moving south-
easterly at 8 knots towards Fécamp at the mouth of the Seine. Their
sparkling snowfall of 'Window' was supported on the sea surface by a
few launches using 'Moonshine', a device that produced multiple
radar images, which gave the same impression of a large assault
convoy to any airborne radar reconnaissance. At the same time, the
Stirling bombers of 218 Squadron created a similar ghost image on the
approaches to Boulogne.

Winston Churchill was not aboard the great armada sailing for
France, but Norman Wilkinson was. The painter who had watched
the Suvla Bay landings at Gallipoli in 1915 was now on the destroyer
HMS *Jervis*, still wearing his old WW1 jacket but astonished by the
thousands of vessels of every imaginable type. Nearly 350 British,
Canadian and US minesweepers led the way, clearing ten approach
channels, closely followed by the bombarding ships, including *Jervis*.
Wilkinson was the only professional artist there on D-Day and he
worked busily as 800 naval guns opened fire at 6.27 a.m. on the
Normandy coast over six miles away.

Off OMAHA beach, Allied rocket ships fired 9,000 explosive
projectiles. More than 300 B-24 bombers swept through grey cloud to
drop 13,000 bombs. All of them missed the German defenders. The
amphibious Sherman tanks were launched too early, and 27 out of 29
foundered in heavy seas and sank with their crews, as did 23 of the 32
howitzers in amphibious 'Ducks'. An 'inhuman wall of fire' met the
first Americans ashore. The photographer Robert Capa reached the

Easy Red sector of OMAHA beach, but got out as quickly as he could. The photo lab accidentally destroyed all but eight of Capa's 'slightly out of focus' pictures of men crawling though bullet-torn surf to shelter behind German beach obstacles. US Rangers who risked life and limb to climb up Point du Hoc found the big guns replaced by wooden dummies.

When the American reporter Ernie Pyle got ashore on the day after D-Day (known as D+1), he found the wreckage of equipment 'vast and startling' and the human litter poignant: 'In the water floated empty life rafts and soldiers' packs and ration boxes, and mysterious oranges.' From a high bluff he overlooked the littered beach and 'the greatest armada man has ever seen. You simply could not believe the gigantic collection of ships that lay out there waiting to unload.' German prisoners also stood watching, on their faces 'the final horrified acceptance of their doom'.

The invasion did achieve surprise. By the end of 'the longest day', 156,000 men had landed by sea in France as well as 23,000 from the air, although none of them had reached their planned objectives. The airborne and seaborne forces met up on 10 June, the beachheads did not link up till the 11th, and chaotic fighting went on for many days. Montgomery did not take Caen for six weeks, and the Americans did not manage to break out to the south-west for two months. In those first days, the Normandy bridgehead was only a toehold; the German Army's resistance was fierce and the *bocage* backcountry of small fields and thick hedges made tank and infantry advance difficult.

The camouflage officer Captain Basil Spence had landed on Sword Beach. On D+2, the day that Montgomery came ashore, he watched British tanks destroy two beautiful Norman churches at Ouistreham and Hermanville by shelling their belfries to kill the German snipers up there. In their dugout that night, a friend asked him what his ambition was. 'To build a cathedral,' said the architect who was to remake Coventry.

Steven Sykes was also a *camoufleur* with No. 5 Beach group, helping to conceal stores from German bombing and shelling. He was putting a belching smokescreen canister into a beached landing-craft when he came across its occupants, a closely packed mass of corpses still pressed together the way they had all died twenty tides before. On D+30 he went to help 6th Airborne Division who had reverted to a static sniper war. He found himself making dummies dressed in Airborne camouflage smocks and demonstrating ghillie hoods, just like Hesketh Prichard in WW1. Mines, booby traps and snipers made progress slow, and cautious.

A huge storm, one of the worst of the century, blew up in the Channel on 19 June and raged for several days, wrecking the American Mulberry harbour and delaying the landing of vital supplies. The storm exposed the vulnerability of the forces ashore: lifelines could be snapped; the cable was fraying. In these early stages, if the Germans had thrown all their forces at it, the D-Day invasion could still have failed. Eisenhower's 'Great Crusade' hung in the balance, and events could have tipped the scale either way. For example, when Churchill visited the Normandy beachhead on 12 June 1944 (see plate 26), he went to Montgomery's HQ at Creully. As senior officers stood outside with the Prime Minister, South African Field Marshal Smuts sniffed the air and said, 'There are some Germans near us now . . . I can always tell!' Two days later two fully armed German paratroopers emerged from a nearby rhododendron bush, where they had been hiding all along. Had they used their guns and grenades on Churchill, everything would have changed.

Now came the culminating moment of all the lies and the spies, the ruses, dupes and lures that make up British military deception in the twentieth century. This is when deception changed the course of history. In the crucial days after the Normandy invasion, the second phase of the deception plan FORTITUDE SOUTH came into play. The genius of Dudley Clarke's pupil David Strangeways revealed itself, because the FUSAG bluff did not evaporate, it continued to grow.

The German Army Group B in France comprised two forces: 7th Army in Normandy and 15th Army away to the east in the Pas de Calais. When the Allied Expeditionary Force landed in Normandy they had to deal with the German 7th Army. 'Just keep the 15th Army out of my hair for the first two days. That's all I ask,' Eisenhower had said to the deceivers months before. He was requesting only hours. But every single day that the German divisions stayed away, fewer Allied soldiers died or were injured, and more Allied men and kit managed to get ashore, building up eventually to a force of nearly two million men.

Two days after D-Day Pujol hosted a fictitious conference of his imaginary agents – including three of Agent SEVEN's sub-agents, DONNY, DICK and DORICK – and, just after midnight, sent his Abwehr masters in Madrid a two-hour-long coded message with a summary of his conclusions, laying out the entire FORTITUDE SOUTH gambit. In essence, he pretended to surmise that the Normandy invasion was part of a two-pronged attack. The landings had just been a feint, a diversionary manoeuvre designed to draw German reinforcements west. If Rommel's 15th Army moved west from the Pas de Calais to reinforce the 7th in Normandy, Pujol warned that they would fall into the trap. The currently inactive FUSAG – with twenty or twenty-five divisions – would cross from south-east England to land the second blow behind them in the Pas de Calais. The implication was that this entirely fictitious second invasion, code-named MARS, would cut the German Armies off in Normandy, leaving the Allies and General Patton free to plunge towards Germany's heartland.

The Spanish message from their trusted agent ARABEL went through several hands and translation into German in the eighteen or so hours it took to travel from London via Madrid to Berlin and arrive by teleprinter in Adolf Hitler's headquarters at Berchtesgaden. Colonel Krummacher, the Ober Kommando Wehrmacht Intelligence chief, read it and handed it to General Jodl, who thought it was important enough to pass to Adolf Hitler himself. 'Diversionary manoeuvre'. . . 'decisive attack in another place' . . . 'probably take place in the Pas de Calais area' . . . 'proximity of air bases'. It all made sense. Cancel the counter-attack on Normandy. Hold back the troops.

Sefton Delmer thought the FUSAG deception was brilliantly tailored to Hitler's psychology, 'his long-displayed lust for self-dramatisation'. Here he was, the hero Führer, confronting many enemies just like the heroic King of Prussia, Frederick the Great, at the end of the Seven Years' War. And just as Frederick II in the eighteenth century was saved at the critical moment by the accession of the pro-Prussian Tsar Peter who pulled his troops back from Berlin, so now a providential spy, Pujol, had appeared like a *deus ex machina* with a message to save him. Hitler would never fall into Eisenhower's trap by moving his forces west to Normandy! The great hero would be ready and waiting to crush the arrogant Patton at Calais. Hitler would still win the war.

And so twenty-one German divisions – two armoured and nineteen infantry and parachute crack troops – were retained in the Pas de Calais area, not for the two days that Eisenhower had asked for, nor for two weeks, but for nearly two months, until the end of July – by which time the Allies had established themselves in north-west France, and the Germans' chance had gone. When the German forces did finally move west, Eisenhower called it 'a belated and fruitless attempt to reinforce the crumbling Normandy front'.

In the conclusion of his *Report by the Supreme Commander to the Combined Chiefs of Staff on the Operations in Europe of the Allied Expeditionary Force*, Dwight Eisenhower wrote that the enemy 'was completely misled by our diversionary operations, holding back until too late the forces in the Pas de Calais which, had they been rushed across the Seine when first we landed, might well have turned the scales against us'. In his history, Winston Churchill wrote, 'Our deception measures both before and after D day had aimed at creating this confused thinking. Their success was admirable and had far-reaching results on the battle.' And Bernard Montgomery wrote in *21 Army Group: Normandy to the Baltic*, 'These deception measures continued, as planned, after D-Day and events were to show that they . . . played a vital part in our successes in Normandy.'

# 28

# V for *Vergeltung*

'We may also ourselves be the object of new forms of attack from the enemy,' Churchill warned, and a week after D-Day came the advent of the 'pilotless plane'. The first of more than 8,000 German V-1 flying bombs was launched at London on 13 June 1944, bringing back the Blitz and prompting, once again, the evacuation of London's children. The 'buzz-bomb' or 'doodlebug' was a twenty-one-foot-long robot plane with stubby wings and a tail that flew at about 360 mph with an orange flame jetting out from the back. When the puttering engine cut out, it fell silently to the ground. V-1s weighed two tons and carried fifteen hundredweight of high explosive. In three months they destroyed 23,000 houses and damaged over a million, killing nearly 6,000 people and wounding over 18,000.

The V-1 flying bomb and the later V-2 rocket, which was twice the size and flew six times faster with a ton of explosive, were called *Vergeltungswaffen*, 'retaliation', 'reprisal' or 'revenge' weapons, because they were a response to the increased Allied firestorm bombing of German cities which killed up to half a million civilians. The V weapons showed that the Germans were neither defeated nor going to give up easily. They worried Churchill, who in July 1944 briefly considered drenching the Ruhr with poison gas.

Juan Pujol had been warned in December 1943 by his Abwehr controller to get out of London, and moved to Taplow in Buckinghamshire. When the V-1s started falling the Germans wanted to know exactly where they landed so the launchers could adjust their aim. Pujol, as ARABEL, replied that his sources in the Ministry of Information told him that the V-1s were falling all over the place, from Harwich to Portsmouth. This was calculated to deceive: by telling the Germans the bombs were overshooting, they might adjust their aim so the bombs would in fact fall short of London.

409

On 22 June, Pujol and Harris composed a long and strictly personal letter to his 'friend and comrade' CARLOS, Karl-Erich Kühlenthal, his Abwehr controller in Madrid. The tone is that of a pompous National Socialist unbuttoning into cautious frankness. He asks himself, has the V-1 a military aim? 'No! Its effect is nil. Is it then intended for propaganda? Possibly, yes!' ARABEL says how much he had looked forward to 'the destruction of this useless town which surrounds me', but confesses the results are very disappointing, both as a military weapon and as propaganda. Life goes on, traffic is the same, and people can see with their own eyes that the German claims that London is burning and Big Ben destroyed are lies. Is this the best use of military resources? he asks. He wonders if the V-bomb could be improved or made faster, and then, fishing, asks if it is really the 'rocket weapon' referred to earlier. His letter ends:

I feel more than ever a sensation of hatred, more than death, for our enemy, and an ever increasing irresistible urge to destroy his entire existence. The arrogance of this rabble can only be conceived when you live among them. Receive a cordial embrace from your comrade and servant, JUAN.

In July 1944, when the real First Canadian Army and the real Third US Army began to move over to Normandy (potentially breaking the spell of the imaginary FUSAG, which was still notionally in Kent and Sussex) MI5 decided to switch off Pujol/ARABEL for a while. This involved another pantomime. On 5 July 1944, imaginary Agent THREE told Madrid he was worried because ARABEL had not arrived for their regular meeting. The next day he reported that ARABEL was missing, and ARABEL's wife was about to go to the police to find out if he had been killed by a bomb, which would be disastrous if ARABEL had in fact gone somewhere prohibited. What should he do?

The fiction culminated in the 'discovery' that ARABEL had been arrested by a plain-clothes police detective while asking questions near a V-1 bomb site in Bethnal Green. Madrid advised the whole ALARIC network to lie low for a while. Agent THREE eventually reported that ARABEL was freed. His forged Spanish republican papers and his friend in the Spanish Section of the Ministry of Information had saved him. Pujol added a further layer of deception. He pretended that he had complained to the Home Secretary about his unlawful detention, and then, having secured a (forged) letter of apology from this luminary, passed it on to Madrid. It all convinced. On 29 July 1944, Madrid was delighted to inform ARABEL that the Führer had awarded him the Iron Cross.

As the Allied ring tightened around Germany in 1945, Sefton Delmer used the giant transmitter Aspidistra as a weapon of psychological warfare. If there was an Allied air raid heading towards a particular German city, genuine German radio transmitters in the locality would be switched off so the bombers could not use the frequency as a homing beacon. Forewarned of the coming raids, Aspidistra could pick up and hold the national signal from another regional transmitter and broadcast it the instant the local one was switched off, so the listeners did not notice any transition. Then Delmer's crew seamlessly inserted their 'black' broadcast or news announcement, while German listeners still thought they were listening to Nazi state radio. Aspidistra thus became a giant pirate radio ship taking over other vessels. This counterfeiting was first used on 24 March 1945, the day Allied forces crossed the Rhine. Delmer's announcer, a PoW who had the perfect official technique from his time in German services radio, gave out evacuation instructions over *Reichsender Cologne*. The following night it was *Reichsender Frankfurt*'s turn for panic measures. Later, Hamburg and Berlin were 'officially' told that the devious Allies were monkeying with Reich communications and that no orders or instructions received by telephone were ever to be obeyed without calling back first to check.

Delmer was caught up with many 'syke-warrior' schemes designed to break the enemy will to resist. Although a small group of aristocratic army officers had attempted to assassinate Hitler in July 1944, there was no organised resistance to Nazism inside Germany. So it seemed like a good idea in 1945 to invent an imaginary resistance and send them some real agents ('turned' German PoWs), wirelesses, stores, explosives and secret messages in order to rattle the Gestapo and make them waste their time persecuting innocent people. Four agents survived and two of them claimed to have contacted a subversive organisation which did not actually exist. (David Garnett says that operation PERIWIG 'resembles the subject of many plays by Pirandello'.) The most cynical duplicity was alleged by Leo Marks in his book *Between Silk and Cyanide* – the deliberate dropping of a German double agent called Schiller with a sabotaged parachute so that a fresh corpse could be found with incriminating evidence. Marks says that this brutal version of MINCEMEAT was fixed by Major General Gerald Templer, but Fredric Boyce in *SOE's Ultimate Deception: Operation PERIWIG* (2005) is more sceptical about whether it happened. Ultimately, Delmer thought that the invented resistance gave camouflage to the wrong people. He called his subsequent book *Black Boomerang* because after the war so many Germans who were committed Nazis or staunch collaborators claimed to have been in the fictional resistance he had partly helped to invent.

As the Third Reich continued to crumble and Allied forces moved deeper into Germany, Sefton Delmer felt the time for the *Soldatensender* was over. He decided to do what he had reported other Germans as doing: abandon ship. At 5.59 a.m. on 14 April 1945 the radio station 'faded from the ether, never to be heard again', and almost the same day, the last edition of his newspaper *Nachrichten* was printed in Luton. The morning after the fancy-dress party in Milton Bryan to celebrate the final broadcast, Delmer performed another ritual and shaved off his beard, to symbolise what he called 'the end of "black".' He was shocked by what he saw in the mirror. 'There, staring at me, was the pallid, flabby-mouthed face of a crook. Was this, I asked myself, what four years of "black" had done to Denis Sefton Delmer?' Clean-shaven, he summoned his team in the canteen to say goodbye. A voice from the back of the room called: *'Der Bart is ab! Der Krieg ist aus!'* – the beard is off, the war is over.

But Delmer also had a serious point in calling the meeting. He told them all that they must resist the temptation to boast about the tricks they had played on the Germans. He said that although the unit had 'within our limits, contributed our bit to the defeat of Hitler', they must remember that their role was 'purely subsidiary' to the work of the fighting services. He was being modest, and advising modesty, for a reason. He reminded his team that after WW1, Lord Northcliffe, 'hungry for public glory', could not restrain himself from boasting how successful his propaganda had been. This had only fed the German propaganda myth that they were beaten 'not by the armies in the field, but by Northcliffe's propaganda', which in turn fostered the illusion that the Germans could have won had they not been deceived – in which case, why not 'have a second go'?

Delmer's fictions ended in the same month, April 1945, that Allied soldiers discovered the realities of Nazism in Germany. They entered the concentration camps of Buchenwald, Bergen-Belsen and Dachau. 'No one will believe us,' a stunned GI said after seeing the ovens and the corpses. Richard Dimbleby was the first British radio reporter into Belsen where 60,000 civilians were dying of hunger and typhus.

I have seen many terrible sights in the last five years, but *nothing, nothing* approaching the dreadful interior of this hut at Belsen. The dead and dying lay close together . . . As we went deeper into the camp, further from the main gate we saw more and more of the horrors of the place . . . Far away in the corner of Belsen camp there is a pit the size of a tennis court. It is fifteen feet deep and piled to the very top with naked bodies that have been tumbled in one on top of the other . . . Our Army doctors, examining some of the bodies, found in their sides a long slit, apparently made by someone with surgical knowledge. They made enquiries and established beyond doubt that in the frenzy of their starvation, the people of Belsen had taken the wasted bodies of their fellow prisoners and removed from them the only remaining flesh – the liver and kidneys – to eat.

The BBC, in disbelief, was initially reluctant to broadcast his dispatch. But Dimbleby telephoned the News Room and said that if they did not put it out he would never broadcast again.

Adolf Hitler in his bunker in devastated Berlin chose not to face a war crimes tribunal. He wept as he poisoned his Alsatian dog, Blondi. Then, after lunch on 30 April, the Führer took out his false teeth and sucked on a cold pistol. Nazi Germany surrendered unconditionally on 7 May 1945.

# Epilogue

The story of British deceiving in WW2 emerged piecemeal, and late. The official history, *Strategic Deception in the Second World War*, by Professor Sir Michael Howard, was published in 1990, a full ten years after its completion. This was because the Prime Minister in 1980, Mrs Margaret Thatcher, wanted no more publicity about the secret services after the revelations that Sir Anthony Blunt had been passing information to the Russians. The MI5 *Summary of the Garbo Case 1941–1945* was not published until 2000. The most comprehensive account of Allied military deception, *The Deceivers*, by Thaddeus Holt, only appeared in 2004.

Why did we have to wait so long to learn about this aspect of WW2? Is it because deception is the most secret thing, or the most shameful? Official Secrecy has weighed heavy on the subject. The 1972 book that revealed the WW2 use of double agents, J. C. Masterman's *The Double-Cross System* (first published by Yale University Press in the USA), excised all mention of 'Most Secret Sources', information derived from deciphered enemy codes, itself code-named ULTRA. This did not come to public knowledge until F. W. Winterbotham's dambusting book *The Ultra Secret* was published in 1974. For three decades, the story of Bletchley Park's great success at breaking the German Enigma machine codes had been strictly *verboten*. Even Winston Churchill kept to the rules, so in his six-volume history, *The Second World War*, the one man who really grasped the importance of SIGINT could say nothing about it at all.

There were two reasons for the official Anglo-American policy of continuing deep secrecy after 1945. The first, clearly, was not to reveal anything about the range and capacity of modern signals intelligence, because the Cold War against Communism had now begun. The second reason was not to give the defeated Axis nations any excuse to complain, as Germany had after WW1, that they only lost because the victors cheated. Books like *The Secrets of Crewe House* by Sir Campbell Stuart, in boasting of British propaganda skills, had only encouraged the Nazis to rewrite history.

Some stories did get out despite the official policy. Books have always been a way of pushing against the various Defence of the Realm and Official Secrets Acts passed since 1914, and in the first years after WW2, secrets emerged not as history but as adventure stories. As we have seen, Duff Cooper's novel *Operation Heartbreak* eventually led to the non-fiction accounts of MINCEMEAT, *The Unknown Courier* and *The Man Who Never Was*, in 1953. The Joint Intelligence Committee probably allowed these accounts of a British deception triumph to come out to counter a story of German success in the deception field published earlier in 1953. The cover of H. J. Giskes's *London Calling North Pole* promised 'An Incredible Disclosure by the Former Chief of German Counter-Espionage in Holland', and Giskes revealed that all British secret agents sent into wartime Holland by SOE were captured and had their radio sets taken over by the Abwehr, in an eighteen-month deception operation called *Nordpol*. Britain could not let Germany win at *Funkspiele* or 'radio games', hence the propaganda need for stories that showed the British secret war in a good light.

Brigadier Dudley Clarke retired from the British Army in 1947, having written the narrative war diary of 'A' Force, 1940–45. From 1948 to 1952, he was head of public opinion research in the Conservative Central Office and for a time a director of Securicor Ltd. He kept in touch with his former deceivers, lending to some of the more penurious moneys which were not returned. After *Seven Assignments* in 1948, Clarke wrote a history of the 11th Hussars 1934–1945, *The Eleventh at War*, and the mild security thriller *Golden Arrow*, which came out in 1955. But he never achieved the fame of his film-writer brother, T. E. B. Clarke, because he was not allowed to publish what he really knew. Dudley Clarke wanted to write *The Secret War* in 1953, but it never got beyond a publisher's proposal because of the Official Secrets Act, which he dutifully obeyed until his death in 1977.

Sir Winston Churchill, meanwhile, who won the Nobel Prize for Literature in 1953, was making millions from his writing, having taken rather more government documents with him than he should have done when he left. 'The way to command history,' he once remarked, 'is to write it yourself.' Re-elected Prime Minister once again in October 1951, a month short of his 77th birthday, Churchill was not immune to deception. After the old man had a massive stroke, the press was squared, the doctors hushed and the political class co-opted so his incapacity was hidden from the people. Churchill was a national icon whose image was not to be tarnished by his own feebleness and occasional petulance. Churchill finally resigned as PM in April 1955 and, after six decades as an MP, retired from Parliament in 1964. He died on 24 January 1965, at the age of 90, and his impressive and moving state funeral brought down the final curtain on the British Empire. In 2002, BBC TV viewers voted the saviour of his country the greatest Briton ever.

Revisionism usually takes a generation. Lytton Strachey's *Eminent Victorians* was not published in the Victorian era or the Edwardian, but the Georgian. Twenty years after his death in 1935, T. E. Lawrence came under attack from Richard Aldington, a writer and critic disillusioned by WW1 and determined to attack any class-bound romanticism that camouflaged its horrors. Richard Aldington's *Lawrence of Arabia: A Biographical Enquiry* indicts T. E. Lawrence as a fraud and fantasist; what it lacks in accuracy, it makes up for in invective. *Lawrence l'Imposteur* (as it was called on first publication in Paris) caused a furore when it came out in England in 1955, and a powerful cabal, headed by Basil Liddell Hart, first tried to suppress, then denigrate it.

One man who read Richard Aldington's book attentively (especially the chapter about Lawrence's sexuality) was Colonel Richard Meinertzhagen, DSO, the WW1 deceiver who by then was a white-bearded old gentleman in his seventies. From 1957 to 1964 Meinertzhagen published four books apparently drawn from the transcriptions of his seventy-six volumes of diaries, including *Kenya Diary*, *Army Diary* and *Middle East Diary*. Just as he was becoming a forgotten figure, he constructed his own larger-than-life legend, full of swaggering violence. Yet he was mostly famed as a bird man. In 1951, the British Ornithologists' Union gave him their Godman Salvin Medal, and in 1957 he was made a CBE for his services to ornithology. The American Ornithologists' Union also made him an honorary fellow, a rare distinction. His lifetime achievements included the amassing of one of the finest private collections in the world, including 25,000 bird skins and half a million *mallophaga* or feather-chewing lice parasites, which he donated to the British Museum (Natural History) in 1954, where it is still kept intact.

But Richard Meinertzhagen was a poacher. To create his collection, he had secretly stolen specimens from other private collectors and from the scientific collections of great institutions around the world. His last ornithology book, *Pirates and Predators,* a detailed ethological study of bird cheats and robbers, was in fact a disguised autobiography.

In 1993, twenty-five years after Richard Meinertzhagen's own death, Alan Knox, then of the Buckinghamshire County Museum, pointed out some blatant anomalies in Meinertzhagen's collection of *Acanthis* finches. Knox was the first person brave enough to suggest quite bluntly in a scholarly article that the colonel had been dishonest about how he got his specimens. Other scholars also uncovered deceptions relating to his 'finds'. 'It was a nuisance,' Alan Knox told me (with typical British understatement) from the University of Aberdeen in August 2005. 'Meinertzhagen corrupted the information which misled many people for a very long time.'

Nor was it only birds. Knox pointed me towards a critique published in 1995 by J. N. Lockman. This focuses on twelve entries relating to T. E. Lawrence in Meinertzhagen's *Middle East Diary 1917–1956.* It shows that those entries were not contemporaneous, but were inserted into the typescript after 1955 when the author had read Aldington's book. These entries were all disparaging to Lawrence's manhood, and are clearly Meinertzhagen's belated revenge for the way Lawrence had described him in *Seven Pillars of Wisdom,* as a man 'who took as blithe a pleasure in deceiving . . . as in spattering the brains of a cornered mob of Germans'.

The first two biographies of Meinertzhagen, *Duty, Honor, Empire* by John Lord and *Warrior* by Peter Hathaway Capstick, were admiring hagiographies; his third, by Mark Cocker in 1989, although edged with doubt, gave him the benefit of it. But his fourth utterly demolishes him: Brian Garfield's *The Meinertzhagen Mystery* (2007) is subtitled *The Life and Times of a Colossal Fraud.* Thus the whirligig of time brings in its revenges. Lawrence's reputation is to a great extent restored, but Meinertzhagen is exposed for what he was.

--

How easy was it to escape from deception once the war was over? The *camoufleur* Basil Spence kept his promise and built his cathedral at Coventry, and David Strangeways took Holy Orders and became a canon in the Church of England. The singular and secular Sefton Delmer wanted to start 'a journalistic revolution in Germany' by establishing a vigorous free press in place of the existing turgid German hackery. But occupied Germany was in the grip of a rigid British control commission and he left disgruntled to return to Fleet Street and his roving job as chief correspondent for the *Daily Express*. He finally quit Lord Beaverbrook's employment after a quarrel about expenses. 'I can only think clearly in a five-star hotel,' Delmer is said to have said. 'Is that it?' he demanded of the apparatchik who fired him. 'After thirty years?' 'Yes, that's it.' 'Well, if I'd known the job was temporary I wouldn't have taken it.'

Delmer broke his own injunction to the *Soldatensender* team not to reveal their secrets and talk about black propaganda when he wrote the second volume of his autobiography, *Black Boomerang*, in 1962, describing his life in 'black' radio. But nine years later, when he tried to go one further by writing a colourful account of strategic deception in WW2, mainly about GARBO and D-Day, he ran into problems because he drew on a secret history of FORTITUDE written for MI5 by Roger Hesketh that had been shown him clandestinely. Eventually a deal was done: Roger Hesketh, John Bevan and even Dudley Clarke were given a hand in editing Sefton Delmer's book *The Counterfeit Spy*. This was published in 1973 as a true story, but with many of the names changed: the code-name GARBO became 'CATO', and the Catalan Juan Pujol became the Basque 'Jorge Antonio'.

On 10 January 1978, the BBC broadcast a memorable television 'Play for Today' that went on to win a top BAFTA award. *Licking Hitler* was the fourth play by David Hare. Written, directed and narrated by him, it located in British WW2 secret activities the seeds of post-war corruption and dishonesty. *Licking Hitler* begins with an unworldly upper-class English girl, Anna Seaton, arriving to work in an English country house where a Black Propaganda Unit is preparing a robust talk about the defection of Rudolf Hess for a 'black' radio station called *Otto Abend Eins*. There is a devious political character from PWE who seems based on Richard Crossman, and a pragmatist who in some ways resembles Ian Fleming. The parallel figure to Sefton Delmer in this 'black' unit is Archie Maclean, a brilliant, savage, alcoholic, working-class Scot who writes lies to fool the Germans by day and has brutal sex with Anna by night. Hare's film is skilfully written, beautifully lit, shot and acted, but to my mind is flawed by a moralistic pseudo-documentary, added at the end, narrating its protagonists' lives in the decades after the war. Their subsequent failures and dishonesties are all laid at the door of their wartime activities in deception: 'The lying, the daily inveterate lying, the thirty-year-old corrosive national habit of lying.'

Tom Delmer himself watched the broadcast of *Licking Hitler* in some confusion. He was unwell, in bed, suffering the after-effects of a stroke, and the scenes of the 'black' broadcasts in German spoke to him, though the deplorable Archie Maclean seemed strange. He knew the BBC had optioned his book *Black Boomerang*, although it had not paid a full fee for adaptation, and now, as the drama unfolded and what was recognisably his work was disparaged, he wondered if 'Auntie' BBC was not somehow getting back at him for those old wartime humiliations.

Delmer died the following year. His last book, *The Counterfeit Spy*, had broken an incredible story and ended with what it claimed actually happened to the real people. 'Jorge Antonio' was said to have gone to Australia and Canada, eventually settling in Portuguese Angola, where he died of malaria in 1959. But that too was another, deliberate, deception.

The real Juan Pujol had asked Tomás Harris to let him vanish, so that vengeful neo-Nazis could not hunt him down. In May 1984, he was traced to Venezuela by the historian of intelligence Nigel West, who persuaded him to return to Europe for the fortieth anniversary of D-Day. Juan Pujol MBE was thanked by the Duke of Edinburgh at Buckingham Palace in London and was finally acknowledged publicly by the British for what he had done.

Dudley Clarke's best hope for deception was that 'The secret war was waged rather to conserve than to destroy . . . it was able to count its gains from the number of casualties it could avert'. Before he died in Caracas in 1988, Pujol in turn also declared that he was proudest to have helped to protect some of the tens of thousands of Allied servicemen fighting to hold the Normandy beachheads: 'Many, many more would have perished had our plan failed.'

Let the author lay some personal cards on the table at the end of the book. My thoughts about camouflage, deception and propaganda are not quite the same as when *A Genius for Deception* was first commissioned. Of course, deception is everywhere and we all practise it to some extent. Only adolescents are indignant at the white lies that politeness demands; there's deception all through nature and in great art as well. I have loved actors' performances all my life, yet we in the audience know they are not real, and willingly suspend our disbelief. Similarly, deception in warfare is not like deception in civilian life. War is an extraordinary state that changes the normal rules: a crime like killing may become a duty. Deception or deceitfulness in ordinary life is wrong because it corrodes trust, the basic glue of human relationships, but deceiving your enemy in wartime is common sense. If the war is just, then deception is also justified, because stratagems increase your chance of victory many times over.

Whenever I told anybody I was writing a book about British military deception they always said, 'That's interesting.' We are interested in deception because we know we all do it when we have to, and we all feel ambivalent about it. The primal warrior/deceiver in Western literature, Odysseus, deviser of the wooden horse gambit at Troy, displays *metis*, cunning or intelligence, in order to get back to protect his wife and child. We understand that, in a fight for survival, his ends justify his means.

But in real life we have to make continuing moral judgements on which ends justify which means. It happened that my last book, *Telegram from Guernica*, was published just three weeks after the 'shock and awe' attacks on Baghdad in March 2003. These were the opening salvoes of a war based on government claims that Iraq had weapons of mass destruction and was prepared to use them against us. Since then, in the years I have been working on *A Genius for Deception*, the word 'deception' has been hanging in the air like a bad smell.

There is a difference between this twenty-first-century deception and those of WW1 and WW2 that you have been reading about. What C.

E. Montague predicted in *Disenchantment* has come true. This time, the deceptive 'story' was aimed at our own people, on the brink of a possibly illegal war, and not at the despotic enemy. Moreover, it did not fool anyone for long.

When a leader says that Gamal Abdel Nasser or Saddam Hussein or someone else is another Adolf Hitler, this does not make the speaker a new Winston Churchill. Other wars are not WW2; metaphors do not confer reality. The USA is not 'our oldest ally' as a British politician stated in 2008 (Portugal is); nor is the Anglo-American alliance the same as six decades ago. We have lost what moral high ground there was in WW2. The three trillion dollar war in Iraq has been a propaganda disaster for 'the good guys'.

Winston Churchill knew from grim experience that 'jaw-jaw was better than war-war' and when toasts were drunk, used to add under his breath '. . . and no war'. He was also a great parliamentarian, 'brought up', as he told his beloved House of Commons on 29 November 1944, 'never to fear the English democracy, to trust the people' – who duly threw him out of office eight months later. Although as a wartime Prime Minister Churchill sometimes bent the truth when addressing the nation, what he and his wizards achieved by deceiving the Axis powers is indisputably justified.

History will judge our generation.

# Source Notes

## PREFACE

Carl von Clausewitz's *Vom Kriege* first appeared in 1832, and the Everyman *On War* was edited and translated by Michael Howard and Peter Paret in 1976. Thomas Hobbes's *Leviathan* was published in 1651. Quotes come from Alan Lascelles's diaries *The End of an Era* (1986), Raymond Seitz's memoir *Over Here* (1998), Richard Eyre's history *Changing Stages: a view of British Theatre in the 20th Century* (2000), Geoffrey Household's novel *Watcher in the Shadows* (1960), Jorge Luis Borges's fiction 'Tlön, Uqbar, Orbis Tertius' (1941) and Sun Tzu's military manual, *The Art of War,* translated by Lionel Giles (1910). The Prophet's *hadith*, cited in *The Encyclopaedia of Islam* in the British Library, is to be found in the *Sahih* of al-Bukhari, vol. 4, book 52, nos 267, 268 and 269. An entire 1971 book by William Woodin Rowe is dedicated to *Nabokov's Deceptive World* and Vladimir Nabokov's autobiography *Speak, Memory* (1966) has a superbly lyrical description of Batesian mimicry, confirming the observation in the writer's 1964 commentary on *Eugene Onegin:* 'Art is a magical deception, as all nature is magic and deception.' Two British historians who have noted the contrast between Britain's size and its standing are Linda Colley and David Cannadine; see their respective *Captives: Britain, Empire and the World 1600–1850* (2002) and *Ornamentalism: how the British saw their Empire* (2001). Among copious Churchilliana, Martin Gilbert's biographical work is indispensable: the paperback *Churchill: a life* was always to hand, Violet Bonham Carter's *Winston Churchill As I Knew Him* (1965) is full of insight, and *Speaking for Themselves: the personal letters of Winston and Clementine Churchill* (1998), edited by their daughter Mary Soames, is fascinating. Winston Churchill's account of the Downing Street tea-party comes from *The World Crisis*, his tremendous history of the First World War from 1911 to 1922, published in five volumes between 1923 and 1932, where the *Most Secret* memo 'The Dummy Fleet' appears in Appendix E of volume 1. There is a picture of SS *Merion* before and after its transformation into a dummy battleship on pp. 168–9 of *The Imperial War Museum Book of the First World War* (1991) by Malcolm Brown.

## I THE WAR OF NERVES

The books by Sir Philip Gibbs I used were *The Soul of the War* (1915), *Realities of War* (1920), *Life's Adventure* (1957) and *The War Dispatches*, edited and introduced by his son Anthony in 1966. Gibbs reported the Western Front for the readers of the *Daily Chronicle* and the *Daily Telegraph* longer than anyone else. John Buchan's twenty-four-volume *Nelson's History of the War* is a remarkable exercise in contemporaneous synthesis. It is of course patriotic and propagandistic but not idiotic, for a first-class intellect is at work, marshalling documentary information. The emotional colouring of Buchan's conservative mind is well caught in *The King's Grace, 1910–1935* from where I drew some personal touches. The Max Aitken observation came from the excellent biography *Beaverbrook: a life* (1992) by Anne Chisholm and Michael Davie. The Richard Harding Davis coverage of Brussels appears in *The Treasury of Great Reporting*, edited by Snyder and Morris, (2nd edition, 1962). Richard Harding Davis's *Notes of a War Correspondent* (1910) covers five wars on four continents and has a splendid essay on the kit required which might have benefited William Boot in Ishmaelia.

Winston Churchill's 'My Spy-Story' is related in his collection of journalism *Thoughts and Adventures* (1932) which is as revealing of the man and as readable as *My Early Life* (1930).

Cable-cutting details came from *The Thin Red Lines* (1946) by Charles Graves and *Gentlemen on Imperial Service* (1994) by R. Bruce Scott, histories of Cable & Wireless and the Pacific Cable Board. The John Keegan quote comes from page 162 of his *Intelligence in War* (2003).

The RFC quotes are taken from volume 2 of *War in the Air: being the story of the part played in the Great War by the Royal Air Force* (1922–37) by Walter Raleigh and H. A. Jones. The Fleet Air Arm Museum published *Warneford, VC* by Mary Gibson in 1979.

## 2 THE NATURE OF CAMOUFLAGE

Eric Partridge's etymology of camouflage appears in 'War as a Word-Maker' in *Words at War: Words at Peace* (1948). The Gertrude Stein stories come from *The Autobiography of Alice B. Toklas* (1933) and were also cited in Roy R. Behren's wonderful *False Colors: art, design and modern camouflage*. Another valuable and pioneering book on this subject is *Camouflage: a history of concealment and deception in war* (1979) by Guy Hartcup. For exhaustive illustration see *DPM: Disruptive Pattern Material: an encyclopaedia of camouflage in nature, warfare and culture* (2004) by Hardy Blechman and Andy Newman.

Many details about Britain's first *camoufleur* come from *Solomon J. Solomon: a memoir of peace and war* by Olga Somech Phillips [1933]. (She also wrote *The Boy Disraeli*, a sympathetic study of the early life of Britain's first Jewish Prime Minister.) Edward Potton edited *A Record of the United Arts Rifles 1914–1919* in 1920. For more on the shift in uniforms see *The British Army on Campaign 4: 1882–1902* by Michael Barthorp and Pierre Turner.

## 3 ENGINEERING OPINION

Frank Lynch helped me via the internet to pin the Dr Johnson quote down to *Idler* No. 30, 11 November 1758. An absolutely invaluable source for much in this chapter is the excellent *British Propaganda and the State in the First World War* (1992) by Gary S. Messinger.

For more on intelligence operations behind enemy lines in WW1, see *The Secrets of Rue St Roch* (2004) by Janet Morgan, and Christopher Andrew's superb history *Secret Service: the making of the British Intelligence community* (1985).

For much more on the *Lusitania*, see www.lusitania.net. Among several books, I found Diana Preston's *Wilful Murder: the sinking of the* Lusitania (2002) the best. Oliver Percy Bernard's remarkable autobiography *Cock Sparrow: a true chronicle* was published in 1936, three years before he died. After WW1, Bernard became the Art Deco designer of the interiors of the Lyons Corner Houses, the Cumberland, the Regent and the Strand Palace Hotels. He also fathered three sons: the poet and translator Oliver Bernard (b. 1925), the picture editor and photographer Bruce Bernard (b. 1928), and the legendary 'Low Life' columnist in the *Spectator*, Jeffrey Bernard (b. 1932). *The Eyes of the Navy: a biographical study of Admiral Sir Reginald Hall* (1955) by Admiral Sir William James is revealing although hagiographical. Margaret FitzHerbert's excellent biography of her grandfather Aubrey Herbert, *The Man Who Was Greenmantle*, was issued in paperback in 1985.

## 4 HIDING AND SNIPING

John Connell's biography to June 1941, *Wavell: scholar and soldier* (1964) is the source of the Archibald Wavell material. Details about Hesketh Vernon Hesketh Prichard come from *Hesketh Prichard D.S.O., M.C., Hunter: explorer: naturalist: cricketer: author: soldier*, a memoir by Eric Parker [1924]. I am grateful to the Librarian of the Marylebone Cricket Club at Lord's for details of H. P.'s cricketing career. H. M. Tomlinson's *All Our Yesterdays* was published by Heinemann in 1930. I quarried books on sniping by Adrian Gilbert, Peter Brookesmith and Andy Dougan as well as Martin Pegler's history of the military sniper, *Out of Nowhere*, published in 2004. The Aubrey Herbert quotes come from his 1919 book *Mons, Anzac and Kut* 'by an M.P.'.

## 5 DECEPTION IN THE DARDANELLES

*The Duff Cooper Diaries 1915–1951*, edited by John Julius Norwich, were published in 2005. *Tell England* by Ernest Raymond, the best-selling novel about naïve public schoolboys going to war, has none of the disenchantment of the Great War poets. The Roger Keyes quote appears on page 363 of Alan Moorehead's *Gallipoli*. John Masefield's sympathetic *Gallipoli* came out in 1916, Henry W. Nevinson's *The Dardanelles Campaign*, with its Greek epigraphs and a frontispiece of Sir Ian Hamilton, was published in 1918, and Compton Mackenzie's *Gallipoli Memories* a decade later, in 1929. Among those who also attended Rupert Brooke's funeral on Skyros were two talented composers who did not live to fulfil their talents. The Australian F. S. Kelly (killed like George Butterworth in the battle of the Somme in 1916) wrote his haunting *Elegy for Strings:'In Memoriam Rupert Brooke'* in hospital in Alexandria in June 1915, the same month that his musical friend W. Denis Browne was killed at Babi Acha. 'Lancashire Landing' is described in volume 1 of *The History of the Lancashire Fusiliers 1914–1918* and in Geoffrey Moorhouse's *Hells' Foundations: a town, its myths and Gallipoli* (1992). Dick

Doughty-Wylie's entry in *ODNB* was written by J. M. Bourne. Albert Barnett Facey (1894–1982) wrote only one book, which was published nine months before he died, the best-selling *A Fortunate Life*. Whether novel or autobiography, it is now deservedly a classic of Australian literature.

The Australian Private Henry Barnes is quoted on page 175 of *Defeat in Gallipoli* (1994) by Nigel Steel and Peter Hart. *The Secret Battle* (1919) is one of the outstanding novels of WW1; its author, A. P. Herbert, supplied additional dialogue for Anthony Asquith's unsuccessful 1931 film *Tell England*, also known as *The Battle of Gallipoli*.

*Painting as a Pastime* was reprinted for its fiftieth anniversary in the Sotheby's catalogue for the 1998 show organised by David Coombs, 'Winston Churchill – his life as a painter'. Norman Wilkinson's autobiography *A Brush with Life* came out in 1969. John Masefield's WW1 writings, including *Gallipoli* and his reviews of Nevinson's book and *Jacka's Mob,* were reprinted in 2007 in the excellent anthology *John Masefield's Great War,* edited by Philip W. Errington. For more on Nevinson, see *War, Journalism and the Shaping of the Twentieth Century* (2006) by Angela V. John. Patrick Beesly's *Room 40: British Naval Intelligence 1914–18* was published in 1982.

## 6 STEEL TREES

Details about Malcolm Wingate, together with much information on early camouflage, come from the Royal Engineers' interesting Library at Chatham. Philip Chetwode's comparison of French and Haig appears in *The Little Field Marshal: a life of Sir John French* (1981) by Richard Holmes. Philip Warner's *Kitchener: the man behind the legend* was published in 1985.

Solomon J. Solomon's correspondence about obtaining tree bark from King George V is in the Royal Archives at Windsor Castle.

I learned a lot about the revival of the steel helmet from Timothy Prus at the Archive of Modern Conflict in London. George Coppard's *With a Machine Gun to Cambrai: the tale of a young tommy in Kitchener's army 1914–1918* was published in 1969.

Some eyewitness details of Churchill's days in the trenches come from *Winston Churchill: his military life 1895–1945* (2005) by Michael Paterson. For more on the first tanks see Liddell Hart *The Tanks: the history of the Royal Tank Regiment and its predecessors* (1959), Patrick Wright *Tank: the progress of a monstrous war machine* (2000), *British Mark I Tank 1916* (2004) by David Fletcher and Tony Bryan, and Christy Campbell's excellent *Band of Brigands: the first men in tanks* (2007).

## 7 GUILE AND GUERRILLA

Figures on animals and lorries employed came from the Animals at War exhibition at the Imperial War Museum and *Statistics of the Military Effort of the British Empire during the Great War 1914–1920*, published by the War Office in 1922. The equine contribution has not been forgotten by British fiction: Michael Morpurgo's well-known *War Horse* (1982) has now been joined by Rosalind Belben's extraordinarily moving *Our Horses in Egypt* (2007).

I found Joshua Teitelbaum's *The Rise and Fall of the Hashimite Kingdom of Arabia* (Hurst, 2001) an invaluable guide to a complicated subject, backed up by *A Peace to End All Peace: the fall of the Ottoman Empire and the creation of the modern Middle East* (1989) by David Fromkin, and *Sowing the Wind: the seeds of conflict in the Middle East* (2003) by John Keay. Details about Ibn Saud come from *Lord of Arabia* (Penguin, 1938) by H. C. Armstrong and *The Kingdom* (1981) by Robert Lacey. Sir Ronald Storrs's *Orientations*, published in 1939, is a window into a vanished world of imperial diplomacy, as is Laurence Grafftey-Smith's autobiography *Bright Levant* (1970).

The literature on T. E. Lawrence is large and still growing. I have basically relied on Jeremy Wilson's authorised biography, *Lawrence of Arabia*, and three books by Malcolm Brown: *A Touch of Genius* (written with Julia Cave in 1988), *Lawrence of Arabia: the life, the legend* (2005) and *T. E. Lawrence in War and Peace: an anthology of the military writings of Lawrence of Arabia* (2005), which includes 'Twenty-Seven Articles'. I have also gleaned interesting details from the *Journal of the T. E. Lawrence Society*. The 'Hindustani fanatics' story comes from *God's Terrorists: the Wahhabi cult and the hidden roots of modern jihad* (2006) by Charles Allen. *Arab Command: the biography of Lieutenant-Colonel Peake Pasha CMG, CBE* by Major C. S. Jarvis was published in 1942. The £11 million figure is in footnote 1 on page 160 of Storrs' *Orientations*. Robert Irwin (author of a defence of Orientalists, *For Lust of Knowing*) reviewed the complete 1922 'Oxford' text of *Seven Pillars of Wisdom* in the *Times Literary Supplement* on 2 April 2004. *Oriental Assembly* (1939) edited by A. W. Lawrence, includes 'The Evolution of a Revolt' as well as many of T. E. Lawrence's photographs. See *Xenophon and the Art of Command* (2000) by Geoffrey Hutchinson for more on the Greek tactics as they fought their way out of Persia.

## 8 THE TWICE-PROMISED LAND

David Lloyd George's attitudes to the Middle East feature in Barbara W. Tuchman's *Bible and Sword: how the British came to Palestine* (1956) and *God, Guns and Israel: Britain, the First World War and the Jews in the Holy City* (2004) by Jill Hamilton.

Richard Meinertzhagen's seventy-six volumes of diaries are in the Rhodes House Library in Oxford, and the forty-two-volume catalogue of his bird collection at the British Museum (Natural History) Tring.

The Trojan horse image is from John Marlowe *Rebellion in Palestine* (1946). The D-Notice on the Balfour declaration is quoted in 'The Last Crusade? British Propaganda and the Palestine Campaign, 1917–18' by Eiten Bar-Yosef, published in the *Journal of Contemporary History*, vol. 36, no. 1 (January 2001).

In his 1959 memoirs, *Not in the Limelight*, Sir Ronald Wingate, the elder son of the Sirdar Sir Reginald Wingate, blamed the 'ill-informed enthusiasm' of T. E. Lawrence and the 'romantic penchant' of Gertrude Bell for the 'continually unstable political situation in Iraq', later made worse by the growing importance of oil. Mahomed bin Abdillah Hassan, the 'Mad Mullah of Somaliland' was actually chased and defeated by ground troops, according to the 1960 *Memoirs of Lord Ismay*, who was there as a young cavalry subaltern in the Somaliland Camel

Corps. Ismay saw all the aeroplane bombs miss, but acknowledges that the RAF did a brilliant political 'snow job' in London which saved them as an independent air force. *Churchill's Bodyguard* (2005) by Tom Hickman is based on the memoirs of Walter H. Thompson of the Metropolitan Police.

### 9 A DAZZLE OF ZEBRAS

*Politics, Press and Propaganda: Lord Northcliffe in the Great War 1914–1919* (1999) is by J. Lee Thompson, who is also the author of the fine biography *Northcliffe: press baron in politics 1865–1922*, published in 2000. Barbara Tuchman's *The Zimmermann Telegram* (1959) remains an exemplary historical study: the number of OB40's wireless operators and clerks comes from there, and the number of German communications they dealt with from p. 278 of *The Codebreakers: the story of secret writing* (1968) by David Kahn.

Norman Wilkinson featured in the Scottish Arts Council touring exhibition 'Camouflage' in 1988, as well as the Camouflage exhibition organised by James Taylor at the Imperial War Museum in 2007. Edward Wadsworth was prominent in the 1974 Hayward Gallery show 'Vorticism and its allies', organised by Richard Cork, also author of *A Bitter Truth: avant-garde art and the Great War* (1994).

'La Guerre Inconnue', a special edition of *Le Crapouillot* published in August 1930, features large-scale camouflage in Paris.

The catalogue of the 1998 show organised by Nicole Zapata-Aubé at the museum at Bernay in France, 'André Mare: Cubisme et camouflage 1914–1918' is richly detailed. *The War the Infantry Knew 1914–1919*, Captain James Churchill Dunn's chronicle of service with the 2nd battalion Royal Welch Fusiliers was first published in 1938.

### 10 LYING FOR LLOYD GEORGE

'Smiling pictures . . .' comes from S. J. Taylor *The Great Outsider: Northcliffe, Rothermere and the 'Daily Mail'* (1996), as quoted in the *ODNB* article on Alfred Harmsworth by D. George Boyce. The Arnold Bennett quote comes from his remarkable novel *Lord Raingo*, based on Beaverbrook's wartime propaganda work.

George Bernard Shaw's WW1 writings, *What I Really Said in the War* were republished in 2006, edited by J. L. Wisenthal and Daniel O'Leary. Shaw's views on his escort at the front appear in *C. E. Montague: a memoir* (1929) by Oliver Elton, and Philip Gibbs's memories of Shaw in *Life's Adventure* (1957).

The Stewart Menzies story comes from *The Secret Servant* (1988) by Anthony Cave Brown. *The Secret Corps: a tale of 'Intelligence' on all fronts* by Captain Ferdinand Tuohy was published in May 1920.

Meinertzhagen's memorandum, AIR 1/1155, appears in chapter 5 of *The British Army and Signals Intelligence in the First World War* (1992), edited by John Ferris. *Secrets of Crewe House: the story of a famous campaign* (1920) and *Opportunity Knocks Once* (1952) by Sir Campbell Stuart tell the story of British propaganda against the Central Powers. *The Inner Circle: the memoirs of Ivone Kirkpatrick* was published in 1959, three years after he retired as head of the Foreign Office, having interpreted for Halifax and Chamberlain with Hitler and interrogated Hess after his bizarre flight. Wickham Steed's *The Fifth Arm* was published in 1940 and G. M. Trevelyan's *Scenes from Italy's War* in 1919. See www.psywarrior.com for 'British Forgeries of the Stamps and Banknotes of the Central Powers' by SGM Herbert A. Friedman (Ret.)

## 11  DECEIVERS DECEIVED

*Strategic Camouflage* by Solomon J. Solomon RA was handsomely published as a demy quarto volume in 1920. Churchill describes 21 March 1918 in chapter 17, vol. 3 of *The World Crisis*. Martin Middlebrook's *The Kaiser's Battle* was first published in 1978, *All The Kaiser's Men: the life and death of the German army on the Western Front 1914–1918* by Ian Passingham in 2003, and Martin Kitchen's *The German Offensives of 1918* in 2005. H. M. Tomlinson's most scathingly angry anti-war book, *Mars His Idiot* (1935), is dedicated to 'Unknown Warriors'.

The correspondence about Solomon J. Solomon's camouflage ideas is in the Royal Engineers' Library at Chatham. Charlie Chaplin's *Shoulder Arms* can be viewed free online at the Internet Archive.

## 12  WIZARDS OF WW2

The extensive papers of Brigadier Dudley Clarke, CBE, CB at the Imperial War Museum, Box 99/2/1–3, include letters and diaries as well as his unpublished memoirs *A Quarter of My Century.*

*Pieces of War* by Lieutenant Colonel A. C. Simonds ('This officer is a pirate; only useful in time of war') is also in the Imperial War Museum. Max Hastings's comments came in a *Sunday Times* review of *Wavell: soldier and statesman* by Victoria Schofield. Bernard Fergusson's affectionate *Wavell: portrait of a soldier* was published in 1961; more details of Wavell's exercises can be found in Part IV of his book *The Good Soldier* (1948).

Denis Sefton Delmer published two volumes of memoirs, *Trail Sinister* (1961) and *Black Boomerang* (1962) and his papers have also been donated to the Imperial War Museum. The personal file (KV/2/2586) held on Sefton Delmer (and his father) by the Security Service is downloadable for a small fee from the National Archive. See also www.seftondelmer.co.uk.

In following the chronology of the rise and fall of the Third Reich, the black and gold volumes 2, 3, 4 and 5 of *Keesing's Contemporary Archives,* from 1 July 1934 to 31 December 1945, were an invaluable resource, continually consulted.

Virginia Cowles wrote about 1937 Madrid in *Looking for Trouble* (1941). Sefton Delmer in Madrid is a notable character in *Single to Spain* (1937) by Keith Scott Watson, later the first British journalist to report the bombing of Guernica. A fine contextual study of the foreign correspondents in Spain is Paul Preston's *Idealistas bajo las balas* (2007) from where the Constancia de la Mora quote comes.

## 13  CURTAIN UP

Clare Hollingworth's autobiography, *Front Line* (1990) recounts her 1939 Polish adventure. See also Esther Addley's profile of her in the *Guardian*, 17 January 2004. The Gleiwitz radio station incident is described in *The Man Who Started The War* (1960) by Gunter Peis and in *Kommando: German Special Forces of World War Two* (1985) by James Lucas. Alfred Naujocks's sworn affidavit about Gleiwitz, Document 2751-PS, dated 20 November 1945, was evidence at the Nuremberg trials. Hugh Trevor-Roper's views on Nazi reading come from the essay 'Admiral Canaris', appended to *The Philby Affair: espionage, treason, and Secret Services* (1968). Gerhard Klein directed an interesting East German feature film about this Nazi deception, *Der Fall Gleiwitz*, in 1961.

The texts of four 'bomphlets' appear on page 68 of the nine-volume history *The Second Great War*, edited by Sir John Hammerton. The rebuffed American journalist was John Gunther; the story was told by Harold Nicolson in a letter to his wife Vita Sackville-West on 14 September 1939. Joan Bright Astley's memoir *The Inner Circle: a view of war at the top*, first published in 1971, is a wonderful book, full of intelligence and insight. Having worked with the founders of British irregular warfare in WW2 and written two regimental histories, she also co-authored, with Peter Wilkinson, *Gubbins and SOE* (1993). *Seven Assignments* was first published in July 1948 and sold a respectable 5,000 copies.

## 14 WINSTON IS BACK

The merchant seaman statistics come from page 383 of *A New History of British Shipping* by Ronald Hope. The Scapa Flow dummy-ship story comes from *Churchill's Bodyguard*. John le Carré's interview, 'The Art of Fiction CXLIX', was in *Paris Review* 39 (1997). Patrick Beesly's *Very Special Intelligence: the story of the Admiralty's Operational Intelligence Centre 1939–1945* (1977) and *Very Special Admiral: the life of Admiral J. H. Godfrey CB* (1980) augment Donald McLachlan's *Room 39: Naval Intelligence in action 1939–45* (1968), and their British Naval Intelligence papers are together at the Churchill Archives Centre. For more on the creator of James Bond, see *The Life of Ian Fleming* (1967) by John Pearson, *17F: the life of Ian Fleming* (1993) by Donald McCormick, *Ian Fleming* (1995) by Andrew Lycett, and *For Your Eyes Only: Ian Fleming and James Bond* (2008) by Ben Macintyre, accompanying the centenary exhibition at the Imperial War Museum. For more on the murkier history of wireless see 'An Improper Use of Broadcasting . . . The British Government and Clandestine Radio Propaganda Operations against Germany during the Munich Crisis and after' by Nicholas Pronay and Philip M. Taylor, in *Journal of Contemporary History*, vol. 19, no. 3, (July 1984), and the interesting and opinionated *Truth Betrayed: radio politics between the wars* (1987) by the late W. J. West.

I have relied on Christopher Andrew's *Secret Service* and the biography of Claude Dansey, *Colonel Z: the life and times of a master of spies* (1984) by Anthony Read and David Fisher, for information about SIS. *The Partisan Leader's Handbook* is Appendix 2 in *SOE in the Low Countries* (2001) by Professor M. R. D. Foot whose *SOE: the Special Operations Executive 1940–1946* is the classic outline history. Details about Section D, Electra House and MI(R) are also in *The Secret History of SOE* (2000) by William Mackenzie and *Special Operations Executive: a new instrument of war* (2006), edited by Mark Seaman.

On GC&CS, see *Thirty Secret Years: A. G. Denniston's work in signals intelligence 1914–1944* (2007) by Robin Denniston. There are many books about Bletchley Park: I consulted *Battle of Wits: the complete story of code-breaking in World War II* (2000) by Stephen Budiansky, *Action This Day* (2001) edited by Michael Smith and Ralph Erskine, and *Station X: the code breakers of Bletchley Park* (2003) by Michael Smith. *The Essential Turing*, edited by B. Jack Copeland, was published by Oxford in 2004. For the repercussions of the Venlo incident see Nigel West's preface to *Invasion 1940: the Nazi invasion plan for Britain by SS General Walter Schellenberg* (2000) introduced by John Erickson. This bizarre volume, which includes the names of those to be arrested, also gives the title to Tom Paulin's interesting cut-up/collage of 1918–1940, *The Invasion Handbook* (2002). See also *Militärgeographische Angaben über England, 1940*, published by the Bodleian Library in 2007 as *German Invasion Plans for the British Isles 1940*.

I learned a great deal about RSS and other matters from *The Secret Wireless War: the story of MI6 Communications 1939–1945* (2006) by Geoffrey Pidgeon. John Masterman's *The Double-Cross System 1939–1945* was originally published by Yale in 1972, but the 1995 Pimlico edition has a useful introduction by Nigel West. MI5's WW2 successes feature in the official history *The Security Service 1908–1945* by John Curry (1999) and *Camp 020: MI5 and the Nazi spies* (2000) edited and introduced by Oliver Hoare. Dusko Popov is the subject of *Codename Tricycle* (2004) by Russell Miller. Professor R. V. Jones's *Most Secret War: British Scientific Intelligence 1939–1945* (1978, called *The Wizard War* in USA) was rightly described by A. J. P. Taylor as 'the most fascinating book on the Second World War that I have ever read.' His *Reflections on Intelligence* (1989) is also most worthwhile.

## 15 HIDING THE SILVER

An excellent, well-illustrated study of the WW2 return of the *camoufleurs* is *Camouflage and Art: design for deception in World War 2* (2007) by Henrietta Goodden of the Royal College of Art. The painter Julian Trevelyan's *Indigo Days* was published in 1957, and details about Roland Penrose came from *Visiting Picasso: the notebooks and letters of Roland Penrose* (2006) by Elizabeth Cowling. The film-maker Geoffrey Barkas's *The Camouflage Story (From Aintree to Alamein)* was published in 1952. A classic study of the visual arms race is *To Fool a Glass Eye: camouflage versus photoreconnaissance in World War II* (1998) by Colonel Roy M. Stanley II, USAF (retd). See also *Eyes of the RAF: a history of photo-reconnaissance* (1996) by Roy Conyers Nesbit. W. Heath Robinson's WW2 cartoons were collected in *Heath Robinson at War* (1942), *The Penguin W. Heath Robinson* (1966), *Inventions* (1973), *The Best of Heath Robinson* (1982) and *Heath Robinson's Helpful Solutions* (2007), which is Simon Heneage's catalogue of the London Cartoon Museum show curated by Anita O'Brien.

## 16 A GREAT BLOW BETWEEN THE EYES

The Wavell quote comes from the introduction to Dudley Clarke's *Seven Assignments*. There is a clear account of the Norwegian campaign by Major

General J. L. Moulton in the 1966 Purnell partwork *History of the Second World War*. Ray Mears's *The Real Heroes of Telemark* (2003) honours the endurance of the Norwegian resistance.

The Guy Liddell diaries (vol. 1: 1939–42; vol. 2: 1942–45) edited by Nigel West, were published by Routledge in 2005. John Colville's diary of his time as Churchill's private secretary, *The Fringes of Power*, was published in 1985. The 'battleship built on land' was General Sir Alan Brooke's description after a visit to the Maginot Line in December 1939. The observations in Marc Bloch's *Strange Defeat: a statement of evidence written in 1940* (1949) are confirmed by the May 1940 diary in *The Rommel Papers* (1953) edited by B. H. Liddell Hart.

Dudley Clarke's experiences are narrated in *Seven Assignments*. Airey Neave's *They Have Their Exits* (1953) is one of the very best WW2 memoirs, edged with the irony of his visits to imprisoned Nazis at the Nuremberg war crimes tribunal. (Neave was murdered by the IRA on 30 March 1979, at Westminster.) Dudley Clarke's mission to Ireland features in Robert Fisk's *In Time of War: Ireland, Ulster and the price of neutrality 1939–1945* (1983). David Mure's *Master of Deception* (1980) followed *Practise to Deceive* (1977). Danchev's comments on Dill are from his entry in *ODNB*.

The most recent books on Dunkirk include Hugh Sebag-Montefiore's excellent *Dunkirk: fight to the last man* (2006), Sean Longden's *Dunkirk: the men they left behind* (2008) and General Julian Thompson's *Dunkirk: retreat to victory* (2008). J. B. Priestley's twenty BBC talks, *Postscripts*, were published at the end of 1940.

## 17 COMMANDO DAGGER

Fifteen-years'-worth of Churchill's speeches are collected in ten volumes edited by his son Randolph. The Dunkirk speech is on page 215 of *Into Battle* (1941). Dudley Clarke was paid 25 guineas for his fifteen-minute talk, which was printed in *The Listener*. The producer was the future historian Ronald Lewin. *The Green Beret: the story of the Commandos 1940–1945* (1949) by Hilary St George Saunders credits Dudley Clarke in chapter 2. Ernest Chappell's account of the first raid is in *Commandos: the inside story of Britain's most elite fighting force* (2000) by John Parker.

## 18 BRITISH RESISTANCE

The classic account of WW2 British internment is 'Enemy Alien' by the Nobel Prize-winning chemist Dr Max F. Perutz, OM, CH originally published in *The New Yorker* in 1985, and reprinted in *Is Science Necessary? Essays on Science and Scientists* (1989). For more on the 'Fifth Column' see *Blackshirt: Sir Oswald Mosley and British Fascism* (2006) by Stephen Dorril.

You get a disturbing sense of what invasion meant in *Occupation: the ordeal of France 1940–1944* (1997) by the late Ian Ousby. William L. Shirer's account of Compiègne (and his global scoop for CBS) is in chapter 16 of *The Nightmare Years 1930–1940*, vol. 2 of *20th Century Journey*, published in 1984. Keitel's speech is from page 1012 of volume 3 of *The Second Great War*. Priestley's comment came from the last *Postscript*, Sunday, 20 October 1940.

George Orwell's 'Patriots and Revolutionaries' first appeared in Victor Gollancz's *Betrayal of the Left* (1941), an indictment of the Communist Party, and was also the final piece in the last non-fiction book published by Gollancz in 1981: *The Left Book Club Anthology*, edited by Paul Laity. Tom Wintringham's version of the Battle of the Jarama in *English Captain* (1939) should be compared with that of Jason Gurney in *Crusade in Spain* (1974). *Picture Post 1938–50*, edited and introduced by Tom Hopkinson, was published in 1970. *Home Guard Socialism: a vision of a People's Army* (2006) by Stephen Cullen gets it all in fifty pages. The Maxwell memo is on pp. 57–8 of the most revealing vol. 4, 'Security and Counter-Intelligence', of *British Intelligence in the Second World War* (1990), written by Professor Sir Harry Hinsley with Anthony Simkins, formerly Deputy Director of MI5, with unrestricted access to the records. The photograph of Lee Miller as a camouflaged nude is on pp. 182–3 of *DPM: Disruptive Pattern Material* and I am grateful to its editor Hardy Blechman for the loan of a copy of Penrose's *Home Guard Manual of Camouflage*.

Kevin Brownlow and Andrew Mollo's fictional film *It Happened Here* (1966) and David Lampe's first class investigation of the Auxiliary Units, *The Last Ditch* (1968), prompted other speculations about German invasion. Norman Longmate wrote the book of the BBC1 television film *If Britain Had Fallen* in 1972, and *The Real Dad's Army: the story of the Home Guard* in 1974, the same year that Duff Hart-Davis published *Peter Fleming: a biography*. Len Deighton's brilliant vision of a Nazi-occupied UK, *SS-GB*, appeared in 1978, paving the way for other novels: Gordon Stevens *And All The King's Men* (1990), Robert Harris *Fatherland* (1992) and Owen Sheers *Resistance* (2007).

*The Home Guard: a military and political history* by S. P. Mackenzie came out in 1995, and further details about the British Resistance can be found in the more excitable *With Britain in Mortal Danger: Britain's most secret army in WWII* (2000) edited by John Warwicker. I am grateful to his daughter Julia Korner, encountered at the Special Forces Club, for a copy of Andrew Croft's autobiography *A Talent for Adventure* (1991). The BFI guide to the film *Went the Day Well?* was written by Penelope Houston in 1992. Michael Korda's memoir *Charmed Lives: the fabulous world of the Korda brothers* was published in 1980.

## 19 FIRE OVER ENGLAND

When *The Big Lie* by John Baker White, first published in 1955, was issued by Pan in 1958, its cover was subtitled *The Art of 'Political Warfare'* with the strapline 'How the Allies Fooled the Nazi High Command'. Dennis Wheatley's War Papers were published as *Stranger than Fiction* in 1959.

Peter Haining's *Where The Eagle Landed: the mystery of the German invasion of Britain, 1940* (dedicated to Dennis Wheatley) is less reliable than James Hayward's excellent *The Bodies on the Beach: Sealion, Shingle Street and the Burning Sea myth of 1940* (2001).

*The Ironside Diaries 137–40* were published in 1962. The Göring boast comes from *Trenchard* (1962) by Andrew Boyle. There is a huge literature on the 1940 air war: Len Deighton's *Fighter: the true story of the Battle of Britain* (1977) and *Battle of Britain* (1980) stand out as clear and vivid.

London bombing details come from *The Night Blitz 1940–41* (1996) by John Ray and the 1942 HMSO publication *Front Line 1940–41: the official story of the Civil Defence of Britain.*

## 20 RADIO PROPAGANDA

*Propaganda in War 1939–1945: organisations, policies and publics in Britain and Germany* by Michael Balfour, published in 1979, is the classic account.

Sefton Delmer features prominently in *The Secret History of PWE: the Political Warfare Executive 1939–1945* written by David Garnett (who edited the letters of T. E. Lawrence) in 1945–6 but which was first published in 2002.

Val Gielgud's memoirs are called *Years of the Locust* (1947). *Stephen Potter at the BBC: 'Features' in war and peace* (2004) by his son Julian Potter throws interesting light on the BBC in wartime. *Lord Haw Haw: the English voice of Nazi Germany* (2003) by Peter Martland, *Germany Calling: a biography of William Joyce, Lord Haw-Haw* (2003) by Mary Kenny, and *Haw-Haw: the tragedy of William and Margaret Joyce* (2005) by Nigel Farndale should sate curiosity about 'Sinister Sam'. *ITMA 1939–1948* by Francis Worsley was published in 1948, and Ted Kavanagh's biography of Tommy Handley the following year.

Orson Welles's Mercury Theatre on the Air had been broadcasting classics of English Literature for CBS like *Dracula* and *Treasure Island* before they came to H. G. Wells. *Broadcasts from the Blitz: how Edward R. Murrow helped lead America into war* by Philip Seib was published in 2006. Kevin Jackson's superb biography *Humphrey Jennings* was published in 2004.

Some of the 'V' campaign material comes from the *BBC Handbook 1942*, volume III of Asa Briggs's great history of broadcasting, *The War of Words 1939–1945*, and Sir John Lawrence's contribution to *Sage Eye: the aesthetic passion of Jonathan Griffin* (1992) edited by Anthony Rudolf. There is a photograph of the BBC 'V' signal and the African drum used to record it in James Blades' autobiography *Drum Roll* (1977). Anthony Rhodes's *Propaganda: the art of persuasion in WW2* (1987) shows visual uses of the 'V'.

Sir Hugh Greene's lectures, speeches and broadcasts are collected in *The Third Floor Front* (1969). For 'black' radio see Delmer's *Black Boomerang*, Garnett's *Secret History of PWE*, as well as *The Black Game: British subversive operations against the Germans during the Second World War* (1982) by Ellic Howe, and *Black Propaganda in the Second World War* (2005) by Stanley Newcourt-Nowodworski.

In 1998 and 2002, the late David Syrett edited two volumes of papers for the Navy Records Society about signals intelligence in the Atlantic battle against the U-boats.

## 21 'A' FORCE: NORTH AFRICA

General Sir Archibald Wavell's address to Australian troops in February 1940 in which he described Middle East Command is in *Generally Speaking* (1946). For more on Bagnold see *Long Range Desert Group* (1945) by W. B. Kennedy Shaw and 'Bagnold's Bluff' by Trevor J. Constable in *The Journal for Historical Review* vol. 8, no. 2 (March/April 1999), also *Bearded Brigands: the diaries of trooper*

Frank Jopling (2002) edited by Brendan O'Carroll and *Desert Raiders: Axis and Allied Special Forces 1940–43* by Andrea Molinari (2007). Alexander Clifford's *Three Against Rommel* was first published in 1943. The Dimbleby quote is from *The Frontiers are Green* (1943). The first book in Alan Moorehead's African trilogy is *Mediterranean Front: the year of Wavell, 1940–41*, and his moving memoir *A Late Education: episodes in a life* (1970) is structured around his friendship with Alexander Clifford.

Brigadier Dudley Clarke's *Seven Assignments* leaves off just where the 'A' Force *Narrative War Diary* (CAB 154/1 in the National Archives) begins. Clarke's meeting with 'Wild Bill' Donovan is recorded on page 34 of *Establishing the Anglo-American Alliance: the Second World War diaries of Brigadier Vivian Dykes* (1990), edited by Alex Danchev.

Barkas's road trip is in *The Camouflage Story*. *With Rommel in the desert* (1951) by Heinz Werner Schmidt and *Rommel's War in Africa* (1981) by Wolf Heckmann are authentically detailed.

On the SAS, *The Phantom Major: the story of David Stirling and the S.A.S. Regiment* (1958) by Virginia Cowles is excellently researched, and Tim Jones has kept up to the mark with two fine books on Stirling's legacy: *SAS: the first secret wars* (2005) and *SAS Zero Hour: the secret origins of the Special Air Service* (2006). *Ghost Force: the secret history of the SAS* (1998) by Ken Connor, who was twenty-three years in the regiment, knows what it is talking about. *The Originals in their Own Words: the secret history of the birth of the SAS* by Gordon Stevens was published in 2006.

Sir Michael Howard's *Strategic Deception in the Second World* (1990), vol.5 of the official history of British Intelligence in WW2, singles out Dudley Clarke's achievements in 'notional' forces, as does Thaddeus Holt's comprehensive *The Deceivers: Allied military deception in the Second World War* (2004).

*I Spied Spies* by Major A. W. Sansom, MBE was published in 1965. Jasper Maskelyne's ghosted *Magic: top secret* (1949) is fantastical; David Fisher's *The War Magician* (1983) is fictional: 'The events depicted in this book are true. Everything Jasper Maskelyne is credited with doing he actually accomplished', says its disclaimer, but see the website run by Richard Stokes, www.maskelynemagic.com, for a more realistic view. Earlier magicians also laid claim to camouflage inventions: see chapter 7 of Horace Goldin's *It's Fun to be Fooled* for a WW1 example.

The Australian Chester Wilmot's book *Tobruk 1941: capture, siege, relief* describes cheerfulness in adversity. Steven Sykes's *Deceivers Ever: the memoirs of a camouflage officer* was published in 1990. For more on WW2 Mesopotamia, see *Five Ventures* (1954) by Christopher Buckley, *Iraq and Syria 1941* (1974) by Geoffrey Warner and *Iraq 1941* (2005) by Robert Lyman. George Steer lent me a copy of *The Road Uphill: episodes in a long life* (1997) by his stepfather Kenyon Jones, together with an undated typed account of 'CHEESE' that KJ sent to Dudley Clarke some time after the war.

## 22 IMPERSONATIONS

If Spain was the Axis neutral, Portugal was the Allied neutral. See *Sympathy for the Devil: neutral Europe and Nazi Germany in WW2* (2001) by Christian Leitz.

On the failure of Washington DC to fit many pieces of the Japanese jigsaw together see the opening chapter of *The Secret War against Hitler* (1989) by William Casey of the CIA.

Juan Pujol wrote his autobiography *GARBO* with Nigel West in 1985. The Far Eastern haversack ruse features in the biography of Peter Fleming by Duff Hart-Davis, in *Wavell: supreme commander* (1969) by John Connell, completed and edited by Michael Roberts, as well as in *The Deceivers* by Thaddeus Holt.

The Cabinet War Rooms are part of the Imperial War Museum, open to the public, and the Churchill Museum now occupies the rooms where the Deception Planners once sat.

The letters and photographs dealing with the consequences of Clarke's arrest in Madrid are in the Churchill Archives Centre, Cambridge (CHAR 20/25/42–52). Clothes, as Virginia Woolf observed in *Orlando*, do more than keep us warm: 'They change our view of the world and the world's view of us.' See *Dressing Up: transvestism and drag – the history of an obsession* (1979) by Peter Ackroyd and *Crossing the Stage: controversies on cross-dressing* (1993) edited by Lesley Ferris.

I am grateful to Dr Paul Adamthwaite of Ontario for supplying me with David Syrett's account 'The Battle for Convoy HG-75, 22–29 Oct 1941' which appeared in *The Northern Mariner,* vol. 9, no 1.

### 23 THE GARDEN OF FORKING PATHS

For a full account of Borges's anti-fascist credentials see *Borges: a reader* (1981) edited by Emir Rodriguez Monegal and Alastair Reid, and *Borges: a life* (2004) by Edwin Williamson. *MI9: escape and evasion 1939–1945* (1979) by M. R. D. Foot and J. M. Langley is the classic account of wartime escaping. Clayton Hutton's story of escape aids, *Official Secret*, was published in America in 1961. Its last three dozen pages are about his fights with British Air Ministry bureaucrats in the 1950s trying to block its publication on grounds of 'security'.

For SOE kit see the PRO's *Secret Agent's Handbook of Special Devices* (2000) and *SOE Syllabus: lessons in ungentlemanly warfare* (2001) as well as *SOE: the scientific secrets* (2003) by Fredric Boyce and Douglas Everett.

Tomás Harris's summary of the Garbo case was published in *GARBO: the spy who saved D-Day* (PRO, 2000) edited by Mark Seaman.

### 24 THE HINGE OF FATE

Dr Hugh B. Cott began a long association with Africa at the Camouflage School, Helwan, Egypt. Later work on the ecology of the Nile crocodile took him thousands of miles through the continent. The second of the 109 marvellous pen drawings in *Uganda in Black and White* (1959) is a Jackson's Chamaeleon whose 'expressionless face' and 'hesitant gait' give it the look of 'a robot'. The undress of soldiers in the Middle East is captured by Cecil Beaton, who arrived in March 1942 on photographic commission from the Ministry of Information, and some of whose photographs can be seen in *Near East* (Batsford, 1943).

Rick Atkinson's *An Army at Dawn: the war in North Africa, 1942–1943* (2003) covers TORCH. Carlo D'Este has written two superb biographies of the key American generals: *A Genius for War* (1995) about George S. Patton, and *Eisenhower: a soldier's life* (2002).

For how Sir Arthur Harris distorted the Casablanca Directive about the bombing of Germany see pp. 201–2 of *The Bomber War* (2001) by Robin Neillands. Macmillan's Greek/Roman analogy is cited in volume 1 of Alastair Horne's official biography, *Macmillan, 1894–1956*; 'Supermac' was still using it as the wise old UK PM when he courted the young US President John F. Kennedy in the early 1960s. Stories of 30 Assault Unit RN/RM Commando are told disjointedly in *Attain by Surprise: capturing top secret Intelligence in WWII* (2003), edited by David Nutting, and the splendid *Arctic Snow to Dust of Germany* (1991) by Patrick Dalzel-Job, which has photographs of the young and handsome Charles Wheeler.

## 25 MINCEMEAT

The detail about Patrick Leigh-Fermor comes from his afterword to the 2001 Folio Society edition of the classic *Ill Met By Moonlight* by W. Stanley Moss, telling how two SOE officers kidnapped General Kreipe on Crete in 1943.

'Deception history is more complicated than we are more normally inclined to believe': Klaus-Jürgen Müller in 'A German Perspective on Allied Deception Operations in the Second World War', in *Strategic and Operational Deception in the Second World War* (1987), edited by Michael I. Handel, warns of the dangers of exaggerating the success of MINCEMEAT.

*The Unknown Courier* by Ian Colvin (with a note on the Axis situation in the Mediterranean in spring 1943 by Field Marshal Kesselring) was published by William Kimber in 1953, after Ewen Montagu's *The Man Who Never Was* (which had an introduction by General the Rt Hon. Lord Ismay, Secretary General of NATO, formerly Churchill's Chief of Staff, 'Pug'.) A combined edition of *Operation Heartbreak* and *The Man Who Never Was*, with an introduction by Duff Cooper's son, John Julius Norwich, was published by Spellmount in 2003.

## 26 THE DOUBLE

*Borges y yo* appeared in *El Hacedor* in 1960 and *Dreamtigers* in 1964. Borges reviewed Victor Fleming's 1941 film of *Jekyll & Hyde* in the magazine *Sur*, saying that two completely different actors would have been preferable to the solo Spencer Tracy overacting. Borges had written the story of another failed impersonation in 'The Implausible Impostor Tom Castro' in 1933. Dennis Wheatley writes about 'the False Montgomery' in chapter 17 of *The Deception Planners: my secret war* (1980). The drunkenness story is on page 140 of Jock Haswell's *The Intelligence and Deception of the D-Day Landings* (1979). He says the Germans paid no attention to COPPERHEAD at all, but a kindlier version of James's treatment and achievement is on page 562 of *The Deceivers* by Thaddeus Holt.

## 27 OVERLORD AND FORTITUDE

Geoffrey Pyke features in *Pyke: the unknown genius* (1959) by David Lampe, the

'Science in War' section of Max Perutz's *Is Science Necessary?* (1989) and Paul Collins's 'The Ozzard of Whizz' in *Fortean Times* 197, June 2005. The ice-shooting incident is recorded in Alanbrooke's *War Diaries 1939–1945* for 19 August 1943. Rommel's report is quoted on page 453 of *The Rommel Papers* (1953).

*Bodyguard of Lies* is also the title of the 1975 book on WW2 deception by Anthony Cave Brown which although pioneering is not always accurate. The modified COSSAC plan is described in Chapter 13 of *Operation Victory* (1947) by Major-General Sir Francis de Guingand. A clear and concise outline of the planning and execution is in *D-Day* (2004) by Martin Gilbert.

Part of the Soviet Russian help to BODYGUARD was a feint towards northern Norway, and a purported amphibious assault from the Black Sea on Rumania. The classified official history, *Fortitude: the D-day deception campaign* was written by Roger Hesketh in 1945–8, but not published until 2000.

*Allied Photo Reconnaissance of World War II* (1998) edited by Chris Staerck shows aerial reconnaissance pictures of the Normandy beaches before and after D-Day. A picture of René Duchez with an account of his activities is on page 94 of *D-Day: June 6, 1944 – the Normandy Landings* (1992) by Richard Collier.

*PLUTO: Pipe-Line Under the Ocean* (2nd edition, 2004) by Adrian Searle tells the definitive story. More engineering ingenuity is displayed in *Churchill's Secret Weapons: the story of Hobart's Funnies* (1998) by Patrick Delaforce and in Gerald Pawle's *The Secret War 1939–1945* (1956) about the 'Wheezers and Dodgers' of DMWD, the Admiralty's Department of Miscellaneous Weapon Development.

Sonic and wireless deceptions around D-Day feature in chapters 6 and 7 of *Trojan Horses: deception operations in the Second World War* (1989) by Martin Young and Robbie Stamp. *King's Counsellor* (2006), the wartime diaries of Sir Alan Lascelles, edited by Duff Hart-Davis, record a visit by 'two MI men' on Friday, 3 March 1944 to explain how the King's visits could help 'to bamboozle the German Intelligence'.

For a critical, pro-Crossman view of *Nachrichten für die Truppen* and Delmer's work see *Sykewar: psychological warfare against Germany, D-Day to VE-Day* (1949) by Daniel Lerner.

TAXABLE and GLIMMER feature in 'Deception, technology and the D-Day invasion' by R. W. Burns in *Engineering Science and Education Journal*, vol. 4, issue 2, April 1995. *The Far Shore* (1960) by Rear Admiral Edward Ellsberg, USN gives a clear account of what went wrong. Robert Capa was not the only one to lose his images: most US motion-picture footage of the landings was lost when the ship holding it was sunk. Ernie Pyle's syndicated dispatches appeared six times a week across the USA and were collected in three books: *Here is Your War: the story of G.I. Joe* (1943), *Brave Men* (1944; the quote here is from chapter 26) and the posthumous *Last Chapter* (1945). The Smuts story comes from chapter 31 of *The War and Colonel Warden* (1963) by Gerald Pawle, based on the recollections of Commander C. R. Thompson.

GARBO's message, as received by teleprinter at Hitler's HQ on 9 June 1944, is reproduced (and translated) at the start of Roger Hesketh's *Fortitude: the D-Day deception campaign*.

## 28 V FOR VERGELTUNG

Chapter 337 in vol. 8 of *The Second Great War* is an illustrated account of the German 'reprisal weapons' in action. Many words have been written about the bombing of Germany, for which 45,000 Allied airmen gave their lives. For a literary response see *On the Natural History of Destruction* (2003) by W. G. Sebald, for a moral inquiry see *Among the Dead Cities* (2006) by A. C. Grayling, and for case studies of two individual cities see *Inferno: the destruction of Hamburg 1943* by Keith Lowe and *Dresden: Tuesday 13 February 1945* by Frederick Taylor.

   *To the Victor the Spoils: D-Day to VE-Day – the reality behind the heroism* (2004) by Sean Longden is a superbly researched account of what soldiering with 21st Army Group into Germany was really like: grim, grimy, and grinning. Richard Dimbleby's account of Belsen was partially printed in *War Report: D-Day to VE-Day* (1946) and more fully in the 1975 biography of him by his son, Jonathan Dimbleby.

# Acknowledgements

'Let thinks be thanks,' wrote Auden. Everyone in the team at Faber and Faber has been very kind. I am grateful to Julian Loose for nurturing and protecting the book, and to his assistant, Kate Murray Browne, who was always a tonic when gloom descended; to Jon Riley for provoking an extension of my original notion; and to Lucy Davey, Gavin Morris, Anne Owen, Anna Pallai, Paula Turner at Palindrome, proofreader Peter McAdie and indexer Alison Worthington for all their careful work.

The Tippexed and typed-over manuscript of my first book, twenty-two years ago, was thickened by literal 'cut-and-paste'. There was no internet for civilians then – you researched in real libraries, posted letters and bought second-hand books in shops you actually visited. This new book, of course, also owes much to Abebooks, Google and Hotmail as well as other online resources like the *Oxford Dictionary of National Biography*, available through my borough library service in Brent. Older institutions I am grateful to include the British Broadcasting Corporation, the National Archives at Kew and the Imperial War Museum in Lambeth, all of whose staffs try to keep the flame of public service flickering. On the private side, it is always a pleasure to prowl the battleship decks of the London Library in St James's Square.

I started writing this book in one writers' retreat and added the final licks of paint in another. I am grateful to Mrs Drue Heinz and the Trustees of Hawthornden Castle in Scotland for the Fellowship that allowed me to stay there in the spring of 2004, and to the Committee of the Fondation Ledig-Rowohlt in Lausanne, Switzerland for inviting me and my wife, in the summer of 2008, to stay at the Château de Lavigny, which once belonged to a great German publisher and his English wife. These places have been small heavens of cool green in a hot and thirsty terrain.

Among those I have talked to in the last four years I am grateful to the following for their help and encouragement: Julia Abel-Smith, Dr Paul Adamthwaite, Oliver Bernard, Anne Bingaman, Hardy Blechman, Jasper Bouverie, Malcolm Brown, Jimmy Burns, Anne Chisholm, Felix Delmer for trusting me with some of his father Sefton Delmer's papers and stories, Aaron Delwiche, Michael Diamond, Moris and Nina Farhi, Roger Fenby, Maggie Fergusson, Professor M. R. D. Foot, Harriett Gilbert, Henrietta Goodden, Professor Barbara Goodwin and her husband Michael Miller QC (my pal, who died before he could read this book), Stephen Gottlieb and Jane Dorner, Colin Grant, Dr Toby Haggith, Roger Hardy, Caroline Herbert, Susannah Herbert, Hesketh Prichard's grandchildren (whose full names are in a tragically lost black Alwych notebook), Sally Higgin, Thaddeus Holt for his friendship and guidance (his own book, *The Deceivers: Allied Military Deception in the Second World War*, was a major resource), Richard Ingrams, Robert Irwin, the brilliant David Jones, P. J. Kavanagh, Dr Douglas Kerr of Hong Kong, Wesley Kerr, Dr Alan Knox, Julia Korner, Shen Litznaisky, Andrew Lycett, Andrew Lownie and Kate Macdonald for help with John Buchan, Hugh MacDougall for 'jiggery-pokery' and Merlin's stone, Margaret Macmillan, Nicholas Mays, Glenn Mitchell for gallantly enduring a first draft, Caroline Moorehead, Ingo Niebel (*bruder im geist*), Bob O'Hara, Drs Andrew R. Parker and Robert Prys-Jones of the Natural History Museum, Hayden Peake, Lawrence Pollard, Timothy Prus, Tom Read, Michael Redley, Zina Rohan, John Ryle, Anthony Rudolf, Dan Shepherd, George Steer for his memories of Dudley Clarke and the loan of Kenyon Jones's autobiography, James Taylor, Claire Tomalin, Nigel West, the late Sir Charles Wheeler, Dr Andrew Whiten, Hugh Whistler, Caroline and Malcolm Winterburn, and Patrick Wright, the son of my old Shrewsbury Headmaster. Forgive me if I have forgotten your name here. Of course, all errors are my own: 'Ignorance, madam, pure ignorance', as Dr Johnson explained.

Most of all I want to thank family. First, my dear siblings: Charles in Malvern, Sarah in Ipswich, Trina in Bury St Edmunds and my elder brother, John, for the long-term loan of Buchanalia and for our battlefield trips to France, to Crécy and Le Cateau, to Ypres and Dunkirk. I particularly wish to remember our late and always beloved great-aunt, Gwynneth Constance Stallard, an Englishwoman of the old school, who died in 2005 at the age of 103. Our grandfather, Colonel Geoffrey Page, was her only brother, and their ashes are scattered on the Ashdown Forest less than a mile away from the secret 'Aspidistra' site. Since Winston Churchill was her hero, it felt apt to write some of *Churchill's Wizards* in her home, Gorse Cottage, where all our family found such happiness. I thank my lovely, witty daughter Rosa Rankin-Gee and, above all, her mother, my wife.

--

The author and the publisher are grateful to the following for use of copyright material. Every effort has been made to trace the copyright holders of material quoted in the text. The publisher would welcome the opportunity to rectify any omissions brought to their attention:

Citation of King George V's personal diary (RA GV/PRIV/GVD/1917: 8 March) and the use of other material from the Royal Archives at Windsor Castle (RA PPTO/PP/WC/MAIN/NS/95) by the gracious permission of Her Majesty Queen Elizabeth II.

Extracts from Sir Winston Churchill's letters, speeches and books, including *My Early Life, Thoughts and Adventures, The World Crisis* and *The Second World War*, are reproduced by permission of Curtis Brown Ltd, London on behalf of The Estate of Winston Churchill, © Winston S. Churchill.

Extracts from the *BBC Handbook 1941* are reproduced by permission of the BBC Written Archives, Caversham; from Oliver Bernard's *Cock Sparrow*, Oliver Bernard; from the writing of John Buchan, including *The Thirty-Nine Steps, Greenmantle, Mr Standfast, The Three Hostages, Nelson's History of the War,*

*John Macnab* and *Memory-hold-the-Door*, by A. P. Watt Ltd on behalf of Jean, Lady Tweedsmuir, and the Executors of the Estate of Lord Tweedsmuir; from Sefton Delmer's autobiographies, *Trail Sinister* and *Black Boomerang*, by Felix Delmer; from Richard Dimbleby's description of Belsen, by Jonathan Dimbleby; from the journalism of Sir Philip Gibbs, by Martin Gibbs and Frances McElwaine; from A. P. Herbert, by A. P. Watt Ltd on behalf of the Executors of the Estate of Jocelyn Herbert, M. T. Perkins and Polly M.V. R. Perkins; from Aubrey Herbert's letters and *Mons, Anzac and Kut*, by Claudia FitzHerbert; from the script of *Desert Victory*, written by James Lansdale Hodson, by the Trustees of the Imperial War Museum, London; from Ted Kavanagh's scripts for *ITMA*, by P. J. Kavanagh; from the letters of T. E. Lawrence, by the Seven Pillars of Wisdom Trust; from *Gallipoli Memories* by Compton Mackenzie, by the Society of Authors as the Literary Representative of his Estate; from *The White Cliffs*, by Pollinger Limited and the Estate of Alice Duer Miller; from the works of George Orwell, *Animal Farm* (Copyright © George Orwell, 1945), *Patriots and Revolutionaries* (Copyright © George Orwell), *Notes on the Way* (Copyright © George Orwell, 1940), *A Review of The Thirties* (Copyright © George Orwell), *The Lion and the Unicorn* (Copyright © George Orwell, 1941), by Bill Hamilton as the Literary Executor of the Estate of the Late Sonia Brownell Orwell and Secker & Warburg Ltd; from Roland Penrose, by A. P. Watt Ltd on behalf of the Executors of the Estate of the late Sir Roland Penrose; from copyright material by J. B. Priestley, by Peters, Fraser and Dunlop (www.pfd.co.uk) on behalf of the Estate of J. B. Priestley; for Crown Copyright material on early camouflage, by the Royal Engineers' Library at Chatham; from 'Joy-Riding at the Front', by the Society of Authors, on behalf of the Bernard Shaw Estate; material © *The Times* 1917 & 1939, by News International Syndication and Times Newspapers; from *The Letters of Evelyn Waugh* © 1980, the Estate of Laura Waugh; from *Put Out More Flags* by Evelyn Waugh (first published by Chapman & Hall 1942, published Penguin Books 1943, reprinted with a new Introduction Penguin Classics 2000), copyright 1942 by Evelyn Waugh, by Penguin Books Ltd; from *A Brush with Life*, by Camilla Wilkinson and the Norman Wilkinson Estate.

# Index

Sparappelhoeck, 167–71

Special Operations Executive (SOE):
agents in Netherlands, 224, 415;
camouflage, gadgets and disguises,
355–7, Plate 15; function and
organisation, 48, 280; in Greece,
374; and Heydrich's assassination,
205; overview, 273

Speer, Albert, 198

Spence, Basil, 399, 405, 418

Spiecker, Dr Carl, 300–1

Spilsbury, Sir Bernard, 375

Spring Offensives (1918), 163–5

Stalin, Joseph, 196, 305, 394

Stalingrad, Battle of (1942–3), 367

Stamp, Robbie, 276

Stanley, Col Oliver, 340–1

Starfish sites *see* SF sites

Steed, H. Wickham, 157

steel wool, 228

Stein, Gertrude, 27–8

Stephens, Pembroke, 196

Stephens, Lt Col Robin 'Tin Eye', 381

Stevens, Gordon, 323–4

Stevens, Maj Richard, 222

Stevens, Tom, 298, 308

Stevenson, Robert Louis, 384

Stimson, Henry L., 126

Stirling, David, 252, 323–4, Plate 19

Stone, Senator, 127

stool pigeons, 151–2

Stopford, Gen, 72

Storrs, Ronald, 98, 103, 116–17, 124

Strachey, Oliver, 223

Stradling, Reginald, 227

Strangeways, Lt Col David: after the
war, 418; and Normandy landings,
395, 398, 401, 406; and TORCH,
364; in Tunisia, 371

Stranks, Mabel, 271

Streicher, Julius, 190

Stroheim, Erich von, 384–5

Strong, Kenneth, 182, 208

Stuart, Lt, 14

Stuart, Sir Campbell, 144, 155–6, 172,
218, 415

submarines *see* U-boats

Sudan, 101–2, 318

Suez Canal, 95–6, 102

Suhren, Reinhard, 350

Sunshields, 330, 366

Sutherland, Graham, 230–1

Suvla Bay, 70–2

Swinton, Col Ernest, 50, 91–2

Swinton, Lord, 257

Switzerland, 235

*Sword of Honour* (Waugh), 252

*Sydney*, HMAS, 16–17

Sykes, Maj, 251

Sykes, Mark, 115

Sykes, Steven, 327, 329–30, 360, 362,
405–6

Sykes–Picot agreement (1917), 119,
121

Symien sniper suits, 91

Symington, Lyndsay D., 83, 85, 91,
167

Syria: WW1, 96–7, 100, 109, 117,
120–1; WW2, 331–2

Taber, Robert, 108

Tallents, Sir Stephen, 217, 285–6, 297

Tanganyika, 101, 112–13

tanks: camouflaged and dummy, 92–3,
313, 321, 330, 360–2, 365, 366,
Plate 3, Plate 18; numbers at El
Alamein, 367; unusual types used in
Normandy, 398; WW1, 92–3

TATE (Wulf Dietrich Schmidt), 400

Tavistock, Lord, 257

Taylor, A. J. P., 73, 268

'The Technique of War' (Shaw), 147

Teheran conference (1943), 394

*Telconia*, 14–15

telegraph, 14–17, 119–20, 211, 220

telephones: interception, 149–50, 211

telescopes, 54

television, 289

*Tell England* (Raymond), 61

Templer, Maj Gen Gerald, 219–20, 412

Thatcher, Margaret, 414

Thayer, Abbott H., 26–7, 129, 161

Thayer, Gerald, 26–7

*The Thirties* (Muggeridge), 260